T0134452

Advances in Computer Vision and Pattern Recognition

Marius Leordeanu

Unsupervised Learning in Space and Time

A Modern Approach for Computer Vision using Graph-based Techniques and Deep Neural Networks

 Springer

Marius Leordeanu
Computer Science and Engineering
Department
Polytechnic University of Bucharest
Bucharest, Romania

ISSN 2191-6586   ISSN 2191-6594   (electronic)
Advances in Computer Vision and Pattern Recognition
ISBN 978-3-030-42130-4   ISBN 978-3-030-42128-1   (eBook)
https://doi.org/10.1007/978-3-030-42128-1

This Springer imprint is published by the registered company Springer Nature Switzerland AG
The registered company address is: Gewerbestrasse 11, 6330 Cham, Switzerland

Dedicated to my parents who let me be a child, my teachers who let me be a student and my students who give everything a purpose.

In memory of Solomon Marcus (1 March 1925–17 March 2016), Romanian mathematician, writer and guardian angel for many of us. He was a member of the Romanian Academy and Professor of mathematics at the University of Bucharest.

Preface

In *Unsupervised Learning in Space and Time* we address one of the most important and still unsolved problems in artificial intelligence, which is that of learning in an unsupervised manner from very large quantities of spatiotemporal visual data that is often freely available. The book covers our main scientific discoveries and results while focusing on the latest advancements in the field from an original and insightful perspective. The text has a coherent structure and it logically connects, in depth, original mathematical formulations and efficient computational solutions for many different unsupervised learning tasks, such as graph and hypergraph matching and clustering, feature selection, classifier learning, object discovery, recognition, and segmentation in video. The tasks are presented with a unified picture in mind, which puts together and relates at many levels different tasks that converge into the more general unsupervised learning problem. We start from a set of intuitive principles of unsupervised learning, and then gradually build the mathematical models and algorithmic tools that are necessary. We eventually reach a general computational model for unsupervised learning, which brings together graphs and deep neural networks. Overall, the book is deeply grounded in the scientific work we have developed together with our professors, colleagues, and doctoral students at the Robotics Institute of Carnegie Mellon University, Intel and Google Research, the Institute of Mathematics "Simion Stoilow" of the Romanian Academy and the University Politehnica of Bucharest. For our work on unsupervised learning, in 2014 we were awarded the "Gigore Moisil Prize", the highest in mathematics given by the Romanian Academy.

Organization and Features

The book is organized into eight chapters, which take the reader from a set of initial intuitions and common sense principles for unsupervised learning, to different tasks, computational models, and solutions which are introduced and integrated together, chapter by chapter, as follows:

Chapter 1: In the first chapter, we introduce seven principles of unsupervised learning, and then make a brief pass through the subjects covered in the next chapters in strong relation to these basic principles and concepts. Chapter 1 gradually builds a big picture of the book, without covering the very last concepts and models, which are presented in the final chapter.

Chapter 2: In the second chapter, we introduce the problems of graph and hypergraph matching, going from initial motivation and intuition to efficient algorithms for optimization and unsupervised learning. In this chapter, we present the Spectral Graph Matching algorithm, which is later related to the method presented in Chap. 6 for unsupervised object segmentation in video. We also present the Integer Projected Fixed Point (IPFP) method, whose clustering extension (Chap. 3) is later used on hypergraph clustering (Chap. 3), unsupervised feature selection and classifier learning (Chap. 4), and descriptor learning and object discovery in video (Chap. 5). We extensively compare our methods to many other approaches for graph and hypergraph matching and demonstrate a significant boost. We also show how unsupervised learning for graph matching can significantly improve performance.

Chapter 3: In the third chapter, we extend the formulations and algorithms from the second chapter to the task of graph and hypergraph clustering. The two problems are strongly related, and similar models and algorithms can address both. We present an efficient clustering algorithm based on the integer projected fixed-point method from the second chapter. The IPFP-clustering method is then applied to the tasks defined in Chaps. 4 and 5.

Chapter 4: In the fourth chapter, we present an efficient approach to linear classifier learning formulated as a clustering problem. The formulation leads to both feature selection and classification, for which we also provide an unsupervised solution. We introduce the idea of a feature sign and show that by knowing this sign we could learn without knowing the samples' labels. The algorithm used for learning is the same as the clustering-IPFP method from Chap. 3. We compare to many linear classification approaches, including Support Vector Machines (SVM) on the task of video classification and show significantly more powerful generalization power from limited training data.

Chapter 5: In this chapter, we put together all the building blocks presented so far. By following the initial unsupervised learning principles from Chap. 1, we create a fully unsupervised system for object segmentation in video, which learns over several generations of classifiers, using features that start from simple pixels to deep features extracted from the whole image. We show in extensive experiments that our approach is more efficient and accurate than previously published methods on several challenging datasets.

Chapter 6: In the sixth chapter, we continue our exploration of unsupervised object discovery in video and present an original formulation of object discovery as segmentation in a space-time graph in which every pixel video is a node. We introduce a novel Feature-Motion matrix, which couples elegantly motion and appearance and demonstrates that the main object of interest can be discovered as a strong cluster in the space-time graph by efficiently computing the eigenvector of the Feature-Motion matrix. The mathematical formulation and solution is thus directly related to the spectral graph matching approach from Chap. 1. Our spectral clustering approach to object discovery in space and time is fast and completely unsupervised, while also capable to accommodate any type of pretrained features, if needed. We test on three challenging datasets and outperform other unsupervised approaches. We also boost the performance of other supervised methods, when including their outputs into the Feature-Motion formulation.

Chapter 7: In the seventh chapter, we move to the next level of unsupervised learning over multiple generations. We introduce a teacher-student system, which learns in a self-supervised manner, such that the population of student ConvNets trained at one iteration form the teacher at the next generation. The ideas build upon the material presented so far, but the system is original and shows how it can learn from video, without any human supervision, to segment objects into single images. While the previous chapters are more focused on classical graph models than on deep neural networks, in Chap. 7 we change the focus to deep learning.

Chapter 8: In the last chapter, we merge the graph and neural networks models into a new recurrent space-time graph neural network (RSTG) model, which leverages the benefits and features of both, including the ability to learn over deep layers of features and multiple scale, as well as the capacity to send messages iteratively across both space and time. The RSTG model takes full advantage of the space-time domain and proves its effectiveness on high-level tasks, such as learning to recognize complex patterns of motion and human activities. In the last part, we introduce the Visual Story Network concept—a universal unsupervised learning machine, which learns through multiple prediction pathways, between different world interpretations, by optimizing its own, self-consensus.

Target Audiences

The book is written especially for people with exploratory, naturally curious, and passionate minds, who are daring to ask the most challenging questions and accept unconventional solutions. They could be young students or experienced researchers, engineers, and professors, who are studying or already working in the fields of computer vision and machine learning. We hope the book will satisfy their curiosity and convey them a unified big picture of unsupervised learning, starting from some basic, common sense principles, and developing towards the creation of a fully universal unsupervised learning machine. However, in order to grasp in sufficient detail the complex material covered by the book, readers are expected to have a solid background in mathematics and computer science, and already be familiar with most computer vision and machine learning concepts. To fully understand the more technical parts, which bring together many graph algorithms and deep neural network models, spread across several computer vision problems, readers are encouraged to master fundamental elements of linear algebra, probability and statistics, optimization, and deep learning. *Unsupervised Learning in Space and Time* is ultimately for people who are determined to find the time and space to learn and discover by themselves.

Bucharest, Romania   Marius Leordeanu
May 2020

Acknowledgements

This book would not be possible without my dear professors, mentors, colleagues, and students who have devoted a considerable amount of effort to our collaborative work. They gave me their time, trust, and support for which I am and will always be grateful. We shared hopes, passion, and values. Above all, we shared our love for knowledge and for the fascinating world of vision. Trying to make computers learn to see the world as we learn to see it, is a really hard problem. But the challenge is fascinating, the intellectual reward could be immense, and it is worth all our focus, passion, and spark of creativity. Unsupervised learning is probably one of the ultimate quests in science and technology today, with the potential to open doors towards a territory that is beyond our imagination now. We can only hope that the way we will understand unsupervised learning in the natural world will help in the way we will create and use resources, communicate with each other, and ultimately live our lives.

We should hope that a better understanding of learning, by establishing bridges between many domains, from mathematics and computer science to neuroscience, psychology, philosophy, and art, will take us to a better understanding of ourselves. At the end, it will lead to developing a clearer vision of our own meaning and purpose. To this noble goal, I dedicate this book and towards this goal I hope to grow and direct my efforts.

My deepest and most sincere thoughts of gratitude go first to my professors and mentors who gave me a chance to find my own way in research and grow over the years.

I thank deeply my dearest Ph.D. advisor Martial Hebert, a true model as a researcher and human being, during my Ph.D. years. He guided me with much kindness, wisdom, and light through my very first steps in becoming a computer vision and robotics researcher, at Carnegie Mellon University (CMU). I learned a lot from him and enjoyed immensely doing research together with him. Many of the core ideas and results presented in this book are created and developed together with Martial, under his wise and caring guidance. I also want to express my deepest gratitude to my other great advisor during my first Ph.D. years, Prof. Robert

Collins, with whom it was such a wonderful pleasure to work, create, and debate ideas, and write my first papers on object tracking and unsupervised learning.

I thank deeply to one of my dearest friends and mentors throughout my entire career, Rahul Sukthankar, who I met during my Ph.D. years at CMU. Our journey of working together has been truly amazing. I have learned so much from him and together with him, at both scientific and spiritual levels. Some of the best papers I wrote were together with Rahul. We met in a magic space, governed by the love of science and discovery, where we ask the most interesting questions, take the most daring intellectual journeys, and find the most surprising solutions.

My deepest gratitude and warmest thoughts also go towards my first professor of computer vision, Ioannis Stamos, at Hunter College of the City University of New York. He welcomed me into this wonderful world of vision and gave me the chance to do, for the first time in my life, what I have always dreamed about. My very first contact with research in computer vision and my very first papers were written together with him. Meeting him was a true blessing and thus, I embarked on the journey to find how vision and mind works. From that moment on, 18 years ago, my dream became my profession.

My deepest thanks and warmest thoughts also go to my dear mentor and friend Cristian Sminchisescu, who gave me the extraordinary chance to do top computer vision in my own country, Romania. Under his guidance and wisdom, I have started my career as an independent researcher. I thank him greatly for our fruitful collaboration together. He has been a true model of diligence, motivation, and success.

I also send my deepest feelings of hope, admiration and love to my extraordinary friend and mentor, Leon Zagrean, Professor of Medicine and Neuroscience at Carol Davila University in Bucharest. He is a true pioneer in neuroscience research in our country and a man of very high moral and human values. His profound love for life and endless search for harmony in space and time, has shaped and influenced the core ideas proposed in this book in surprising and wonderful ways.

I also thank deeply my dear colleagues, mentors and friends at the University Politehnica of Bucharest, Institute of Mathematics of the Romanian Academy and University of Bucharest, who actively participated in the creation of this book though numerous discussions, exchange of ideas, projects, and papers that we worked on together. Adina Florea, Ioan Dumitrache, Stefan Trausan Matu, Emil Slusanchi, Traian Rebedea, Elena Ovreiu, Nirvana Popescu, Mihai Dascalu, Lucian Beznea, Vasile Brinzanescu, Dan Timotin, Cezar Joita, Cristodor Ionescu, Sergiu Moroianu, Liviu Ignat, Ionel Popescu, Bogdan Ichim, Bogdan Alexe, Radu Ionescu, Viorica Patraucean, Razvan Pascanu, and Gheorghe Stefanescu are the people together with whom we participate every day in growing a young community of AI researchers, professionals, and engineers in Romania and Eastern Europe and bring a wind of hope and change in this part of the world. I must thank them, from the bottom of my mind and heart for being strong and staying together in our endeavor.

Last, but not least, I thank my dear Ph.D. and graduate students. They are the main purpose of my work as a professor and researcher. This book is made possible by them and ultimately dedicated to them and to the young generation, which they represent. A large part of the material presented in this book is the result of their effort and passion for knowledge and discovery. Ioana Croitoru, Vlad Bogolin, Andrei Zanfir, Mihai Zanfir, Emanuela Haller, Otilia Stretcu, Alexandra Vieru, Iulia Duta, Andrei Nicolicioiu, Elena Burceanu, Alina Marcu, Dragos Costea, Nicolae Cudlenco, Mihai Pirvu, Iulia Paraicu, and Cristina Lazar participated substantially in the research results reported in this book and also gave me constant feedback during the writing process. I must thank them for their full support and great work. They give me one more strong reason to believe that the next generation will surpass the last and the world will be in good hands.

Source Materials: The material presented in most chapters is based in large part on a number of published works, as follows:

- Chapter 2:
  Leordeanu, Marius, and Martial Hebert. "A spectral technique for correspondence problems using pairwise constraints." IEEE International Conference on Computer Vision (ICCV), 2005.
  Leordeanu, Marius, Martial Hebert, and Rahul Sukthankar. "An integer projected fixed point method for graph matching and map inference." In *Advances in Neural Information Processing Systems* (NIPS) 2009.
  Leordeanu, Marius, Rahul Sukthankar, and Martial Hebert. "Unsupervised learning for graph matching." *International Journal of Computer Vision* 96, no. 1 (2012): 28–45.
  Leordeanu, Marius, Andrei Zanfir, and Cristian Sminchisescu. "Semi-supervised learning and optimization for hypergraph matching." *IEEE International Conference on Computer Vision*, 2011.
- Chapter 3:
  Leordeanu, Marius, and Cristian Sminchisescu. "Efficient hypergraph clustering." In *Artificial Intelligence and Statistics* (AISTATS), 2012.
- Chapter 4:
  Leordeanu, Marius, Alexandra Radu, Shumeet Baluja, and Rahul Sukthankar. "Labeling the features not the samples: Efficient video classification with minimal supervision." In Thirtieth AAAI conference on artificial intelligence (AAAI). 2016.
- Chapter 5:
  Stretcu, Otilia, and Marius Leordeanu. "Multiple Frames Matching for Object Discovery in Video." British Machine Vision Conference (BMVC) 2015.
  Haller, Emanuela, and Marius Leordeanu. "Unsupervised object segmentation in video by efficient selection of highly probable positive features." IEEE International Conference on Computer Vision (ICCV) 2017.
- Chapter 6:
  Haller, Emanuela, Adina Magda Florea, and Marius Leordeanu. "Spacetime Graph Optimization for Video Object Segmentation." arXiv preprint arXiv: 1907.03326 (2019).

- Chapter 7:
  Croitoru, Ioana, Simion-Vlad Bogolin, and Marius Leordeanu. "Unsupervised Learning of Foreground Object Segmentation." *International Journal of Computer Vision* (IJCV) 2019: 1–24.
  Croitoru, Ioana, Simion-Vlad Bogolin, and Marius Leordeanu. "Unsupervised learning from video to detect foreground objects in single images." IEEE International Conference on Computer Vision (ICCV) 2017.
- Chapter 8:
  Nicolicioiu, Andrei, Iulia Duta, and Marius Leordeanu. "Recurrent Space-time Graph Neural Networks." In Advances in Neural Information Processing Systems (NeurIPS) 2019.

Funding Sources: Writing of the book and a good part of the research work presented was supported through UEFISCDI, from EEA Grant EEA-RO-2018-0496 and projects PN-III-P1-1.1-TE-2016-2182 and PN-III-P1-1.2-PCCDI-2017-0734.

Endorsements

The book is a pleasure to read. It is timely, in the all-important quest for effective strategies for unsupervised learning. The author describes his work on the past decade, with many new additions, and an interesting philosophical outlook in the last chapter through his Visual Story Graph Neural Network. A key mechanism in the book is the use of his "Integer Projected Fixed Point" (IPFP) method with first-order Taylor series expansion, so the problem can be solved in a cascade of linear programs. Another returning key mechanism is the spectral clustering by "learning" the fit to the principal eigenvector of the adjacency matrix or feature-motion matrix. Many books on deep learning and the quest for unsupervised strategies lack a focus on video analysis, and this book fully compensates this. The author has for many years been a pioneer in this spatiotemporal AI domain. His work has significantly influenced many other works and patents. The chapters are built up in a logical order, increasing in complexity from graph/hypergraph matching and clustering to appearance and motion, and to exploiting large numbers of networks forming "students" creating "teacher" over a number of generations. A realistic approach is offered by allowing a little bit of supervised information at the start of the process, like assigning a small number of "Highly Probable Positive Features" (HPP) and "feature signs". It's impressive to learn how excellent the implementations of the described theories almost always outperform the current state of the art, often by a significant margin, and very often explicitly in the speed domain. The author takes great care in comparing the proposed methods with many current models and implementations. Also, each chapter gives a deep and complete overview of the current literature. The theory is described well, with both a solid mathematical theory, an intuitive story, and with critical discussions, many parameter variations, discussion of pitfalls, and extensive quantitative results. As the author remarks, there is a lot of work to do. But this book is a significant step forward, with proven effectivity of the many ideas. All in all, an important new book on stage.

Bart M. ter Haar Romeny
Emeritus Professor in Computer Vision and Biomedical Image Analysis
Department of Biomedical Engineering, Eindhoven University
of Technology
Eindhoven, The Netherlands

I very much enjoyed the systematization and the logical sequencing of the questions posed and addressed, in analyzing the unsupervised learning process. Marius Leordeanu is a wonderful critical thinker, who takes us on an exciting journey, sharing original insights and drawing us into the story of a beautiful adventure from imaging to seeing things in images and videos. I found his approach appealing and inviting, his enthusiasm contagious, and his arguments solid and well presented and I am sure this book will become a standard reference for researchers in the field of Computer Vision.

Alfred M. Bruckstein
Technion Ollendorff Chair in Science
Technion Israel Institute of Technology
Haifa, Israel

I thoroughly enjoyed reading *Unsupervised Learning in Space and Time*! This is a complex topic of very active research and is challenging to capture in a single volume. Rather than presenting a dry review of recent work, Marius Leordeanu guides the reader along a fascinating journey using a handful of well-formulated but intuitive principles that motivate this perspective on the research space. The authors' infectious love for the subject is evident throughout the book—it will inspire the next generation of researchers to dream bigger dreams!

Rahul Sukthankar
Distinguished Scientist, Google Research
Mountain View, California, USA

Unsupervised Learning in Space and Time outlines a pathway to solving one of the most complex open issues in today's Computer Vision. Just as the completion of big puzzle game should start by sorting the pieces by color, form, size, and orientation before actually trying to fit the pieces together, Marius puts into perspective all the complementary tasks and processing steps involved in unsupervised learning before letting us enjoy the fully assembled picture. We are then left to dream with our minds' eye of the possibilities just opened by the acquired insights.

Emil Slusanschi
Professor of Computer Science
University Politehnica of Bucharest
Bucharest, Romania

In a captivating storytelling fashion, Marius captures the reader's attention by displaying the pieces of a puzzle whose completion leads to a result that seems to be the answer to a key question in the field of Computer Vision: how can we learn in an unsupervised manner.

Bogdan Alexe
Pro-Dean and Associate Professor of Computer Vision
University of Bucharest
Bucharest, Romania

Unsupervised learning is such an important topic in machine learning and computer vision, which has not been fully explored yet. The book takes us in an in-depth exploration of recent techniques and methods proposed for unsupervised learning for several tasks in computer vision. And it is a wonderful learning experience, not only for more experienced researchers, but also for beginners in machine learning. The various proposed models and theories converge at the end, when Marius introduces a new model for unsupervised learning in computer vision, the Visual Story Graph Neural Network, which makes use of classifiers based on weak signals, trained in a teacher-student fashion, and reinforcing each other on several layers of interpretation for an image in time. This model also opens new research opportunities for visual-language tasks, but this will probably be the topic of a different book!

Traian Rebedea
Associate Professor of Computer Science
University Politehnica of Bucharest
Bucharest, Romania

Marius Leordeanu, one of the most prolific and creative researchers of his generation, describes in this book fundamental elements of unsupervised learning for vision, spanning topics from graph matching, clustering, feature selection, and applications, to neural networks. These topics are a necessary read for all who want to acquire a deep understanding of the field. The book follows Leordeanu's research path, making it not only current, but also essential for researchers. Marius' passion, enthusiasm, along with his intuition and insights are all reflected here. This is an important book for computer vision.

Ioannis Stamos
Professor of Computer Science
Hunter College
City University of New York
New York City, USA

Contents

Chapter 1
Unsupervised Visual Learning: From Pixels to Seeing

1.1 What Does It Mean to See?

I am trying to imagine how the world looked like the first time I opened my eyes. What did I see? I am pretty sure that I saw every pixel of color very clearly. However, did I see objects or people? When I looked at my mother, what did I see? Of course, there was that warm, bright presence keeping me close and making me feel good and safe, but what did I know about her? Could I see her beautiful, deep eyes or her long dark hair and her most wonderful smile? I am afraid I might have missed all that, since I did not really know back then what "eyes", "nose", "mouth", and "hair" were. How could I see her as a person, a human being just like myself when I did not even know what I am or what a human being is.

As I could not relate anything to past experiences, there was nothing that I could "recognize" or see as something. There were no "things to see" yet because there was no relationship yet built between the different parts of the image. Pixels were just pixels, and I was probably blind to everything else. By just observing that some groups of pixels have similar colors did not mean that I could see them as being something. There was no past behind anything, to connect things to other things and give them meaning and their "reality" as things.

Now when I look at my mother I can see who she is: there are so many experiences which bind us together. There are many images of her at different stages of my life as well as many images of myself at those times. All those memories are strongly connected together and interact within a story that is coherent in both space and time. All those images of mother, linked through her unique trajectory in my life, make her who she is in my universe. They make her what I see when I look at her.

Now, when I see a single pixel on her face, in fact I see so many things at the same time. I see a skin pixel, a face, and a human pixel. I also see a pixel of my own mother—that special and uniquely important person in my life. Then, at a higher level in space and time, that pixel is also part of us, myself and her, mother and son, which connects all the way back to that very first moment. I also see a pixel belonging to the human race and to life and Earth, as it travels around the Sun. It is only now that I can see all those layers of reality, which required years to form.

My vision is now deeper than it was then, at the very first moment, when it barely existed. All these visual layers containing objects and parts of objects, interactions,

© Springer Nature Switzerland AG 2020
M. Leordeanu, *Unsupervised Learning in Space and Time*,
Advances in Computer Vision and Pattern Recognition,
https://doi.org/10.1007/978-3-030-42128-1_1

and activities exist now and are real at the same time. Back when I opened my eyes for the very first time, the world just started to move. I must have felt a deep, strong urge pulling me towards the unexpected lights, with their surprising and seductive dance of patterns. But what did I know then about what would follow? Everything I see now is our story. It is the visual story of myself as part of the world, as I am growing to see it even better and hopefully become able to imagine how might have been then and what followed next.

1.2   What Is Unsupervised Visual Learning?

While my mother has always been next to me when it mattered most and thought me some of the most important lessons that I needed to know, during the first years of my life, she was definitely not the one who taught me how to see. She did not take every pixel in my visual field to give it meaning at so many levels. That would have been simply impossible. It was my brain who taught me how to see, after learning, mostly in a natural and unsupervised way from many experiences I had in the world. It is my brain that takes every pixel that I perceive and gives it meaning and value. It is my brain that makes it all real for me, copying all these pixels and arranging them on different higher or lower, simultaneous space and time layers of seeing. All those layers, present and past, find consensus into a coherent story. It is in that story that I give a meaning to here and now, it is in that space and time world that I see.

From this fundamental point of view, understanding unsupervised visual learning will help us understand that there is so much more to learn about the world and so much more to see in order to better take care of our world and improve our lives. Unsupervised visual learning in the natural world is, for the same reasons, also fundamental for understanding intelligence, how the mind works, and how we can build truly intelligent systems that will learn to see as humans do and then learn to be in harmony with what we are.

Unsupervised learning is one of the most intriguing problems that a researcher in artificial intelligence might ever address. How is it possible to learn about the world, with all its properties and so many different types of objects interacting in simple or very complex ways? And, on top of that, how is it possible to learn all this without access to the truth? Is that even possible? We do not have an answer to that yet, but what we do know is that children can learn about the world with minimal interventions from their parents. And when parents or teachers do intervene in our educations, they do not explicitly tell us everything. They definitely do not start marking every pixel in our visual field with a semantic label. From the first months of our lives, our brain starts learning, by itself, to arrange and group pixels into regions, to which we also begin to give meanings as we continuously gain experience. At least, at the very beginning, our experiences do not involve complex physical interactions with the world and our visual learning is mostly based on observation.

While interacting with the world is crucial for learning, in this book we are mainly interested in what we can learn *by watching* the world over space and time, as it reveals itself in front of our eyes. While we briefly touch the topic of interaction in final chapter and discuss how we could take actions and then learn from the outcome, we believe that we should first focus on the more limiting case of learning from space-time data on which we do not have influence over to better understand what are the fundamental limits of unsupervised learning. What assumptions should we make? What type of data do we need to have access to and how much of it is required? What type of classes and concepts can we learn about? What types of computational models could solve unsupervised learning?

As we show throughout the book, these essential questions can reveal universal answers, which could become practical tools for making possible many real-world applications in computer vision and robotics. In the final part of the book, we will adventure ourselves more into the world of imagination and dare to envision a universal computer vision system that could learn in space and time by itself. From the beginning, we will establish a set of general principles for unsupervised learning, which we will demonstrate chapter by chapter with specific tasks, algorithms, and extensive experimental evaluations. At the end, we will show how one could use these basic principles to build a general system that learns by itself multiple layers of interpretation of the space-time world within a single, unified Visual Story Network.

By the end of the book, we will better understand that unsupervised learning is ultimately about learning in the natural world and it cannot happen by itself without input from the vast ocean of data, which obeys physical and empirical statistical laws. Unsupervised learning is not just about fast and efficient algorithms or the architecture of a certain computational model, it is also very much about the world in which that model operates. At the end, we have to reach the level of learning in which we interact with the outside world, of which the learning system itself is part of. From this perspective, unsupervised learning in computer vision becomes a wonderful chance to learn about learning and about ourselves and how we came to discover and "see" the world around us.

1.3 Visual Learning in Space and Time

Vision is so rich that a picture can tell a thousand words. Vision is also our first window into the world and our most important sense. Vision happens in space and time, creating everything we see, from an object at rest or in motion to an activity that takes place in the scene and the whole story that puts all actors and their intricate interactions together. Vision almost has it all, from the simplest pixels of color and common physical objects to the wildest and most profound imagination, in an attempt to create and reflect the world in which we live in. As a consequence, vision must take into consideration the physical and empirical statistical laws, which give the natural world coherence and consistency in space and time. Therefore, it has to build upon certain grouping properties and statistical principles, which reflect such consistency. We must understand these principles and use them in building computational models of learning if we want to have a real chance to learn, in an unsupervised way, in the wild.

There is a certain advantage in thinking in space and time. Everything that exists and happens around us is in both space and time. There is nothing truly static. Today, however, computer vision research is largely dominated by methods that focus on single image processing and recognition. There are very few approaches that start directly with videos in mind. That is due more to historical reasons: it is less expensive to process single images and humans show that recognition in a single image is possible. So, if humans do it and it is less expensive, why not make programs do the same thing? We argue that single-image tasks make more sense in the supervised setting. If we want to learn unsupervised, then we should better consider how things are in the real world: objects and higher level entities, actions and interactions, complex activities and full stories, all exist in both space and time, and the two are deeply linked. Every object changes a little bit, in appearance or position, from one moment to the next. In every single place, there is an element of change, a vibration in both time and space, which we should learn from. Objects usually move differently than their surroundings. They also look different than their background. Thus, changes co-exist in both temporal and spatial dimensions from the start. At the same time, objects are consistent and coherent in both space and time. Their movements usually vary smoothly and their interactions follow certain patterns. Their appearance also varies smoothly and follows certain geometric and color patterns of symmetry. Therefore, considering videos as input at the beginning of our journey seems to be a must.

All these properties of the physical world could be taken advantage of only if we consider the space-time reality. We should definitely take advantage of every piece of information and property that universally applies in the natural world if we want to solve in a fundamental and general way the hard problem of unsupervised learning.

There is a second, practical advantage in considering videos at input and approaching space and time, together, from the beginning. There is a tremendous amount of unlabeled video data freely available and being able to label it automatically would give any solution a huge advantage over the strictly supervised setting that requires very expensive manual annotation.

Thus, the ability to perform unsupervised learning is extremely valuable for both research and industry. Moreover, the increased amount of available video data, as compared to single images, has the potential to greatly improve generalization and would allow learning of objects and their interactions together, from the beginning. Object classes are often defined by their actions and roles they play in the larger story. There should definitely be an agreement between the properties of an object at its local level and its role played in the global spatiotemporal context and we could only take advantage of such agreements if we consider the spatiotemporal domain from the start.

1.3.1 Current Trends in Unsupervised Learning

The interest in unsupervised learning is steadily increasing in the machine learning, computer vision, and robotics research. Classical works are based on the observation that real-world data naturally groups into certain classes based on certain core, innate properties, related to color, texture, form, or shape. Thus, elements that are similar based on such properties should belong to the same group or cluster, while those that are dissimilar should be put in different clusters. Consequently, the very vast research field of clustering in machine learning was born, with a plethora of algorithms being proposed during the last fifty years Gan et al. [1], which could be grouped into several main classes: (1) methods related to K-means algorithm Lloyd [2] and Expectation-Maximization (EM) Dempster et al. [3], which have an explicit probabilistic formulation and attempt to maximize the data likelihood conditioned on the class assignments; (2) methods that directly optimize the density of clusters, such as the Mean Shift algorithm Comaniciu and Meer [4], Fukunaga and Hostetler [5] and Density-Based Spatial Clustering (DBSCAN) Ester et al. [6]; (3) hierarchical approaches that form clusters from smaller sub-clusters in a greedy agglomerative fashion Day and Edelsbrunner [7], Ward Jr [8], Sibson [9], Johnson [10] or divisive clustering methods (DIANA) Kaufman and Rousseeuw [11], which start from a large cluster and iteratively divide the larger clusters into smaller ones; (4) spectral clustering algorithms, which are based on the eigenvectors and eigenvalues of the adjacency matrix or the Laplacian of the graph associated with the data points Cheeger [12], Donath and Hoffman [13], Meila and Shi [14], Shi and Malik [15], Ng et al. [16]. The clustering algorithms discussed in the present book and applied to different computer vision problems are mostly related to the class of spectral clustering methods.

Until not so long ago, most unsupervised learning research was focusing on proposing and studying various kinds of clustering algorithms Duda et al. [17] and for a good reason. Most unsupervised learning tasks require, implicitly or explicitly, some sort of clustering. We all researchers in machine learning would hope, even without saying it, that the full structure of the world, with its entities moving, relating, and acting in different ways and being grouped into specific classes, should emerge naturally in a pure unsupervised learning setup. The discovery of such structure with well-formed entities and relations immediately implies some sort of data clustering. Also, the insightful reader will surely observe throughout the book, that the methods proposed here are also based, at their core, on clustering principles.

Current research in unsupervised learning is much more versatile, diversified, and complex than the more general clustering approaches from 20 years ago. However, unsupervised learning is still in its infancy with many pieces missing in the still mysterious unsolved jigsaw puzzle. There are still a lot of unanswered questions and some other questions that have not been even asked yet. At this point, we begin to realize that the space-time domain offers a great advantage over the single-image case, as the temporal dimension brings an important piece of information when it comes to clustering. Objects differ from each other not only in the way they look but also in the way they move. Things that belong together look alike but also move together, whereas those which do not, separate in time and space. The time dimension, which enforces additional consistency and coherence of the world structure, suddenly becomes a crucial player in the unsupervised learning puzzle.

Therefore, there it comes as no surprise the fact that the initial modern techniques specific to computer vision for unsupervised learning were dedicated to the spatiotemporal, video domain. For example, in a classic pioneering paper Sivic and Zisserman [18], authors propose an algorithm for retrieving objects in videos which is based on discovering a given object in video based on matching keypoints that are stable with respect to appearance and geometry between subsequent frames. Then, such stable clusters are associated with individual physical objects. While the paper is not specifically dedicated to unsupervised learning and clustering, it is in fact heavily relying on it for the task of object retrieval from videos. In our earlier work on object discovery in videos Leordeanu et al. [19], we took a similar approach and discovered objects as clusters of keypoints, matched between video frames, that are geometrically stable for a sufficient amount of time. We noticed an interesting fact in our experiments: when a group of keypoints is geometrically stable for a specific amount of time, the probability that they indeed belong to a single object increases, suddenly from almost 0 to almost 1—this indicates, again, that time can provide very strong cues for what should and what should not belong together in the unsupervised learning game.

Since the first methods that discover objects in videos in an unsupervised manner, other researchers have started to look into that research direction as well Kwak et al. [20], Liu and Chen [21], Wang et al. [22]. The task of object discovery in video is gaining momentum and nowadays, most approaches are formulated in the context of deep learning. There seem to be several directions of research related to learning about objects from video in an unsupervised fashion. One direction crystallizes around the task of discovering object masks from videos. There are already several popular benchmarks for video object segmentation (e.g., DAVIS dataset [23]) with methods that vary from the fully unsupervised case [24–28] to models that are still unsupervised with respect to the given dataset but are heavily pretrained on other datasets [29–42] or having access to the human annotation for the first video frame [23]. While the case when a method is allowed to use powerful pretrained features on different datasets in order to learn on a new dataset is very interesting and has an important insight to give in the future of unsupervised learning, the case itself is definitely not unsupervised. However, once we have powerful pretrained features it really does not matter whether they have been trained in a supervised or unsupervised manner. We should therefore consider this situation, when we are allowed to use pretrained features as a very important one, since in practice it is always the case that we already have a huge number of pretrained models and features available and we should find the best and most efficient way in order to use them, within an unsupervised learning scheme, on novel tasks and datasets, which keep growing in number, diversity, and size.

Another direction of research on the task of learning about objects from video is that of discovering specific keypoints and features that belong to single objects, along with modeling the dynamics of such object keypoints or object parts

[19, 43–45]. On a complementary side, we also have a limited number of papers that address the problem of unsupervised learning for matching such keypoints of object parts, while taking into consideration both the local appearance of parts and their geometric relationships [46–51]. The core general idea is to optimize the model parameters such that matching process will find the most consistent group of assignments between keypoints or dense object regions, which yield the strongest consistency (or cluster) in terms of both local appearance and topology or geometric relationships.

So far, the approaches discussed have been limited to discovering objects as they are seen in images or videos, without taking into consideration the consistent spatial structure of the 3D world. We should keep in mind that it is precisely this stable structure that yields consistent video, depth, or RGB-D sequences. Once we discover the keypoints or regions that belong to a certain object, we could leverage the geometric and camera constraints, even if only implicitly, in order to improve the object discovery in the image and also infer its 3D structure in the world [52–54]. Moreover, once we make the connection between the image of the world and the 3D world itself, we could also take into consideration the static 3D world, the moving objects as well as the camera motion. In fact, we could consider them simultaneously and make them constrain each other during learning, such that a system that predicts motion could be constrained and "taught" by the one that predicts depth and vice versa, alternatively. We then reach a new level in unsupervised learning, which is currently receiving a growing attention, in which complex systems composed of complementary and collaborative pathways are put together to reinforce and also to constrain each other. Thus, we can learn in an unsupervised way, simultaneously, to predict the depth, the camera motion, and its intrinsic parameters from simple RGB video [55–62].

While the concept and art of combining multiple pathways into a global unsupervised learning system still has to be developed, it immediately leads to our novel concept of a universal unsupervised learning machine, the Visual Story Network (VSN), which we propose in the last chapter. This unsupervised learning machine would learn through self-supervision and consensus among many prediction pathways, such that a unified and coherent story is obtained about "everything" that it can sense, interact with, and predict. For more details regarding our novel VSN concept, we refer the reader to Chap. 8.

Before we discuss the Visual Story Network, we should also bring to the reader's attention a novel trend in unsupervised learning that is focusing on putting together multiple modalities and senses. Once we get the idea of using multiple sources of information as self-supervisory signal and observe that the more such sources we have the better, we immediately want to cross the barrier of vision and include touch, auditory, equilibrium, temperature, smell, or any other type of sense into the unsupervised learning equation. This research direction of unsupervised learning by cross-modal prediction, while it is not new [63] it is currently generating an increasing interest [64–69] and it directly relates to more general principles, which we aim to lay down and substantiate in this book. The more diverse and independent types of input and predictions the better, because the harder it will be to find consistency and consensus, but also the more reliable and robust the final results will be when that will happen. We could start imagining how, in fact, the unsupervised learning problem could begin solving by itself as we keep adding information and constraints into the game. However, we should keep in mind that we might not be able to learn everything from the start and begin by learning first simpler and fewer tasks, with a limited set of data and predictive pathways. This relates to the idea of curriculum learning [70], which has been researched in machine learning in the past decade. At the same time, we should also expect that learning could take several generations of students and teachers, which become stronger from one iteration to the next as they explore a continuously growing world. However, at this point, we should not jump too far ahead and instead return to the more basic ideas and principles, which we will use to build a stronger case for an unsupervised learning system in the final chapters of the book.

First, we propose to go back to some of the earliest ideas ever proposed for unsupervised learning and grouping in humans. We should be prepared to go back and deep in time if we want to be able to see far ahead into the future.

1.3.2   Relation to Gestalt Psychology

Many of the current approaches in unsupervised learning are strongly related and some even inspired by the ideas laid down by the Gestalt school of psychology which was established in Austria and Germany at the very beginning of the twentieth century Koffka [71], Rock and Palmer [72]. The German word *gestalt* used in Gestalt psychology refers to configurations of elements or patterns. This school of thought introduced and studied the main idea that objects are wholes that emerge from parts and are something else than the simple sum of their parts. That is where the saying "the whole is greater than the sum of its parts" comes from. Therefore, the Gestalt cognitive scientists proposed and studied several "laws of grouping," which the brain uses to form such "whole" entities. Such grouping laws include, for example, the *law of proximity* which states that elements that are nearby are likely to belong together, or the *law of similarity* that states that similar things should also be grouped together. In the same way, elements that display symmetry, geometric continuity, or a common fate (similar motions) are also likely to belong together. Besides these principles of combining smaller things into greater ones, Gestalt psychology studied the way we consciously see things as whole entities and interpret them in relation to past experiences.

Below we will present our own key observations regarding unsupervised learning, which we group into a set of principles, which we expect to be true only in a statistical sense, not necessarily all the time. Our principles are strongly related to the initial Gestalt laws and in some sense they could be seen as a modern re-interpretation of those laws in the context of modern machine learning and computer vision. While the Gestalt principles talk mostly about the initial stages of grouping, we go further and present principles from a computational point of view in order to eventually build artificial systems that learn to see by themselves.

1.4 Principles of Unsupervised Learning

Based on our motivations and intuitions presented above, it is now useful in our theoretical and experimental journey to first lay down succinctly the core principles by which we guide our work. As we show throughout the book, the results obtained and the computational models designed are based on such elementary ideas that aim to capture the essence of our intuition and general approach to unsupervised learning. While these *unsupervised learning principles* are not described in precise mathematical terms, they are sufficiently clear and intuitive. Then, in the next chapters, we will transform these common sense rules into specific mathematical formulations and algorithms. For now, let us express them in concise natural language. The principles enumerated below are strongly related. We could condense them into a smaller set of essential ideas, but we let them evolve gradually one from the other in order to better relate to each book chapter and follow more naturally the unsupervised learning process.

Principle 1
Objects are local outliers in the global scene, of small size and with a different appearance and movement than their larger background.

It makes sense that objects look generally smaller than their scene, which contains them. More often than not, they have a different appearance, in contrast with their immediate background. Let us imagine a flower on a green field, with its standout colors, or a red fruit in a tree, waiting to be picked up. Even the big Sun in the sky has a very strong yellow light, in contrast with the even larger blue sky. It is true that the perceived color of objects, as they are seen, is a product of our visual system. However, it is no accident that the visual part of the mind created perceptions of colors that put these objects in contrast with their surroundings. It can discover them, learn about them, and then recognize them easier. There are, of course, exceptions to this statistical observation. The famous chameleon with his amazing ability to camouflage is just one such case. But if all objects were always camouflaged it would be very hard for intelligent vision and the capacity for object recognition to develop.

Objects also have a different movement relative to the scene. They are usually more agile, moving in more complex and faster ways than their more static background. In fact, in the case of living things, the smaller they are the faster and more complex their behavior is. Smaller birds change direction faster than larger ones. It is not by accident that biological visual systems are endowed with the capacity to detect motion. The estimation of optical flow, that is, the apparent motion of pixels in an image, is a fundamental problem in computer vision and also an essential capability of any robotic vision system. For virtually every task in which motion is involved the usage of optical flow is often a must.

Principle 2
It is often possible to pick with high precision (not necessarily high recall) data samples that belong to a single object or category.

Indeed, many times we can spot parts of the scene or a stream of data that belong to a single object or entity. While the object category or entity type might be still unknown, we could be almost sure that the samples (which could be pixels, patches, entire regions, or other groups of data samples) do belong to the same class, based on certain grouping properties (which we will discuss in more detail in the book) that make them unlikely to be otherwise.

Let us think of the case when studying a new kind of animal, which we see for the first time. When we learn about the new properties of the animal in order to understand its unique features, we first apply the knowledge that we already know. We might first break the image of the animal into pieces that we can recognize. We might recognize where the legs are, its fur, head, and other parts of the body. By putting them all back together at the end, we might learn for the first time its unique overall look. Similarly, by decomposing its pattern of behavior into recognizable pieces, we might eventually manage to learn its unique behavior pattern. Based on its appearance, physical features and behavior we might later decide, if possible, in which larger class it belongs to. While at first we know nothing about this new animal, by recognizing several smaller pieces that were all together at the same time it triggers the idea that in that region of space there is an animal—a single animal. Let us just think about it: what is the chance of seeing together legs, body, and a head, very close in space and seemingly grouped together by some fur, when they all come from different animals or objects? We can be sure that all those parts make a single whole, while their unique appearance and arrangement could make us realize that we might see a new kind of animal for the first time.

In the book, we will see many cases, when, as in the example above, the consistent co-occurrence of several known classes in the same region of space and time is indication of the presence of a new class and can be reliably used, in an unsupervised way, as a positive case for training a new classifier.

Principle 3
Objects display symmetric, coherent, and consistent properties in space and time. Grouping cues based on appearance, motion, and behavior can be used to collect positive object samples with high precision. Such cues, which are very likely to belong to a single object or category, are called Highly Probable Positive (HPP) features.

The grouping cues that our human perception uses to form objects in the visual field are in fact Highly Probable Positive (HPP) features—they bring together, with high probability, parts of the scene that belong together. Pixels in those parts of the scene have similar color distribution, similar textures, and symmetric shapes. It is very unlikely that such agreements in color, shape, and appearance all happen accidentally. It is also very unlikely that such a group of similarly looking pixels start moving together smoothly, changing directions at the same time and clearly standing out against the background scene. Grouping in space and time might not give us everything, every single piece of information about an object, but it could give us enough and with high-precision (not necessarily high recall) positive training cases to start the learning process when background negative training cases are usually very easy to collect.

> **Principle 4**
> Objects form strong clusters of motion trajectories and appearance patterns in their space-time neighborhood.

This principle follows immediately from the previous ones. It also makes a lot of intuitive, common sense. An object should be something consistent and coherent in space and time. If we add a fourth temporal dimension to space, then the physical parts of an object through this 4D world would be very tightly connected. Consequently, the same tight connection should be expected in the sequence of images, in the 2D (image space) + 1D (time) world of video frames—since each frame is a projection of what is in the 3D world. We should be able to describe objects as clusters of points in space and time, strongly connected through similar trajectories and appearance features. It is clear that these clusters should be seen relative to a given neighborhood in scale and time. For example, a car forms such a strong cluster. But so does the whole Earth with all its people, animals, and objects, but at a much larger scale. Earth is an object too in the Solar system.

> **Principle 5**
> Accidental alignments are rare. When they happen they usually indicate correct alignments between a model and an image. Alignments, which could be geometric or appearance based, rare as they are, when they take place form a strong cluster of agreements that re-enforce each other in multiple ways.

This principle is at the core of unsupervised learning. By alignments, we refer to complex symmetries, fine geometric alignments of shapes, co-occurrence of many independent classifier outputs or agreements in patterns of motion, and appearance in a given local space-time neighborhood. Such alignments may happen accidentally, but they do not happen consistently. They are rare events in space-time, unless there is an entity there which causes them. That is why such alignments are HPP features which should be used for starting to learn about new objects and other visual categories.

Principle 6
When several weak independent classifiers consistently fire together, they signal the presence of a higher level class for which a stronger classifier can be learned.

Strongly related to the previous principle, it is very unlikely for many independent classifiers to fire together unless they fire at the same thing, potentially unknown class at a higher level of semantics or abstraction. For example, different classifiers could fire at different parts of an object, without existing a single classifier that can "see" the whole object. One can fire at its legs, another at its trunk, yet another at its head or fur, but none at the whole animal. When that happens, a stronger classifier could be learned to see the whole animal, by exploiting the co-firing of the group of classifiers.

Principle 7
To improve or learn new classes, we need to increase the quantity and level of difficulty of the training data as well as the power and diversity of the classifiers. In this way, we could learn in an unsupervised way over several generations, by using the agreements between the existing classifiers as a teacher supervisory signal for the new ones.

New classes could be concepts at higher semantic levels than the classes we already know. They could also be just different categories that we did not know about. Either way, in order to learn about a new class we can only use the classifiers that we already have so we will rely on their consistent (and unlikely) co-firing in order to spot space-time regions where we might catch the presence of a new object or class, that is, up to this point unseen. That is because we did not learn yet how to "see" it. Learning a new class requires the co-firing of the existing classifiers but it also requires the sufficient amount of data that contains the presence of that new class. It is therefore mandatory that new and more challenging data (containing difficult unseen cases or new classes) is added. Also, as the incoming unlabeled training data becomes more challenging it is important to match that with an increased level of classifier power and diversity. Thus, we could learn over several generations, with the old classifiers co-firing together providing supervision to the new but more powerfully equipped ones. Generation by generation the society of classifiers can evolve, from recognizing only very simple and local objects to develop an almost complete, round, and coherent sense of the surrounding visual world—to become a universal unsupervised learning system, such as the Visual Story Network concept which we discuss in the last chapter.

1.4.1 Object Versus Context

It is time now to exemplify how to use the first principle of unsupervised learning presented in Sect. 1.4. We succinctly present a relatively simple method for rapidly discovering the object masks in videos, which is detailed further in Chap. 4. The first principle for unsupervised learning states the following:

Principle 1
Objects are local outliers in the global scene, of small size and with a different appearance and movement than their larger background.

The method we are about to present is based on this observation that objects are usually smaller than the scene that contains them, and they have different appearance and movement patterns than the surrounding scene. We could regard them as small elements of change, outliers, separated, and in complementary contrast with the rest of the scene.

The frames in a video sequence contain a lot of redundant information. It is clear that the frames live in a much smaller subspace than their size (the number of pixels in a frame). We could also expect that the subspace captures mainly information from the background scene, since that covers the largest part and dominates the video. Consequently, if we model the video with Principal Component Analysis, such that each frame is a data point, the first principal components should model this large background. Objects are small and change more from one frame to the next so they could be seen mostly as noise with respect to the background and are expected to be represented by the last, higher frequency eigenvectors.

Our simple algorithm, which is termed VideoPCA, was first published in [73], exploits this observation, and, based on Principal Component Analysis applied to the video frames, rapidly estimates the frame pixels that are more likely to belong to the foreground object.[1]

According to the video PCA model, the background is represented by the image reconstructed in the reduced subspace. Let the principal components be $\mathbf{u}_i$, $i \in [1 \ldots n_u]$ (we used $n_u = 8$) and the reconstructed frame $\mathbf{f}$ be $\mathbf{f}_r \approx \mathbf{f}_0 + \sum_{i=1}^{n_u}((\mathbf{f} - \mathbf{f}_0)^\top \mathbf{u}_i)\mathbf{u}_i$. We obtained the error image $f_{diff} = |\mathbf{f} - \mathbf{f}_r|$. As expected, we notice that the difference image enhances the pixels belonging to the foreground object or to occlusions caused by the movement of this object (Fig. 1.1). By smoothing these regions with a large enough Gaussian and then thresholding, we obtain masks whose pixels tend to belong to objects rather than to background.

In the next section, we develop the idea further and present a more mature and also effective method for obtaining the full object masks starting from the pixels that belong to the object with high probability and are discovered using VideoPCA.

---

[1]Code available at: https://sites.google.com/site/multipleframesmatching/.

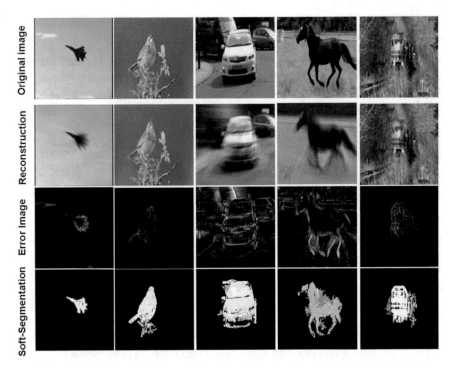

Fig. 1.1 First row: original images. Second row: reconstructed images with PCA and the first 8 principal components chosen. Third row: error image between the original and the reconstructed. Note that the error images enhance pixels that belong to the foreground object. Fourth row: foreground segmentation as the posterior pixelwise probability of foreground computed from the empirical color distributions of foreground and background estimated from the error images in the third row

1.4.2 Learning with Highly Probable Positive Features

In the previous Section, we showed a relatively simple but effective method for discovering pixels that belong to the foreground object in a given video sequence, which is based on the observation (given by the first principle) that objects are in contrast to their background and could be regarded as outliers relative to the background model. In this section, we further explore such general grouping cues to discover objects and present a method that takes into consideration the second principle as well.

Principle 2
It is often possible to pick with high-precision (not necessarily high recall) data samples that belong to a single object or category.

We refer to samples or features that are highly likely to be positive (associated with the class of interest that we would like to discover) as highly probable positive features. These features or samples do not need to cover the whole set of positives. We should keep in mind that we need to discover them in an unsupervised manner, so it would be unreasonable to think that we can discover all of them with high precision. If that was the case, the unsupervised task would be solved and completely fall under the supervised scenario where all negatives and positives are known. What we claim here is that we can discover with high precision at least a small percentage of the positives (therefore the recall could be low) and that these positives constitute the basis for learning our first classifiers and kick start the unsupervised learning process.

> **Highly Probable Positive (HPP) features**: HPP features constitute data or feature samples that are very likely to belong to the positive class, which is the class of interest that we want to learn. Our proposed principles of unsupervised learning state that such samples are often easy to detect with high precision, but not necessarily high recall. The high recall is ultimately achieved through a repeated learning process of discovery over several generations of classifiers.

Next, we present briefly an effective approach for using HPP samples in order to discover the object masks. We first present a method that is a simplified version of a more general algorithm that will follow. The complete approach will be discussed in detail in Chap. 5.

In Fig. 1.2, we show a toy example of an object for which we assume a small, but incorrect bounding box. This bounding box represents the set of considered positive examples, which are highly likely to be correctly labeled as positive. Notice that if we know a single point on the object then a small box around it probably contains positive samples with a high probability. We will then use the pixels inside the box as positive pixels and the ones outside of it as negative ones and build color models for the foreground and background. Without getting into the mathematical details, the basic idea is simple. Once we learn these color models based on the assumed box as if it was the true object mask, we could use these models as if they were the true ones and obtain essentially the same result. This is in fact true mathematically if a few assumptions hold—assumptions that make intuitive sense and are often met in practice. We provide a full proof of this claim in Chap. 5.

Below we state the main assumptions considered by our simple object segmentation using color based on learning with HPP features (Fig. 1.2). The assumptions refer to both the specific case of using pixelwise color and the more general case of unsupervised learning with HPP samples:

1. The first assumption states that the object is generally smaller than the background scene in the image. In the general case, this translates into the fact that the set of true positives is a small percentage of the whole set of samples.

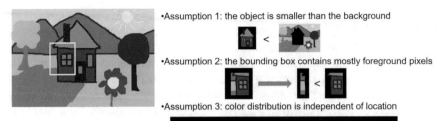

Fig. 1.2 A visual representation of the assumptions behind our simple segmentation algorithm. These assumptions could also be applied in the general case, where instead of colors we could consider any visual property or more abstract feature and instead of the selected bounding box we would have a set of considered HPP samples

2. The second assumption is particular to the bounding box: We require the bounding box to contain mostly foreground pixels, without necessarily being the true box. This case could be often achieved once we know a single point inside the object region and then take a small bounding box around it. In the general case, this means that the set of considered HPP samples contains mostly positive samples (high precision). It is often possible in practice to select a small set of (mostly) uncontaminated positive samples.

3. The third and strongest assumption is that if a particular color is more often found on the object than in the background, then this property is true everywhere in the image, independent of location. This idea intuitively makes sense and could be used effectively in practice, even though it is only an approximation. It is based on the observation that very often objects have particular, distinctive color patterns with respect to their background scene, and that the patterns display some symmetry—the distribution of colors is often stable across the object surface. For the general case, we could replace colors with any other appearance property or higher level feature. This third assumption requires that the small set of HPP positives is representative for the whole set: that is, if a specific property is more often found in the positive set, it is also more often found in the considered HPP set.

In Fig. 1.3, we show how to apply these assumptions and obtain an efficient method for foreground object segmentation. For a given point selected on the object of interest, different bounding boxes of different sizes (scales) are selected. Note that the true object scale is not known so we take a multi-scale approach and combine the results. The final mask, shown on the right, is computed as the average over the masks computed for different scales. Note how robust the masks are when the reference object point is taken at different locations. The masks seem to be stable across scales as well, indicating that our assumptions are statistically valid in this case.

This relatively simple idea reminds us of the next principle proposed for unsupervised learning which states the following:

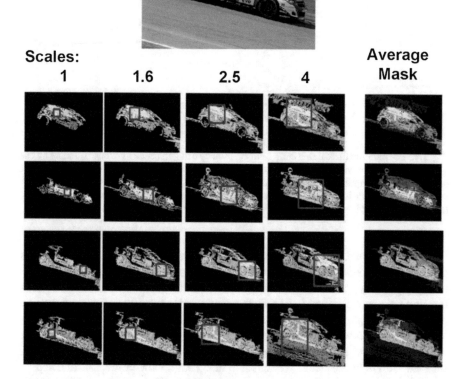

Fig. 1.3 A very simple foreground segmentation approach, using the empirical distributions of pixel colors for background and foreground. Distributions are computed by considering the pixels inside the red box as foreground and the remaining ones as background. Then the posterior probability of foreground offers an object mask that is much closer to be correct than the initial box. The final mask is the average over all masks from multiple scales. Notice how robust the foreground mask is relative to the box location and size (far from the true bounding box or mask)

Principle 3
Objects display symmetric, coherent, and consistent properties in space and time. Grouping cues based on appearance, motion, and behavior can be used to collect positive object samples with high precision. Such cues, which are very likely to belong to a single object or category, are called Highly Probable Positive (HPP) features.

Learning with Highly Probable Positive (HPP) Features

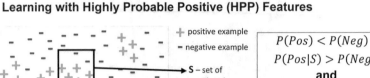

Fig. 1.4 The intuition for learning color segmentation by using a small and wrong bounding box as a positive training set is extended to the general case for learning with HPP features. We show that under certain assumptions, which generally hold in practice, we can use a set of positive training samples that has very low recall (most true positive training samples are left outside), as long as it has high precision (it contains mostly correct positive samples) reliably to train a classifier

The idea above relates directly to Gestalt principles and states that such properties could be used reliably as HPP features when learning in an unsupervised fashion. We have consistently used this observation when creating our models for unsupervised learning for many different tasks and observed that it generally holds in practice.

Our toy example can be immediately extended to the general case and the assumptions generalize as well, as discussed. All we need to do is to replace the idea of pixels with data samples and the bounding box with a set S, from which we use samples selected with high precision. We illustrate the general view in Fig. 1.4. In Chap. 5, we present in detail a method that generalizes the idea of learning from HPP features over several generations, starting from single pixels as shown already and moving to patches and then to whole images. The basic idea is the following: at the first iteration, we select positive pixels with high precision (using VideoPCA) and build a pixel-level classifier of foreground regions. At the second iteration, we select such foreground regions that more are probable to be correct and learn patch-level foreground object classifiers. Then, we obtain novel and more accurate object masks using these patch-based classifiers. Then, we combine the second iterations segmentations with motion cues that separate the foreground and the background even more to obtain even better foreground masks. At the very final stage, we train a deep neural network (U-Net [74]) for semantic segmentation on previous output (as supervisory ground truth), chosen based on the same HPP idea such that only the masks that have stronger average response and thus have high precision are given as training cases to the network. In Fig. 1.5, we show some representative visual results.

| original image | initial soft-seg | UNet soft-seg | original image | initial soft-seg | UNet soft-seg |

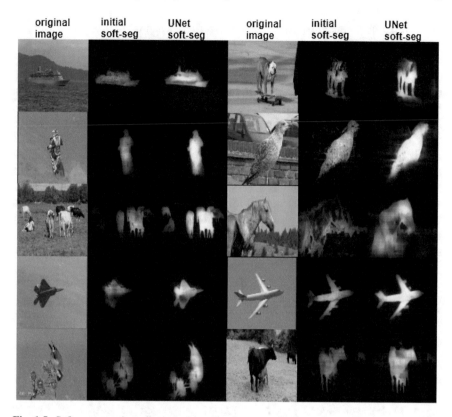

Fig. 1.5 Soft segmentations discovered in video in a completely unsupervised way. We present some qualitative results on representative video frames. Left column: input image. Middle column: masks discovered after several learning iterations with HPP. Right column: final masks produced by the U-Net, which is trained using as supervisory masks some of the output provided by the previous stage, automatically selected using an HPP strategy

1.5   Unsupervised Learning for Graph Matching

In this section, we start developing the optimization tools which we will use for efficiently solving the mathematical problems tackled in the following chapters. We start by introducing the problem of graph matching from a modern computer vision perspective and offer solutions to this NP-hard problem that are approximate but very efficient in practice. In vision graph, matching addresses the task of finding correspondence between two images or between two shapes or views of the same object in different images. The problem of finding correspondences is very general in computer vision and has a long list of applications and already a long history. Modern graph matching in computer vision was initiated in 2005 by our original spectral matching formulation, which showed that such a highly computationally expensive problem can be solved fast if we take advantage of the natural statistics of geometric alignments. It was then, for the first time, when we realized the rarity and power of geometric alignments. We understood that accidental alignments are rare and when they happen, it is due to a correct set of correspondences between a template object and the object in the image. Moreover, when geometric alignments take place, they usually align in multiple ways and form an abundance of geometric agreements that manifest themselves as a cluster of candidate correspondences which reinforce each other. This idea became the core of the spectral matching algorithm, which finds a matching solution as cluster reflected by the principal eigenvector of a special matrix of candidate assignments. The likely agreements between correct candidate correspondences versus the unlikely agreements between wrong ones constitute the statistical basis for our unsupervised learning method developed in 2008 Leordeanu et al. [47]. Below, we succinctly present the mathematical formulations and solutions of our graph matching optimization and learning approaches. In the next chapter, we expand the material and provide more in-depth analysis along with extensive experimental evaluation.

1.5.1 Graph Matching: Problem Formulation

We define the graph matching problem in its most general form, as follows: we want to match two attributed graphs $G^P = (V^P, E^P, A^P)$ and $G^Q = (V^Q, E^Q, A^Q)$. For each node $i \in V^P$, there is an associated feature vector $A_i^P \in A^P$. This feature usually describes the local appearance at node i. Similarly, for each node $a \in V^Q$, we have $A_a^Q \in A^Q$. For each edge $(i, j) \in E^P$, we have an associated vector $A_{ij}^P \in A^P$, usually describing the pairwise geometric relationship between nodes i and j. Similarly, we have $A_{ab}^Q \in A^Q$ for each edge $(a, b) \in E^Q$. For each pair of edges $(i, j) \in E^P$ and $(a, b) \in E^Q$, there is a pairwise score function $M_{ia;jb} = f(A_i^P, A_{ij}^P, A_a^Q, A_{ab}^Q)$ that measures the agreement of the first-order local features (described by A_i^P and A_a^Q) and second-order relationships (defined by A_{ij}^P and A_{ab}^Q) between the pair of candidate correspondences (i, a) and (j, b). We can similarly define unary-only score functions $M_{ia;ia} = f(A_i^P, A_a^Q)$, which, in matrix form, could be stored on the diagonal of $\mathbf{M}$. Then, the graph matching problem can be formulated as an integer quadratic program (IQP) and consists of finding the indicator vector $\mathbf{x}^*$ that respects certain mapping constraints (such as one-to-one or many-to-one) and maximizes the quadratic score function $S(\mathbf{x})$ (Fig. 1.9a). This function could also be interpreted as an intra-cluster score, which captures how well the group of candidate assignments that are considered to be correct (by the solution indicator vector $\mathbf{x}^*$) agree with each other in terms of local appearance and pairwise geometric or other types of second-order relationships:

$$\mathbf{x}^* = \arg\max_{\mathbf{x}} S(\mathbf{x}) = \arg\max_{\mathbf{x}} (\mathbf{x}^T \mathbf{M} \mathbf{x}) \text{ s. t. } \mathbf{A}\mathbf{x} = \mathbf{1}, \ \mathbf{x} \in \{0, 1\}^n, \tag{1.1}$$

given the one-to-one/many-to-one constraints $\mathbf{Ax} = \mathbf{1}$, $\mathbf{x} \in \{0, 1\}^n$, which require that $\mathbf{x}$ is an indicator vector such that $x_{ia} = 1$ if feature i from one image is matched to feature a from the other image and zero otherwise. Usually, one-to-one constraints are imposed on $\mathbf{x}$ such that one feature from one image can be matched to at most one other feature from the other image. The quadratic formulation is a recent, generalized definition of graph matching in computer vision that can also accommodate the earlier, classical formulations, such as *exact, inexact, and weighted graph matching*, which focused mainly on finding the mapping between the nodes of two graphs such that the edges are preserved as well as possible.

Equation 1.1 incorporates, in a general formulation, most cases of graph matching. In this book, we discuss both how to design and learn, with varying degree of supervision powerful second-order terms $M_{ia;jb}$, useful for different computer vision applications, and how to approximately optimize the matching objective function efficiently, since finding the true optimum of Eq. 1.1 is NP-hard. We will show that one of the principles introduced in this book, which states that in nature "accidental alignments are rare," is also the key element which makes unsupervised learning for graph matching possible.

1.5.2 Spectral Graph Matching

Principle 5
Accidental alignments are rare. When they happen they usually indicate correct alignments between a model and an image. Alignments, which could be geometric or appearance based, rare as they are, when they take place form a strong cluster of agreements that re-enforce each other in multiple ways.

Here, we briefly present our spectral matching algorithm, published in 2005 Leordeanu and Hebert [75] designed to give an approximate solution to the graph matching problem. This algorithm can be used for a wide variety of applications because it handles the graph matching problem in its most general formulation, as an Integer Quadratic Program, and it does not impose a specific mathematical formulation on the unary and second-order terms. The only constraint we require is that the unary and pairwise scores should be non-negative and they should increase with the quality of the match, that is, decrease with the deformation errors between the candidate correspondences. These scores can capture any type of deformations/changes/errors at the levels of both appearance and geometric relationships. Our key insight into the graph matching problem, as applied to computer vision, is that the correct assignments will have strong, large valued second-order scores between them, while such large valued scores between incorrect ones are unlikely because accidental geometric alignments are very rare events.

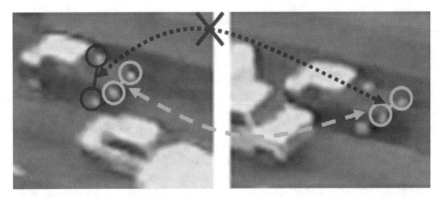

Fig. 1.6 Pairs of wrong correspondences (red) are unlikely to preserve geometry, thus having a low pairwise score. Pairs or correct assignments will preserve geometry and have a high pairwise score. Second-order geometric relationships are often more powerful than unary terms based on local appearance. In this example, correct matching is not possible by considering only local information. The pairwise geometry is preserved only by the correct matches, thus using such type of second-order geometric information is encouraged for robust matching

These probabilistic properties of the second-order scores give the match matrix M, containing the second-order terms/scores, a very specific structure: a strong block with large values formed by the correct assignments, and mostly zero values everywhere else. This allows us to drop the integer constraints on the solution during the optimization step and impose them only after, as a binarization procedure applied to the leading eigenvector of this matrix M. In Fig. 1.6, we present an example that illustrates the intuition behind our algorithm. The images of the red car have a very low resolution. The local appearance of each feature is therefore not discriminative enough for reliable matching between two consecutive frames. However, the pairwise geometry is well preserved between frames, so it is very likely that the pairs of assignments that preserve this geometry will be correct. This is just an illustrative example, but using pairwise geometric constraints for robust matching is suitable for a lot of computer vision applications.

We can think of M as the weighted adjacency matrix of the graph of candidate assignments. Each candidate assignment (or correspondence) (i, a) can be seen as a node in this graph, containing information regarding how well the local appearance between the candidate matches agrees. The links between the nodes can contain information regarding how well the pairwise geometric information between candidate assignments is preserved. Figure 1.7 shows a possible instance of such a graph of candidate assignments. Larger nodes correspond to stronger unary scores (better agreements at the level of local appearance) and thicker edges correspond to stronger pairwise scores reflecting stronger agreements at the second-order, geometric level. We expect that the correct assignments will form a strongly connected cluster, which can be found by analyzing the leading eigenvector of the weighted adjacency matrix M of the assignments graph. The elements of the eigenvector can be interpreted as confidences that each candidate assignment is correct, and the integer constraints can

Fig. 1.7 Correct
assignments (shown in
green) are likely to form a
strong cluster by establishing
stronger pairwise
second-order agreements
(many more thicker edges)
and preserving better the
local appearance of features
(larger nodes)

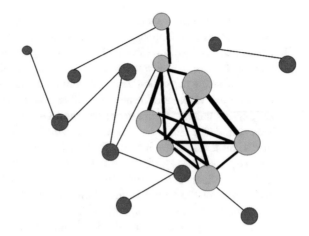

Fig. 1.8 The structure of the
matrix **M**: correct
assignments will form a
strong block in **M** with large
pairwise elements, while the
pairwise scores between
incorrect assignments will be
mostly zero. This statistical
property of **M** will be
reflected in its principal
eigenvector

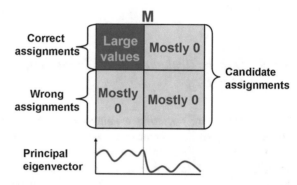

be applied to binarize the eigenvector and get an approximate solution to the graph matching problem.

In Fig. 1.8, we show the likely structure of the matrix **M**. The correct assignments will form a strong block in this matrix with large pairwise elements, while the pairwise scores between incorrect assignments will be mostly zero. This will be reflected in the leading eigenvector of **M**.

The **Spectral Matching** algorithm is summarized as follows:

1. **Create the graph of candidate assignments**: Build the matrix **M** of candidate assignments ia that puts in correspondence feature i from the model with feature a from the image, such that the off-diagonal elements $M_{ia;jb}$ capture the level of agreement between candidate assignments ia and jb: they effectively measure how well the geometry and appearance of model pair (i, j) match the geometry and appearance of image pair (a, b). The matrix should have non-negative elements, with values increasing with the quality of the match (agreement) between candidate assignments ia and jb. Thus, each candidate assignment ia becomes

a node in a graph of candidate assignments. The values $M_{ia;jb}$ on the edges (ia, jb) measure the agreement between the candidate assignments ia and jb.

2. **Efficient optimization**: Compute the leading eigenvector $\mathbf{v}$ of $\mathbf{M}$ using the power iteration algorithm. A few iterations (10–20) would suffice in practice. The eigenvector will have the same size as the number of candidate assignments and only non-negative values. Each value v_{ia} will represent the degree to which candidate assignment belongs to the main, strongest cluster in the graph of assignments. That confidence value is a measure of how likely the match ia is to be correct.

3. **Find a final discrete solution**: The final discrete solution is quickly obtained in a greedy fashion: (1) We choose as correct the candidate assignment ia^* with highest confidence v_{ia} value, which hasn't been chosen nor eliminated yet. (2) We then eliminate all remaining candidate assignments that are in conflict with ia^* (might have the same source or destination feature). Then we return to step 1. The process continues until all assignments are either picked as correct or eliminated.

1.5.3   Integer Projected Fixed Point Method for Graph Matching

After spectral graph matching, we introduced another efficient graph matching algorithm (IPFP) [76] that outperforms spectral matching and many other state-of-the-art methods. Since its publication, modified versions of IPFP have been applied to different computer vision tasks Brendel and Todorovic [77], Jain et al. [78], Semenovich [79], Monroy et al. [80]. IPFP can be used as a standalone algorithm, or as a discretization procedure for other graph matching algorithms, such as spectral matching. As we show in this book, we apply IPFP and the spectral approach to different problems that share very similar mathematical formulations, in Chaps. 3–6. IPFP can be used effectively, with minimal modification, to problems such as Maximum A Posteriori (MAP) inference in Markov Random Fields (MRFs) Leordeanu et al. [76], graph and hypergraph clustering Leordeanu and Sminchisescu [81] (and Chap. 3), unsupervised feature selection and classification Leordeanu et al. [82] (and Chap. 4), unsupervised descriptor learning ([83] and Chap. 5), and unsupervised object segmentation in video Haller et al. [84] (and Chap. 6).

IPFP solves efficiently a tighter relaxation of Problem 1.1, also known to be NP-hard:

$$\mathbf{x}^* = \arg \max_{\mathbf{x}} S(\mathbf{x}) = \arg \max_{\mathbf{x}} (\mathbf{x}^T \mathbf{M} \mathbf{x}) \quad \text{s.t. } \mathbf{A}\mathbf{x} = 1, \; x \geq 0. \qquad (1.2)$$

The only difference between Problems 1.1 and 1.2 is that in the latter the solution is allowed to take continuous values. IPFP takes as input any initial solution, soft or discrete, and rapidly computes a solution that satisfies the initial discrete constraints of Problem 1.1 with a score that is most often significantly higher than the initial one:

Algorithm 1.1 Integer Projected Fixed Point Method for Graph Matching

---

Initialize $\mathbf{x}^* = \mathbf{x_0}$, $S^* = \mathbf{x}_0^T \mathbf{M} \mathbf{x}_0$, $k = 0$, where $x_i \geq 0$ and $\mathbf{x} \neq \mathbf{0}$;
repeat
    1) Let $\mathbf{b}_{k+1} = P_d(\mathbf{M} \mathbf{x}_k)$, $C = \mathbf{x}_k^T \mathbf{M}(\mathbf{b}_{k+1} - \mathbf{x}_k)$, $D = (\mathbf{b}_{k+1} - \mathbf{x}_k)^T \mathbf{M}(\mathbf{b}_{k+1} - \mathbf{x}_k)$;
    2) If $D \geq 0$, set $\mathbf{x}_{k+1} = \mathbf{b}_{k+1}$. Else let $r = \min(-C/D, 1)$ and set $\mathbf{x}_{k+1} = \mathbf{x}_k + r(\mathbf{b}_{k+1} - \mathbf{x}_k)$;

    3) If $\mathbf{b}_{k+1}^T \mathbf{M} \mathbf{b}_{k+1} \geq S^*$ then set $S^* = \mathbf{b}_{k+1}^T \mathbf{M} \mathbf{b}_{k+1}$ and $\mathbf{x}^* = \mathbf{b}_{k+1}$;
    4) Set $k = k + 1$
until $\mathbf{x}_{k-1} = \mathbf{x}_k$
return $\mathbf{x}^*$

Here, the $P_d(.)$ in Step 1 denotes a projection onto the discrete domain. IPFP is a general method and can be applied to many different tasks, the main difference between them, in terms of the mathematical formulation, being the actual discrete domain on which we project with $P_d(.)$. This domain of valid discrete solutions is defined by matrix of constraints $\mathbf{A}$. In Chap. 2, we analyze in more detail the theoretical properties of IPFP along with its experimental validation on various feature matching tasks. While the algorithm as presented above might seem a bit more technical, it can be easily interpreted in terms of two main stages, which are repeated until convergence.

1. **Stage 1**: At step 1, $P_d(.)$ is a projection on the discrete original domain of valid solutions, as defined by $\mathbf{A}\mathbf{x} = 1$, $x \in \{0, 1\}^n$. While the power iteration used by spectral matching performs, at each iteration, a similar projection but on the unit sphere (by normalizing $\mathbf{x}$ to have unit norm) and optimizes in the continuous domain, IPFP aims to maximize the objective score in the desired discrete domain. The projection is meant to find the closest possible vector in the discrete domain in terms of the Euclidean distance. Since all possible discrete solutions have the same norm (we expect to have the same number of valid assignments), $P_d(.)$ boils down to finding the discrete vector $\mathbf{b}_{k+1} = \arg\max_{\mathbf{b}}(\mathbf{b}^T \mathbf{M} \mathbf{x}_k)$, which maximizes the dot product with $\mathbf{M} \mathbf{x}_k$. For one-to-one constraints, this can be efficiently accomplished using the Hungarian method; for many-to-one constraints, the projection is achieved in linear time. For the case of graph clustering, as shown in Sect. 1.6.1 and Chap. 3, the projection could also be rapidly performed in linear time.
2. **Stage 2**: The next steps could be seen as a single stage in which we perform a maximization of the score on the line between the current solution $\mathbf{x}_k$ and the discrete solution $\mathbf{b}_{k+1}$. Note that $\mathbf{x}_k$ could have soft values but it obeys the slightly relaxed constraints $\mathbf{A}\mathbf{x} = 1$, $x \geq 0$ of Problem 1.2, which are also satisfied by $\mathbf{b}_{k+1}$. Therefore, any point on the line between the two will also satisfy the constraints. The line maximization step can be optimally solved in closed form, since the line search boils down to optimizing a quadratic function in a single parameter r. The Algorithm 1.1 provides the exact equations for finding the optimal $\mathbf{x}_{k+1}$ on the line between $\mathbf{x}_k$ and $\mathbf{b}_{k+1}$. The method repeats until convergence, and the best discrete solution $\mathbf{x}^*$ found along the way is returned. The algorithm

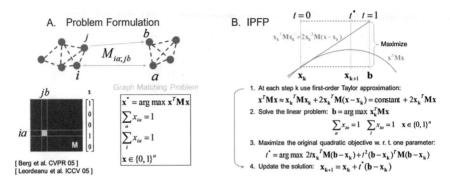

Fig. 1.9 **a** Visual representation of graph matching in matrix form, with IQP formulation. **b** IPFP algorithm steps based on first-order Taylor approximation of the original quadratic score

often converges to discrete solutions by itself as it passes mostly through or very near discrete solutions. It usually improves over the initial starting point by a significant margin.

Relation to Spectral Matching and Power Iteration Algorithm: IPFP is inspired by and strongly related to the power iteration method for eigenvectors, used by spectral matching. It replaces the projection on the unit sphere with a projection $P_d(.)$ on the discrete domain followed by line optimization solved in closed form. As the discrete domain could be any domain of valid solutions, as required by the specific task, IPFP is very general and can be applied to many vision tasks such as graph matching, MAP inference in graphical models as well as graph clustering. In Chaps. 2 and 3, we show how IPFP can be easily adapted to be efficiently applied to optimization tasks on hypergraphs, in which hyperedges connect together three or more nodes.

Relation to first-order Taylor approximation: Another intuition behind IPFP is that, at every iteration the quadratic score $\mathbf{x}^T\mathbf{M}\mathbf{x}$ is approximated by the first-order Taylor expansion around the current solution $\mathbf{x}_k$: $\mathbf{x}^T\mathbf{M}\mathbf{x} \approx \mathbf{x}_k^T\mathbf{M}\mathbf{x}_k + 2\mathbf{x}_k^T\mathbf{M}(\mathbf{x} - \mathbf{x}_k)$. The first-order approximation is then maximized within the discrete domain of Problem 1.1, in step 1, where $\mathbf{b}_{k+1}$ is found. Next, the maximization of the original quadratic score on the line between the current and the discrete solution of the first-order approximation is possible in closed form, since it boils down to maximizing a quadratic function of a single parameter. In Fig. 1.9b, we present IPFP as a sequence of first-order optimization steps in the original discrete domain, each followed by a quadratic line optimization step in the continuous domain.

Relation to Iterative Conditional Modes Algorithm: IPFP can also be seen as an extension to the popular Iterated Conditional Modes (ICM) algorithm [85], having the advantage of updating the solution for all nodes in parallel, while retaining the optimality and convergence properties. While parallelizing ICM as it is does not guarantee improving from one iteration to the next, IPFP by allowing the solutions to take soft values and move in between discrete solutions, it guarantees improvement and convergence by updating nodes in parallel. This gives it a clear computational advantage over the classic Iterative Conditional Modes algorithm.

Relation to Frank–Wolfe Algorithm: Last by not least, IPFP is also related to the Frank-Wolfe method (FW) Frank and Wolfe [86], a classical optimization algorithm from 1956 most often used in operations research. The Frank-Wolfe method is applied to convex programming problems with linear constraints and in form, at a high level, is similar to IPFP, with two main differences: FW algorithm optimizes convex problems not concave ones as IPFP does. For that matter, FW is slow at convergence (as convex functions become quadratic close to the optimum), while IPFP is very fast, but sub-optimal. FW performs a line search between the initial solution and the optimum of the first-order function approximation, whereas IPFP does it in closed form. IPFP is a method specifically designed for solving problems that require discrete solutions and takes advantage of that in several ways, depending on the specific tasks, some detailed in Chaps. 2 and 3. FW is very general and high level and is mostly used in the continuous domain of convex optimization problems.

1.5.4 Learning Graph Matching

Spectral Matching (SM) requires $\mathbf{M}$ to have non-negative elements but the Integer Projected Fixed Point (IPFP) method does not. The unary terms of candidate assignments can be stored on the diagonal of $\mathbf{M}$, but in practice it works better if we consider the local information directly on the off-diagonal, pairwise score $M_{ia;jb}$, and leaving zeros on the diagonal. $M_{ia;jb}$ is essentially a function that is defined by a certain parameter vector $\mathbf{w}$. The type of pairwise scores $M_{ia;jb}$ that we use is

$$M_{ia;jb} = \exp\left(-\mathbf{w}^T \mathbf{g}_{ia;jb}\right), \tag{1.3}$$

where $\mathbf{w}$ is a vector of learned parameter weights, and $\mathbf{g}_{ia;jb}$ is a vector that contains non-negative errors and deformations, to describe the changes in geometry and appearance when we match the pair of features (i, j) to the pair (a, b).

1.5.5 Supervised Learning for Graph Matching

We aim to learn the geometric and appearance parameters $\mathbf{w}$ that maximize the expected correlation between the principal eigenvector of $\mathbf{M}$ and the ground truth $\mathbf{t}$, which is empirically proportional to the sum of dot products between the two, over all training image pairs:

$$J(\mathbf{w}) = \sum_{i=1}^{N} \mathbf{v}^{(i)}(\mathbf{w})^{T}\mathbf{t}^{(i)}, \qquad (1.4)$$

here $\mathbf{t}^{(i)}$ is the ground truth indicator vector for the ith training image pair. We maximize $J(\mathbf{w})$ by gradient ascent:

$$w_j^{(k+1)} = w_j^{(k)} + \eta \sum_{i=1}^{N} \mathbf{t}_i^T \frac{\partial \mathbf{v}_i^{(k)}(\mathbf{w})}{\partial w_j}. \qquad (1.5)$$

To simplify notations, we will use $\mathbf{F}'$ to denote the vector or matrix of derivatives of any vector or matrix $\mathbf{F}$ with respect to some element of $\mathbf{w}$. One way of taking partial derivatives of an eigenvector of a symmetric matrix (when λ has order 1) is given in Magnus and Neudecker [87], Cour et al. [88], but that method applies to the derivatives of any eigenvector and does not take advantage of the fact that in this case $\mathbf{v}$ is the principal eigenvector of a matrix with large eigengap. Our idea is to use the power method to compute the partial derivatives of the approximate principal eigenvector: $\mathbf{v} = \frac{\mathbf{M}^n\mathbf{1}}{\sqrt{(\mathbf{M}^n\mathbf{1})^T(\mathbf{M}^n\mathbf{1})}}$. Here $\mathbf{M}^n\mathbf{1}$ is computed recursively by $\mathbf{M}^{k+1}\mathbf{1} = \mathbf{M}(\mathbf{M}^k\mathbf{1})$. Since the power method is the preferred choice for computing the principal eigenvector, $\mathbf{v}$, the derivatives of $\mathbf{v}$ estimated based on the power method, are not an approximation, but actually the exact ones. Thus, the partial derivatives of $\mathbf{v}$ are computed as follows:

$$\mathbf{v}' = \frac{\|\mathbf{M}^n\mathbf{1}\|^2(\mathbf{M}^n\mathbf{1})' - ((\mathbf{M}^n\mathbf{1})^T(\mathbf{M}^n\mathbf{1})')(\mathbf{M}^n\mathbf{1})}{\|\mathbf{M}^n\mathbf{1}\|^3}. \qquad (1.6)$$

The derivative of $\mathbf{M}^n\mathbf{1}$, which is needed above, can be obtained recursively:

$$(\mathbf{M}^n\mathbf{1})' = \mathbf{M}'(\mathbf{M}^{n-1}\mathbf{1}) + \mathbf{M}(\mathbf{M}^{n-1}\mathbf{1})'. \qquad (1.7)$$

We showed in Leordeanu and Hebert [75] that $\mathbf{M}$ has a large eigengap (difference between the first and the second eigenvalues), which makes the method above stable and efficient in practice.

1.5.6   Unsupervised Learning for Graph Matching

The idea for unsupervised learning in the case of graph matching is to maximize v_r^* instead of v_r, which could be achieved through the maximization of the dot product between the eigenvector and the binary solution obtained from the eigenvector, instead of the ground truth. Thus, during unsupervised training, we maximize the following function:

$$J(\mathbf{w}) = \sum_{i=1}^{N} \mathbf{v}^{(i)}(\mathbf{w})^T\mathbf{b}(\mathbf{v}^{(i)}(\mathbf{w})). \qquad (1.8)$$

In Chap. 2, we provide a theoretical and statistical justification for the choice of the current discrete matching solution as pseudo-ground truth to be used during unsupervised learning. The key idea is that for any sensible starting point (initialization of parameters $\mathbf{w}$, which we aim to learn), the correct assignments are expected to appear on the current discrete solution $\mathbf{b}$ more than the wrong ones. Such a sensible starting point could be found, for example, picking by hand a few examples of positive and negative pairs of candidate assignments and learning $\mathbf{w}$ which best separates (classifies) the positive and the negative pairs. Then, we show that as long as a given correct assignment has a better chance to be chosen as correct than a negative one, the few correct assignments chosen will immediately start forming a strong cluster of agreements in the sub-graph of candidate assignments chosen which will appear on $\mathbf{b}$. This cluster, much stronger than the noisy links formed by the incorrect assignments, will pull the parameter vector towards the correct solution, through gradient ascent with respect to the objective function (1.8).

The numerical difficulty here could be that $\mathbf{b}(\mathbf{v}^{(i)}(\mathbf{w}))$ is not a continuous function and also it may be impossible to express explicitly as a function of $\mathbf{w}$, since $\mathbf{b}(\mathbf{v}^{(i)}(\mathbf{w}))$ is the result of an iterative discretization procedure. However, it is important that $\mathbf{b}(\mathbf{v}^{(i)}(\mathbf{w}))$ is piecewise constant and has zero derivatives everywhere except for a finite set of discontinuity points. We can expect that we will evaluate the gradient only at points where $\mathbf{b}$ is constant and has zero derivatives; therefore, we should not worry about the discontinuities of $\mathbf{b}$, which were also not an impediment in practice. Again, we learn by gradient ascent:

$$w_j^{(k+1)} = w_j^{(k)} + \eta \sum_{i=1}^{N} \mathbf{b}(\mathbf{v}_i^{(k)}(\mathbf{w}))^T \frac{\partial \mathbf{v}_i^{(k)}(\mathbf{w})}{\partial w_j}. \tag{1.9}$$

Note that the supervised and unsupervised cases could be naturally combined, if some correct assignments are known, by simply setting and keeping constant the corresponding values in the discrete solution $\mathbf{b}$ for which the ground truth is known. As mentioned, we provide more theoretical justifications in Chap. 2 for the statistical reasons behind the efficiency of our unsupervised approach, for which we also present thorough experiments. Nevertheless, it is worth emphasizing here that these statistical observations are based entirely on the same power of alignments— only a few candidate N_c assignments need to be correct in order to start forming a strong cluster, through their $O(N_c^2)$ pairwise geometric agreements, which will pull the learning into the right direction. Once that process started and a first step is made into the right direction, there is no way back. As the vector of parameters $\mathbf{w}$ that defines the elements of matrix $\mathbf{M}$ gets closer to the true one, the number of true correct candidate assignments labeled as correct increases at the next learning iteration. Thus, $\mathbf{b}$ gets closer to the ground truth, and the next parameter update will be even more efficient, bringing both $(\mathbf{b}(\mathbf{w}), \mathbf{w})$, iteration by iteration, closer and closer to the true values.

1.6   Unsupervised Clustering Meets Classifier Learning

In the following section, we first show how the IPFP method can be applied to the case of graph clustering with minimal modifications. In Chap. 3, we will also show that IPFP could be easily adapted to the case of clustering with higher order edges (hypergraph clustering as well).

Then, we show that we can formulate the task of learning a linear classifier as a clustering problem for which IPFP is immediately applicable. Thus, we could use IPFP to learn a linear classifier, which has two important properties: (1) the classifier is very robust to shortage of labeled training data and could even be trained in the unsupervised case if we have a minimal piece of knowledge about its input features, which we refer to as *feature signs* as explained later in the section. Second, the optimal classifier found with IPFP will have the property that it is the average of a sparse group of automatically selected features. The classifier will thus be a feature selector, an average of features (or ensemble of lower level classifiers), and the optimum of a linear classifier learning task at the same time.

1.6.1   Integer Projected Fixed Point Method for Graph Clustering

In Chap. 3, we present in detail the more general case of applying IPFP to hypergraph clustering. Here, for simplicity and to better connect to the previous material, we will limit our case to graphs with pairwise relationships between nodes. We have a set of N nodes, forming pairwise similarity links in a graph G, whose symmetric adjacency matrix M captures these pairwise affinities. Similarly to the graph matching case, we have a quadratic clustering score of the type $S(\mathbf{x}) = \mathbf{x}^T M \mathbf{x}$, which we aim to maximize under specific clustering constraints on the solution $\mathbf{x}$. Again, here $\mathbf{x}$ is an indicator vector, with a unique index for each node i such that $x_i = 1$ if i is part of the final cluster and 0 otherwise. $M_{i;j}$ represents the pairwise similarity between the nodes (i, j), the stronger their edge $M_{i;j}$, the more likely (i, j) are to belong to the same cluster.

Finding a good cluster requires finding a subset of features with a high cluster score S. Even if we knew the number of elements in the cluster, maximizing S optimally would be an expensive combinatorial problem. For practical applications, we want to find a good approximate solution rapidly. Again, we relax the problem and allow x to take values in the continuous domain $[0, \varepsilon]$; ε acts as an upper bound of the cluster membership probability:

$$\mathbf{x}^* = \arg\max\ S(\mathbf{x}) \text{ s.t. } \sum x_i = 1,\ \mathbf{x} \in [0, \varepsilon]^n. \tag{1.10}$$

In the case of $\varepsilon = 1$ and pairwise affinities in $\{0, 1\}$ (corresponding to unweighted graphs), the problem is identical to the classical computation of maximal cliques [89, 90].

The L1 norm constraint in Problem 1.10 favors sparse solutions and biases towards categorical values for the membership assignments. This makes it easier to discretize and find actual clusters. As it is formulated above, the solution will produce a single, strong cluster in the graph. In order to find multiple clusters, one idea [91] is to remove the points belonging to a cluster that was found and restart the problem on the remaining points. Another idea [92] is to start from different initial solutions that are close to different clusters and locally maximize the clustering score (1.10).

We immediately observe that Problem 1.10 suits IPFP very well and that the only thing that we need to change is the projection function $P_d(.)$, since the discrete domain in the case of clustering is different than in the case of graph matching. Let us first define $\mathbf{d(x)} = \mathbf{Mx}$, to simplify notation. In order to better understand IPFP, we will now look at it from the perspective of a first-order Taylor expansion of the clustering score, around the current solution $\mathbf{x}_t$ at a given iteration t:

$$S(\mathbf{x}) \approx 2\mathbf{d(x_t)}^\top \mathbf{x} - S(\mathbf{x_t}). \qquad (1.11)$$

Maximizing the first-order approximation in the continuous domain of (1.10) results in the following linear program defined using the current solution $\mathbf{x}_t$:

$$\mathbf{y}^* = \arg\max \ \mathbf{d(x_t)}^\top \mathbf{y} \ \text{ s.t. } \sum y_i = 1, \ \mathbf{y} \in [0, \varepsilon]^n. \qquad (1.12)$$

Note that this linear problem is the same as the projection $P_d(.)$ presented in Sect. 2.4.2, where we introduced IPFP. In other words, at each iteration t, $\mathbf{y}^* = \mathbf{P_d(d(x_t))} = \mathbf{P_d(Mx)}$. Here, the discrete $\mathbf{y}^*$ represents a possible clustering solution and plays the same role as the discrete $\mathbf{b}$ in the case of graph matching in Sect. 2.4.2. If in the case of graph matching $\mathbf{b}$ can be efficiently found by the Hungarian algorithm, here, in the case of graph clustering it is even easier to find the optimal $\mathbf{y}^*$ of the linear program (1.12) by the following algorithm:

Algorithm 1.2 Optimize Problem 1.12.

$\mathbf{d} \leftarrow \mathbf{d(x_t)}$
$c \leftarrow \lfloor 1/\varepsilon \rfloor$
Sort $\mathbf{d}$ in decreasing order $d_{i_1} \geq \cdots \geq d_{i_c} \geq \cdots \geq d_{i_n}$
$y_{i_l} \leftarrow \varepsilon$, for all $l \leq c$
$y_{i_{c+1}} \leftarrow 1 - c\varepsilon$
$y_{i_l} \leftarrow 0$, for all $l > c + 1$
return $\mathbf{y}$

It is relatively easy to show that the above method returns a global optimum of (1.12). Once we know how to find $\mathbf{y}^*$, we can immediately apply IPFP to the case of graph clustering.

Algorithm 1.3 Integer Projected Fixed Point Method for Graph Clustering

---

Initialize $\mathbf{x_0}, t \leftarrow 0$
repeat
   Step 1: $\mathbf{y}^* \leftarrow \arg\max \, \mathbf{d(x_t)}^\top \mathbf{y}$ s.t. $\sum y_i = 1$, $\mathbf{y} \in [0, \varepsilon]^n$. If $\mathbf{d(x_t)}^\top (\mathbf{y} - \mathbf{x_t}) = 0$ stop.
   Step 2: $\alpha^* \leftarrow \arg\max \, S((1-\alpha)\mathbf{x_t} + \alpha\mathbf{y}^*)$, $\alpha \in [0, 1]$.
   Step 3: $\mathbf{x_{t+1}} \leftarrow (1 - \alpha^*)\mathbf{x_t} + \alpha^*\mathbf{y}$, $t \leftarrow t + 1$
until convergence
return $\mathbf{x_t}$

We presented IPFP in the context of two problems, that of graph matching and the other of graph clustering. We also presented it in two algorithmic forms (Algorithm 1.1 and 1.3) and gave two main theoretical justifications, one based on a projection that relates IPFP to the power iteration algorithm used in spectral matching and the other using the first-order Taylor expansion of the objective score, which relates IPFP more to the Frank–Wolfe method. We believe that all these different views are important to understand and to have in mind, in order to better understand this relatively simple method that has a large area of applications in modern computer vision in different problems that can be formulated in similar ways. The general approach is the following:

IPFP in a nutshell:
1. Start with a first-order Taylor approximation of the original objective score.
2. Optimize the first-order approximation, which usually boils down to a projection on the original discrete domain. This is usually equivalent to the maximization of a dot product between the linear approximation of the objective and the solution in the original discrete domain.
3. Optimize, usually in closed form, the original objective score between the current solution and the discrete solution found at the previous step. Update the solution with the new one found at this step.
4. Iterate steps 2 and 3 until convergence.

1.6.2 Feature Selection as a Graph Clustering Problem

Now we show how to formulate the well-known problem of feature selection as a graph clustering problem and present supervised and unsupervised quadratic formulation, to which IPFP as used on the clustering task can be applied without any modification. Classes are often *triggered* by only a few key input features (which could be other classifier outputs), as we show in Fig. 1.10, which presents a case that has been revealed in our experiments detailed in Sect. 4.4. Each such feature acts as a weak classifier that is related to the class of interest but only weakly. The

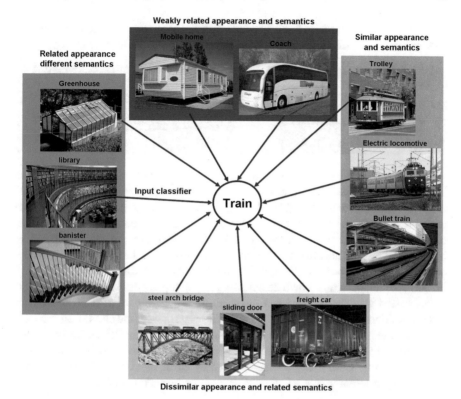

Fig. 1.10 What classes can *trigger* the idea of a "train"? Many classes have similar appearance but are semantically less related (blue box); others are semantically closer but visually less similar (green box). There is a complex discrete-continuum that relates appearance, context, and semantics. Can we find a group of classifiers that combined signal the presence of a specific, higher level concept, which are together robust to outliers, over-fitting, and missing features? Here, we show classifiers (e.g., library, steel arch bridge, freight car, mobile home, trolley, etc.) that are consistently selected by our method from limited training data as giving valuable input to the class "train"

presence of a single such feature or related classifier's positive output is not sufficient for determining the presence of our class of interest. However, when many such key outputs fire at the same time, their combined presence is a strong signal and their average output could become a strong classifier itself. Now we show that we could formulate classifier learning such as a feature selection task, in which we select automatically the subset of classifiers that are related to a given classification problem. We also show that the average response of the selected features (or classifiers) is the optimal solution of our problem formulation and it becomes equivalent to the graph clustering formulation presented before. IPFP becomes immediately a method of choice for solving this feature selection task, for which we present supervised and unsupervised approaches. The work introduced here and presented in detail in

Chap. 4, which was initially published in Leordeanu et al. [82], reveals a specific use case of one of the principles of unsupervised learning that we have proposed:

Principle 6
When several weak independent classifiers consistently fire together, they signal the presence of a higher level class for which a stronger classifier can be learned.

In Fig. 1.10, we present a result that is representative of our approach. We use image-level CNN classifiers Jia et al. [93], pretrained on ImageNet, to recognize trains in video frames from the YouTube-Objects dataset Prest et al. [94]. All we know, in terms of supervision signal, is how are the CNN classifiers correlated with the class of interest. More specifically, do they have on average larger output values on positive train images than on negative ones? This is the only piece of information required by our unsupervised approach. We do not need to know, during training, which training images contain trains and which do not. Our method rapidly discovers, in this almost unsupervised way, the relevant feature group from a large pool of 6000 pretrained CNN classifiers which together form a powerful classifier for the class "train". Since each classifier corresponds to one ImageNet concept (which often turns out to be very different from the concept "train"), we show a few of the classifiers that have been consistently (every single time) selected by our method over 30 experimental trials, using different unlabeled training sets of positive and negative cases.

1.6.2.1   The Blind Men and the Elephant

The experiment presented in Fig. 1.11 reminds us of an ancient parable originated in the Indian subcontinent, which tells about six blind men who want to learn about the properties of a new strange animal, an elephant, which came to their town. They go and touch the animal and based on their limited perception of touch they form an image of the new animal. One person whose hand touches the elephant's trunk says that the animal is like a thick snake. Another man, who touches the ear of the elephant thinks the elephant is like some type of fan. Another blind man, whose hand lands upon the leg of the elephant, says the animal seems like a tree trunk, while another, who touches its side, feels the animal is like a wall. The one who touches its tail thinks the elephant is like a rope, while the last one, touching its tusk, says the elephant is smooth and hard like a spear (Fig. 1.11).

The parable could have many interpretations and indeed applies to a very broad range of cases in which we learn and form an image about the world. Having a limited capacity of perception (due to our limited senses) and understanding (based on limited experience), we inherently form a very subjective and limited image of the world. However, we can immediately see that the parable perfectly describes the situation addressed by our sixth principle, as it is also exemplified by the result in Fig. 1.11. We notice that each classifier responds in its own way at the view of the

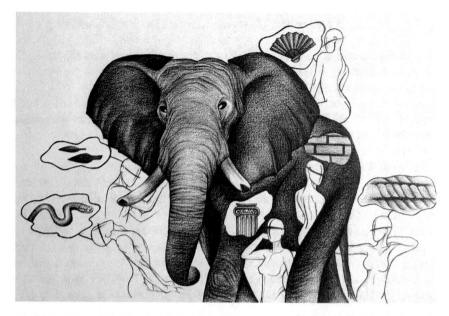

Fig. 1.11 The parable of the six blind men and the elephant. Each blind man sees the elephant completely differently. Our interpretation of the parable in the context of unsupervised learning is that when their weak perceptions take place at the same time (an unlikely event in a random case) then their combined perceptions could strongly indicate the presence of an elephant. From this point of view, the co-occurrence of what they "see" as elephant could be a Highly Probable Positive (HPP) event. Original drawing by artist Cristina Lazar (Cristina Lazar, Nicolae Rosia, Petru Lucaci and Marius Leordeanu, *Smile Project: Deep Immersive Art with Human-AI Interaction* https://sites. google.com/view/smile-Project)

train: one sees it as a library, another as a steel arch bridge, or a freight car, or a mobile home, or a trolley. None sees it as a "train" because they were not trained to see "trains". Their limited experience and capacity to perceive beyond what they know prevent each classifier to respond to trains and only to trains. Instead, they respond to classes that are often as different from trains as there are the "categories" seen by the blind men different from the elephant.

However, what our different fresh look at this parable brings new, in the context of unsupervised learning, is the following: even though none of the blind men see the elephant in isolation, when they all see the objects they claim to see at the same time, it is very likely that they have an elephant and not some other animal in front of them. As Principle 6 says, the combined co-occurrence of their weak classifiers (that output a wrong class but are positively correlated with the elephant) becomes, jointly, a very strong classifier, which could accurately detect the elephant.

1.6.2.2 Feature Selection and Classifier Learning as Graph Clustering

We address the case of binary classification and apply the one versus all strategy to the multi-class scenario. We have a set of N samples, with each ith sample expressed as a column vector $\mathbf{f}_i$ of n features our outputs of classifiers with values in $[0, 1]$. We want to find vector $\mathbf{w}$, with elements in $[0, 1/k]$ and unit l^1-norm (these constraints basically enforce the number of selected features will be k), such that $\mathbf{w}^T \mathbf{f}_i \approx \mu_P$ when the ith sample is from the positive class and $\mathbf{w}^T \mathbf{f}_i \approx \mu_N$ otherwise, with $0 \leq \mu_N < \mu_P \leq 1$.

For a labeled training sample i, we fix the ground truth target $t_i = \mu_P = 1$ if positive and $t_i = \mu_N = 0$ otherwise. We impose clustering-style constraints on $\mathbf{w}$ limit the impact of each individual feature f_j, encouraging the selection of features that are powerful in combination, with no single one strongly dominating. The solution we look for is a weighted feature average with an ensemble response that is stronger on positives than on negatives. For that, we want any feature f_j to have expected value $E_P(f_j)$ over positive samples greater than its expected value $E_N(f_j)$ over negatives. We estimate a features' sign $sign(f_j) = E_P(f_j) - E_N(f_j)$ from labeled samples (or consider it as given) and if it is negative we simply *flip* the feature: $f_j \leftarrow 1 - f_j$.

Supervised Learning: We formulate the supervised learning task as a least squares constrained minimization problem. Given the $N \times n$ feature matrix $\mathbf{F}$ with $\mathbf{f}_i^\top$ on its ith row and the ground truth vector $\mathbf{t}$, we look for $\mathbf{w}^*$ that minimizes $\|\mathbf{F}\mathbf{w} - \mathbf{t}\|^2 = \mathbf{w}^\top(\mathbf{F}^\top\mathbf{F})\mathbf{w} - 2(\mathbf{F}^\top\mathbf{t})^\top\mathbf{w} + \mathbf{t}^\top\mathbf{t}$ and obeys the required constraints. We drop the last constant term $\mathbf{t}^\top\mathbf{t}$ and obtain the following convex minimization problem:

$$\mathbf{w}^* = \arg\min_{w} J(\mathbf{w}) \tag{1.13}$$
$$= \arg\min_{w} \mathbf{w}^\top(\mathbf{F}^\top\mathbf{F})\mathbf{w} - 2(\mathbf{F}^\top\mathbf{t})^\top\mathbf{w}$$
$$s.t. \sum_j w_j = 1 , \; w_j \in [0, 1/k].$$

The least squares formulation is related to Lasso, Elastic Net, and other regularized approaches, with the distinction that in our case individual elements of $\mathbf{w}$ are restricted to $[0, 1/k]$. This leads to important properties regarding sparsity and directly impacts generalization power. This constraint on $\mathbf{w}$ is in fact the key, the most important element that transforms our problem into a case of feature selection and graph clustering.

The Almost Unsupervised case: Consider a pool of signed features correctly flipped according to their features signs, which could be given or estimated from a few labeled samples. We make the assumption that the signed features' expected values for positive and negative samples, respectively, are close to the ground truth target values (μ_P, μ_N). The assumption could be realized in practice, in an approximation sense, by appropriate normalization, estimated from the small supervised set.

Then, for a given sample i, and any $\mathbf{w}$ obeying the constraints, the expected value of the weighted average $\mathbf{w}^\top\mathbf{f}_i$ is also close to the ground truth target t_i: $E(\mathbf{w}^\top\mathbf{f}_i) =$

$\sum_j w_j E(\mathbf{f}_i(j)) \approx (\sum_j w_j) t_i = t_i$. Then, for all samples, we have the expectation $E(\mathbf{Fw}) \approx \mathbf{t}$, such that any feasible solution will produce, on average, approximately correct answers. Thus, we can regard the supervised learning scheme as attempting to reduce the variance of the feature ensemble output, as their expected value is close to the ground truth target. If we approximate $E(\mathbf{Fw}) \approx \mathbf{t}$ into the objective $J(\mathbf{w})$, we get a new ground-truth-free objective $J_u(\mathbf{w})$ with the following learning scheme, which is unsupervised once the feature signs are known. Here $\mathbf{M} = \mathbf{F}^\top \mathbf{F}$:

$$\mathbf{w}^* = \arg \min_w J_u(\mathbf{w}) \qquad (1.14)$$
$$= \arg \min_w \mathbf{w}^\top (\mathbf{F}^\top \mathbf{F}) \mathbf{w} - 2(\mathbf{F}^\top (\mathbf{Fw}))^\top \mathbf{w}$$
$$= \arg \min_w (-\mathbf{w}^\top (\mathbf{F}^\top \mathbf{Fw})) = \arg \max_w \mathbf{w}^\top \mathbf{M} \mathbf{w}$$
$$\text{s.t.} \sum_j w_j = 1 \;,\; w_j \in [0, 1/k].$$

Note that while the supervised case is a convex minimization problem, the unsupervised learning scheme is a concave minimization problem, which is NP-hard. This is due to the change in sign of the matrix $\mathbf{M}$. As $\mathbf{M}$ in the (almost) unsupervised case could be created from larger quantities of unlabeled data, $J_u(\mathbf{w})$ could in fact be less noisy than $J(\mathbf{w})$ and produce significantly better local optimal solutions—a fact confirmed by experiments.

Intuition and overall approach: In Chap. 4, we discuss in more detail the insights revealed by the two formulations. From the start, we notice that both formulations suit IPFP perfectly for optimization and are identical to the clustering approach presented before. The relation to clustering becomes more clear when we look at the more interesting (almost unsupervised case), which has a quadratic term formed by the positive (or semi-positive) definite matrix $\mathbf{F}^\top \mathbf{F}$, which contains the dot products between pairs of feature responses over the samples. This matrix is strongly related to the feature's covariance matrix which reflects how related their values are at the pairwise level. It becomes clear that maximizing the quadratic term could be interpreted as a clustering task with pairwise constraints. As we will show in Chap. 3, under the clustering constraints an optimum of the problem is guaranteed to be sparse with all the non-zero weights being equal to $1/k$. Both from an intuitive and theoretical point of view, we can say that our classifier learning task is nothing but a clustering problem in the (almost) unsupervised case. We solve the task by finding the group of features with strongest intra-cluster score, which is the largest amount of covariance. They strongly fire or not fire together, and that agreement, which is unlikely in the random case, is a robust supervisory signal, which indicates the presence of a new class.

In the absence of ground truth labels, if we assume that features in the pool are, in general, correctly signed and not redundant, then the maximum covariance is attained by those whose collective average varies the most as the hidden class labels also vary.

Thus, we hope that the (almost) unsupervised variant seeks features that respond in a united, combined manner, which is sensitive to the natural variation into the distributions of the two classes. In other words, the variation in the total second-order signal (captured by the covariance matrix) of the good feature responses, caused by the natural differences between the two classes, should be the strongest when compared to noisy variations that are more or less independent and often cancel each other out.

The overall algorithm for classifier learning as a feature selection and clustering problem can be summarized as follows:

Algorithm 1.4 Feature Selection and Classifier Learning by Clustering

---

Learn feature signs from a small set of labeled samples.
Create $\mathbf{F}$ with flipped features from unlabeled data.
Set $\mathbf{M} \leftarrow \mathbf{F}^\top \mathbf{F}$,
Find $\mathbf{w}^* = \arg\max\limits_{w} \mathbf{w}^\top \mathbf{M} \mathbf{w}$
  s.t. $\sum_j w_j = 1$, $w_j \in [0, 1/k]$.
return $\mathbf{w}^*$

As discussed, IPFP can be immediately applied, as we can easily notice the similarity of the problem at Step 4 of Algorithm 1.4 with the clustering Problem 1.10 previously discussed: $\mathbf{w}$ plays the role of the indicator vector $\mathbf{x}$ and $1/k$ plays the role of ε. The formulation is of a clustering task and IPFP directly applies.

1.7 Unsupervised Learning for Object Segmentation in Video

So far, we have established the basic elements, ideas, and algorithms for making the fundamental step towards unsupervised learning in space and time. Discovering novel objects in videos without human supervision, as they move in space, is one of the most interesting and challenging problems in visual learning. How much could we learn just by watching videos, what is the limit for learning when huge amounts of video data are available but they do not have any ground truth labels and we are only passive observers. This case seems to be the hardest and that is exactly why we want to study it, in order to better understand the limits of learning in the space-time domain. The work we introduce in this section, which we then present in detail in Chap. 6, is based at its core on the observation that was captured by one of the key principles of unsupervised visual learning, which we presented at the beginning:

Principle 4
Objects form strong clusters of motion trajectories and appearance patterns in
their space-time neighborhood.

While people tend to agree on which is the main object in a given video shot,
which describes a certain neighborhood in space and time, it is not yet clear how
we do it. There are many interesting questions waiting for an answer: which are the
particular features that make a group of pixels stand out as a single, primary object
in a given sequence? Is it the pattern of point trajectories, the common appearance or
the contrast with the background that make an object stand out as a single entity? Or
is it a combination of all these factors and what is the best way to combine them? Do
we need extensive pre-training or some innate, higher level "objectness" Alexe et al.
[95] features in order to begin learning or we could in principle start from scratch?

We make the observation that the world is consistent and coherent in space and
time, being well structured into objects that share many grouping properties as the
early psychologists have also recognized [71]. Since videos are projections of the
4D world in space and time, they should inherit many of its properties and likewise,
the mind should be well equipped to exploit them. Next, we briefly present our
original approach to object discovery in space and time, initially published in Haller
et al. [84], which proves, by extensive experimental validation (Chap. 6) that is
possible to discover objects as clusters in space and time using very simple motion and
appearance features. The approach is general enough and can be naturally extended
to incorporate more powerful pretrained features.

Our method is based on a dual view of objects that couples the spatial and temporal
dimensions in complementary ways: (1) the same point of an object, captured by
different pixels at different moments in time, connects those pixels through long-
range motion chains—trajectories through space as functions of time; (2) different
points of an object that occupy distinct spatial locations share similar patterns of
motion and appearance; and (3) motion and appearance patterns of an object can often
predict one another and are in agreement regarding the shape and presence of the
object in space-time—what could be estimated from motion should also be estimated
from appearance. Thus appearance and motion could offer supervisory signal to
each other and function in tandem. The same idea of looking for agreements along
different pathways of prediction, based on different types of information (e.g., motion
versus appearance) is further extended in the last part of the book (Chap. 8), where
we propose a novel Visual Story concept, which aims to offer a general approach to
unsupervised learning. Again we understand that it is only through mutual agreements
that we could eventually achieve unsupervised learning in the wild, where such
agreements naturally happen due to the structure, coherence, and consistency of
space and time.

Next, we propose our novel graph structure in space and time, in which segmentation is formulated as a spectral clustering problem with motion and appearance constraints. Segmentation is then found as the leading eigenvector of a Feature-Motion

matrix. Mathematically, the approach relates immediately to our earlier formulations for graph matching, which also exploited the fact that the principal eigenvector of a symmetric matrix with non-negative elements naturally reflects the main, strongest cluster of its associated graph. Conceptually, every pixel in the video is a node in our graph, but we never explicitly compute the matrix, as that would be impossible. Instead, we offer an efficient method which finds the leading eigenvector of the matrix without having to actually construct it.

Given a sequence of m video frames, we want to extract m soft segmentations, one per frame, that define the masks of the main object in the sequence. Conceptually, the entire video is a graph with one node per pixel.

1.8 Space-Time Graph

Before going deeper into the mathematical formulation and optimization, we first need to define and understand some basic terms in our representation:

Graph of pixels in space-time: We define the space-time graph $G = (V, E)$: each node $i \in V$ is associated with a pixel of the video sequence.

Optical flow chains: The trajectories defined by the optical flow that starts from a given pixel in both directions and moves, frame by frame, through the space-time volume, define *optical flow chains* (Fig.1.12a). Two nodes (pixels) are connected if and only if there is at least one chain between them. These pixels are expected to represent the same physical point on the object.

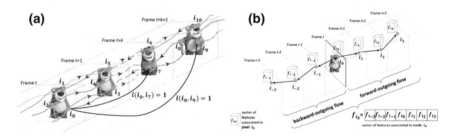

Fig. 1.12 Visual representation of how the space-time graph is formed. **a** i_0 to i_{10} are graph nodes, corresponding to pixels from different frames. **Colored lines** represent optical flow chains, formed by following the optical flow, forward or backward, sequentially from one frame to the next. Thus, two nodes are connected if there exists a chain of optical flow vectors that link them through intermediate consecutive frames. From a given node, there will always be two flows going out, one in each direction, but there could be none or multiple chains coming in (e.g., node i_7). **Black lines** correspond to graph edges, defined between nodes that are connected by at least one flow chain. **b** Flow chains are also used to build node features. For a given node i_0, the features forming feature vector $\mathbf{f}_{i_0}$ are collected along the outgoing flow chains, representing motion and appearance patterns along the chains

Motion matrix: $\mathbf{M} \in \mathbb{R}^{n \times n}$, with $\mathbf{M}_{i,j} = l(i, j) \cdot k(i, j)$, where $k(i, j)$ is a Gaussian kernel as a function of the temporal distance between nodes i and j, while $l(i, j) = 1$ if nodes i and j are connected as explained above and zero otherwise. $\mathbf{M}$ is symmetric, semi-positive definite, has non-negative elements, and is expected to be very sparse.

Node features: Each node i is also described by its appearance features, represented by vectors $\mathbf{f}_i \in \mathbb{R}^{1 \times d}$, collected along the outgoing motion chains starting in i, one per direction (Fig.1.12b). We stack all appearance features into a features matrix $\mathbf{F} \in \mathbb{R}^{n \times d}$.

Node labels: Each node i has an associated (soft) mask label $x_i \in [0, 1]$, which represents our belief that the node is part of the object of interest. Thus, we can represent a segmentation solution in both space and time, over the whole video, as an indicator vector $\mathbf{x} \in \mathbb{R}^{n \times 1}$, with a label $\mathbf{x}_i$, for each ith video pixel (graph node). Note the similarities in notation with the previous graph matching and clustering formulations.

We find the main object of interest as the strongest cluster in the space-time graph, such that pixels that belong to the main object are strongly connected through motion chains but are also predictable from the object appearance features. As a side note, we do not expect the motion chains to be always accurate but we do expect the pixels that indeed belong to the same object to form a strong motion cluster in $\mathbf{M}$. As in the case of graph matching and clustering discussed before, we represent the space-time cluster by the labels indicator vector $\mathbf{x}$, with $\mathbf{x}_i = 1$ if node i belongs to the cluster and $\mathbf{x}_i = 0$, otherwise. Then, the intra-cluster score is $S_C = \mathbf{x}^T \mathbf{M} \mathbf{x}$, with the property that labels $\mathbf{x}$ are allowed to take soft values, have unit norm $\|\mathbf{x}\|_2 = 1$ (only their relative values matter for segmenting the object and the background) and lives in the features subspace: it is a linear combination of the columns of the feature matrix $\mathbf{F}$. The condition that $\mathbf{x}$ is a linear combination of the columns of $\mathbf{F}$ can be enforced by requiring that $\mathbf{x} = \mathbf{P}\mathbf{x}$, where $\mathbf{P}$ is the feature projection matrix $\mathbf{P} = \mathbf{P}^T = \mathbf{F}(\mathbf{F}^T\mathbf{F})^{-1}\mathbf{F}^T$, which projects a vector onto the subspace spanned by the columns of $\mathbf{F}$. We obtain the following result, which we prove in Chap. 6:

Proposition 1.1 $\mathbf{x}^*$ *that maximizes* $\mathbf{x}^T\mathbf{M}\mathbf{x}$ *under constraints* $\mathbf{x} = \mathbf{P}\mathbf{x}$ *and* $\|\mathbf{x}\|_2 = 1$, *also maximizes* $\mathbf{x}^T\mathbf{P}\mathbf{M}\mathbf{P}\mathbf{x}$ *under constraint* $\|\mathbf{x}\|_2$.

We can now define our final optimization problem as

$$\mathbf{x}^* = \arg \max_{\mathbf{x}} \mathbf{x}^T \mathbf{P}\mathbf{M}\mathbf{P}\mathbf{x} \quad \text{s.t.} \quad \|\mathbf{x}\|_2 = 1 \tag{1.15}$$

Feature-Motion Matrix: We refer to $\mathbf{P}\mathbf{M}\mathbf{P}$ as the Feature-Motion matrix, which couples the motion through space-time of an object and its appearance features $\mathbf{F}$ through the projection $\mathbf{P}$ on the features space.

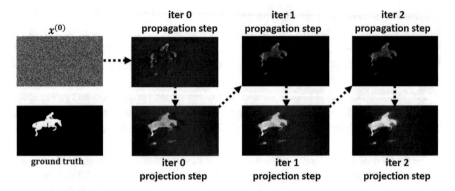

Fig. 1.13 Qualitative results of our method, over three iterations, when initialized with a random soft segmentation mask and using only unsupervised features (i.e., color a motion along flow chains—Sect. 6.4.2). Note that the main object emerges with each iteration of our algorithm

1.8.1 Optimization Algorithm

The problem (Eq. 6.1) is optimally solved by the leading eigenvector $\mathbf{x}^*$ of $\mathbf{PMP}$. It is a classical case of spectral clustering [96] and also related to spectral approaches in graph matching Leordeanu et al. [97]. Note that $\mathbf{x}^*$ must have non-negative values by Perron–Frobenius theorem, since $\mathbf{PMP}$ has non-negative elements. Our efficient algorithm (Algorithm 6.1) which converges to $\mathbf{x}^*$ by the Power Iteration Method, cleverly exploits the space-time video graph, without explicitly computing $\mathbf{PMP}$. The steps of algorithm (GO-VOS) alternate between propagating the labels $\mathbf{x}$ according to their motion links (defined by optical flow chains in $\mathbf{M}$) and projecting $\mathbf{x}$ through $\mathbf{P}$ onto the features columns space of $\mathbf{F}$ (Fig. 1.13), while also normalizing $\|\mathbf{x}\|_2 = 1$.

Algorithm 1.5 GO-VOS: Unsupervised Video Object Segmentation. Below we present the three main steps of our spectral approach to unsupervised object segmentation, which compute by power iteration the leading eigenvector. Note our implementation efficiently computes the three steps below, without constructing matrix $\mathbf{M}$ explicitly. it denotes the iteration number.

1: **Propagation:** $\mathbf{x}^{(it+1)} \leftarrow \mathbf{M}\mathbf{x}^{(it)}$
2: **Projection:** $\mathbf{x}^{(it+1)} \leftarrow \mathbf{P}\mathbf{x}^{(it+1)}$
3: **Normalization:** $\mathbf{x}^{(it+1)} \leftarrow \mathbf{x}^{(it+1)}/\|\mathbf{x}^{(it+1)}\|_2$

Note the immediate relation between GO-VOS Algorithm 1.5 and the spectral method for graph matching. The algorithm reduces to finding the principal eigenvector of the Feature-Motion matrix $\mathbf{P}^T\mathbf{MP}$. Interestingly enough, the algorithm is also related to IPFP, since after the propagation step it is followed by a projection, which in this case is on the features subspace. The second projection is directly on the unit sphere. It is relatively easy to show that the three steps of the method are equivalent with power iteration applied to matrix $\mathbf{P^T MP}$. They could be interpreted as follows:

The propagation step: This step propagates motion information only performing on power iteration for matrix $\mathbf{M}$ which encodes motion only. This step alone is not sufficient for discovering the objects as appearance features are also important to re-enforce the grouping information transmitted by motion. The step can be efficiently implemented by a vote casting procedure that is linear in the number of pixels in the video (when optical flow chains are limited to a fixed neighborhood around each node) and does not need to explicit construction of $\mathbf{M}$.

The feature projection step: This step ensures that the solution obtained lives in the column space of the feature matrix $\mathbf{F}$, which means that the pixel labels can be linearly regressed on the features subspace: each soft label is basically a linear combination of the feature values at that node. Through this projection step, the motion and the appearance information are coupled in complementary way such that they re-enforce each other and make the object discovery process more robust. The step can be more efficiently applied separately frame by frame, with improved speed and accuracy.

The normalization step: The last normalization step keeps the L_2 norm of $\mathbf{x}$ and thus does not let it grow or shrink in unpredictable ways. Note that in the case of segmentation only the relative soft pixel label values matter for separating the soft masks into object and background regions.

The algorithm has excellent accuracy and speed in practice, producing state-of-the-art results on several current benchmarks such as DAVIS, SegTrack, and YouTube-Object Datasets, which we present in detail in Chap. 6. In Fig. 1.14, we present some qualitative results on the SegTrack dataset.

1.8.2 Learning Unsupervised Segmentation over Multiple Teacher-Student Generations

Now, at the end of our introductory chapter, we combined the ideas presented so far and introduce a system that learns, in an unsupervised way, to segment objects in single images over multiple teacher-student generations. At its core, the approach presented below and in more detail in Chap. 7, Croitoru et al. [98, 99] combines many key ideas discussed in this chapter, especially the last, Principle 7, proposed here, for unsupervised learning.

Principle 7
To improve or learn new classes, we need to increase the quantity and level of difficulty of the training data as well as the classifiers' power and diversity. In this way, we could learn in an unsupervised way over several generations, by

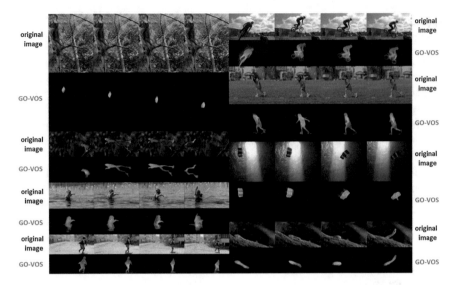

Fig. 1.14 Visual qualitative results on SegTrack dataset with our unsupervised GO-VOS formulation

using the agreements between the existing classifiers as a teacher supervisory signal for the new ones.

The system introduced here is composed of two main pathways, one that performs unsupervised object discovery in videos or large image collections—the teacher branch, and the other—the student branch, which learns from the teacher to segment foreground objects in single images. The unsupervised learning process then continues over several generations of students and teachers. In Algorithm 1.6, we present the high-level description of our method. The key aspects of our approach, which ensure improvement in performance from one generation to the next, are (1) the existence of an unsupervised selection module that is able to pick up good quality masks generated by the teacher and pass them for training to the next-generation students; (2) training of multiple students with different architectures, able through their diversity to help train a better selection module for the next iteration and form together with the selection a more powerful teacher pathway at the next iteration; and (3) access to larger quantities of, and potentially more complex, unlabeled data, which becomes more useful as the generations become stronger.

Our approach is general in the sense that the student or teacher pathways do not depend on a specific neural network architecture or implementation. Through many experiments and comparisons to state-of-the-art methods, we also show (Chap. 7) that it is applicable to different tasks in computer vision, such as object discovery in video, unsupervised image segmentation, saliency detection, and transfer learning.

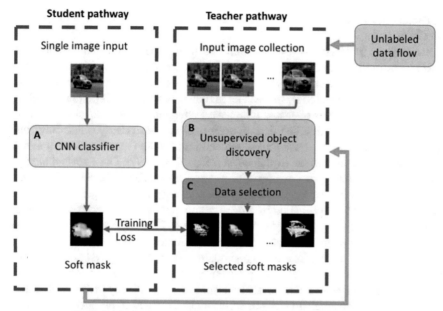

Fig. 1.15 The dual student-teacher system proposed for unsupervised learning to segment foreground objects in images, functioning as presented in Algorithm 1.6. It has two pathways: along the teacher branch, an object discoverer in videos or large image collections (module B) detects foreground objects. The resulting soft masks are then filtered based on an unsupervised data selection procedure (module C). The resulting final set of pairs—input image (or video frame) and soft mask for that particular frame (which acts as an unsupervised label)—is used to train the student pathway (module A). The whole process can be repeated over several generations. At each generation, several student CNNs are trained, and then they collectively contribute to train a more powerful selection module C (modeled by a deep neural network) and form an overall more powerful teacher pathway at the next iteration of the overall algorithm

In Fig. 1.15, we present a graphic overview of our full system. In the unsupervised training stage, the student network (module A) learns, frame by frame, from an unsupervised teacher pathway (modules B and C) to produce similar object masks in single images. Module B discovers objects in images or videos. Module C selects which masks produced by module B are sufficiently good to be passed to module A for training. Thus, the student branch tries to imitate the output of module B for the frames selected by module C. Its input is only a single image—the current frame—while the teacher has access to an entire video sequence.

The strength of the trained student (module A) depends on the performance of the module B. However, as we see in experiments (Chap. 7), the power of the selection module C contributes to the fact that the new student will outperform its initial teacher module B. Therefore, throughout the paper, we refer to B as the initial "teacher" and to both B and C together, as the full "teacher pathway". The method presented in Algorithm 1.6 follows the main steps of the system as it learns from one iteration (generation) to the next. The full system and the steps are discussed in more detail in Chap. 7.

During the first iteration of Algorithm 1.6, the unsupervised teacher (module B) has access to information over time—a video. In contrast, the student is deeper in structure, but it has access only to a single image—the current video frame. Thus, the information discovered by the teacher in time is captured by the student in added depth, over neural layers of abstraction. Several student nets with different architectures are trained at the first iteration. In order to use as supervisory signal only good quality masks, an unsupervised mask selection procedure (very simple at Iteration 1) is applied (module C), as explained in Chap. 7, Sect. 7.4. Once several student nets are trained, they can form (in various ways, as explained in Sects. 7.4.1 and 7.5.1) the teacher pathway at the next iteration, along with a stronger unsupervised selection module C, represented by a deep neural network, EvalSeg-Net, trained as explained in detail in Chap. 7, Sect. 7.4.3. In short, EvalSeg-Net learns to predict the output mask agreement among the generally diverse students, which statistically takes place when the masks are of good quality. Thus, EvalSeg-Net could be used as an unsupervised mask evaluation procedure and a strong selection module. Then, we run, at the next generation, the newly formed teacher pathway (modules B and C) on a larger set of unlabeled videos or collections of images, to produce supervisory signal for the next-generation students. In experiments, we show that the improvement of both modules B and C at the next iterations, together with the increase in the amount of data, are all important, while not all necessary, for increasing accuracy at the next generation.

Note that, while at the first iteration the teacher pathway is required to receive video sequences as input, from the second generation onwards, it could receive as input large image collections, as well. Due to the very high computational and storage costs, required during training time, we limit our experiments to learning over two generations, but our algorithm is general and could run over many iterations. We show in extensive experiments that even two generations are sufficient to outperform the current state of the art on object discovery in videos and images. We also demonstrate experimentally a solid improvement from one generation to the next for each component involved, the individual students (module A), the teacher (module B) as well as the selection module C.

In Fig. 1.16, we present visual comparative results between the different models used as student nets, including their ensemble, at different iterations. The qualitative improvement from one iteration to the next is clearly noticeable and also reflected by a 5% increase in accuracy (F-measure) on average. We invite the reader to take a closer look at the material in Chap. 7 where we provide, in detail, the description and experimental analysis of our proposed system. Below, we iterate here some concluding remarks and observations which we made at the end of our analysis.

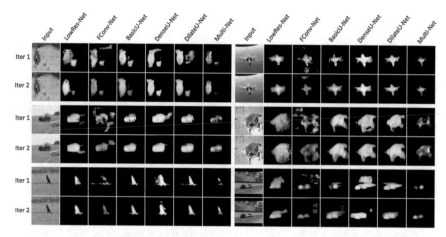

Fig. 1.16 Visual comparison between models at each iteration (generation). The Multi-Net, shown for comparison, represents the pixelwise multiplication between the five models. Note the superior masks at the second-generation students, with better shapes, fewer holes, and sharper edges. Also note the relatively poorer recall of the ensemble Multi-Net, which produces smaller, eroded masks

---

Algorithm 1.6 Unsupervised learning of foreground object segmentation

---

Step 1: perform unsupervised object discovery in unlabeled videos (or image collections, at later iterations), along the teacher pathway (module B in Fig. 7.1).

Step 2: automatically filter out poor soft masks produced at the previous step (module C in Fig. 7.1).

Step 3: use the remaining masks as supervisory signal for training one or more student nets, along the student pathway (module A in Fig. 7.1).

Step 4: use as new teacher one or several student nets from the current generation (a new module B) and learn a more powerful soft mask selector (a new module C), for the next iteration.

Step 5: extend the unlabeled video or image dataset and return to Step 1 to train the next generation (Note: from the first iteration forward, the training dataset can also be extended with collections of unlabeled images, not just videos).

1.8.3 Concluding Remarks

One of the interesting conclusions in our analysis done in Chap. 7 is that the system is able to improve its performance from iteration 1 to iteration 2. There are several factors that are involved in this outcome, which we believe are related through the following relationship: (1) Multiple students of diverse structures ensure diversity and somewhat independent mistakes; (2) in turn, point (1) makes possible the unsupervised training of a mask selection module that learns to predict agreements; (3) thus, the selection module at (2) becomes a good mask evaluation network; (4) once that evaluation network (from 3) is available, we can then add larger and potentially more complex data to select a larger set with good object masks of more interesting cases at the next iteration; (5) finally, (4) ensures the improvement at the next iteration and now we could return to point (1).

1.9   Next Steps

In this introductory chapter, we laid down the foundations of our approach to unsupervised learning in space and time. We started with some intuitive insights and motivations and presented a set of concise and effective principles for unsupervised learning. They are based on theoretical and empirical observations which we reached during our research on the topic, covering various problems, from graph matching and clustering, to unsupervised classifier learning, object discovery in video, and unsupervised learning over multiple teacher-student generations of classifiers.

We related these tasks with respect to the main unifying principles and followed closely the development of theoretical formulations, optimization algorithms, graph representations, and solutions which started from relatively simple and reached more general approaches and algorithms, such as learning with HPP features, spectral techniques applied to different tasks, and the Integer Projected Fixed Point Method. All the approaches are presented next, during the following chapters, in detail at both theoretical and experimental levels. We hope that by now the reader is familiar with most of these concepts and will have an easier time exploring in more depth the material presented next.

In the final chapter, we reiterate and synthesize even more the ideas covered in the book, but from the perspective of taking unsupervised learning into the future. Our final destination is to imagine a universal machine, the Visual Story Network, which could learn to see by itself. What would be required to make such a system possible? Are the mathematical tools and neural nets of today, combined with our practical experience enough to realistically think about such a system?

We invite, especially the young readers, who are passionate about artificial intelligence and learning, to explore the book all the way to the end. Their creative minds will take an unexpected journey of imagination, growth, and discovery, with many interesting ideas and intellectual satisfactions coming along the way. We encourage them to take the time to read, think, and then create in the world of unsupervised learning through space and time. In this way, the new generation will manage to push our knowledge beyond what we now think it is possible, towards wonderful and surprising future territories.

References

1. Gan G, Ma C, Wu J (2007) Data clustering: theory, algorithms, and applications, vol 20. Siam
2. Lloyd S (1982) Least squares quantization in pcm. IEEE Trans Inf Theory 28(2):129–137
3. Dempster AP, Laird NM, Rubin DB (1977) Maximum likelihood from incomplete data via the em algorithm. J Roy Stat Soc: Ser B (Methodol) 39(1):1–22
4. Comaniciu D, Meer P (2002) Mean shift: a robust approach toward feature space analysis. Pattern Anal Mac Intell 24(5):
5. Fukunaga K, Hostetler L (1975) The estimation of the gradient of a density function, with applications in pattern recognition. IEEE Trans Inf Theory 21(1):32–40

6. Ester M, Kriegel HP, Sander J, Xu X et al (1996) A density-based algorithm for discovering clusters in large spatial databases with noise. In: Kdd, vol 96, pp 226–231
7. Day WH, Edelsbrunner H (1984) Efficient algorithms for agglomerative hierarchical clustering methods. J Classif 1(1):7–24
8. Ward JH Jr (1963) Hierarchical grouping to optimize an objective function. J Am Stat Assoc 58(301):236–244
9. Sibson R (1973) Slink: an optimally efficient algorithm for the single-link cluster method. Comput J 16(1):30–34
10. Johnson SC (1967) Hierarchical clustering schemes. Psychometrika 32(3):241–254
11. Kaufman L, Rousseeuw PJ (2009) Finding groups in data: an introduction to cluster analysis, vol 344. Wiley
12. Cheeger J (1969) A lower bound for the smallest eigenvalue of the laplacian. In: Proceedings of the Princeton conference in honor of Professor S. Bochner, pp 195–199
13. Donath WE, Hoffman AJ (1972) Algorithms for partitioning of graphs and computer logic based on eigenvectors of connection matrices. IBM Tech Discl Bull 15(3):938–944
14. Meila M, Shi J (2001) Learning segmentation by random walks. In: Advances in neural information processing systems, pp 873–879
15. Shi J, Malik J (2000) Normalized cuts and image segmentation. PAMI 22(8)
16. Ng A, Jordan M, Weiss Y (2002) On spectral clustering: analysis and an algorithm. In: NIPS
17. Duda RO, Hart PE, Stork DG (2012) Pattern classification. Wiley
18. Sivic J, Zisserman A (2003) Video google: a text retrieval approach to object matching in videos. In: CVPR
19. Leordeanu M, Collins R, Hebert M (2005) Unsupervised learning of object features from video sequences. In: IEEE computer society conference on computer vision and pattern recognition, IEEE computer society; 1999, vol 1, p 1142
20. Kwak S, Cho M, Laptev I, Ponce J, Schmid C (2015) Unsupervised object discovery and tracking in video collections. In: Proceedings of the IEEE international conference on computer vision, pp 3173–3181
21. Liu D, Chen T (2007) A topic-motion model for unsupervised video object discovery. In: CVPR
22. Wang L, Hua G, Sukthankar R, Xue J, Niu Z, Zheng N (2016) Video object discovery and co-segmentation with extremely weak supervision. IEEE transactions on pattern analysis and machine intelligence
23. Perazzi F, Pont-Tuset J, McWilliams B, Van Gool L, Gross M, Sorkine-Hornung A (2016) A benchmark dataset and evaluation methodology for video object segmentation. In: Computer vision and pattern recognition
24. Lao D, Sundaramoorthi G (2018) Extending layered models to 3d motion. In: Proceedings of the European conference on computer vision (ECCV), pp 435–451
25. Papazoglou A, Ferrari V (2013) Fast object segmentation in unconstrained video. In: Proceedings of the IEEE international conference on computer vision, pp 1777–1784
26. Keuper M, Andres B, Brox T (2015) Motion trajectory segmentation via minimum cost multicuts. In: Proceedings of the IEEE international conference on computer vision, pp 3271–3279
27. Faktor A, Irani M (2014) Video segmentation by non-local consensus voting. In: BMVC, vol 2, p 8
28. Haller E, Leordeanu M (2017) Unsupervised object segmentation in video by efficient selection of highly probable positive features. In: Proceedings of the IEEE international conference on computer vision, pp 5085–5093
29. Luiten J, Voigtlaender P, Leibe B (2018) Premvos: proposal-generation, refinement and merging for the davis challenge on video object segmentation 2018. In: The 2018 DAVIS challenge on video object segmentation-CVPR workshops
30. Maninis KK, Caelles S, Chen Y, Pont-Tuset J, Leal-Taixé L, Cremers D, Van Gool L (2017) Video object segmentation without temporal information. arXiv preprint arXiv:170906031
31. Voigtlaender P, Leibe B (2017) Online adaptation of convolutional neural networks for the 2017 davis challenge on video object segmentation. In: The 2017 DAVIS challenge on video object segmentation-CVPR workshops, vol 5

32. Bao L, Wu B, Liu W (2018) Cnn in mrf: video object segmentation via inference in a cnn-based higher-order spatio-temporal mrf. In: Proceedings of the IEEE conference on computer vision and pattern recognition, pp 5977–5986
33. Wug Oh S, Lee JY, Sunkavalli K, Joo Kim S (2018) Fast video object segmentation by reference-guided mask propagation. In: Proceedings of the IEEE conference on computer vision and pattern recognition, pp 7376–7385
34. Cheng J, Tsai YH, Hung WC, Wang S, Yang MH (2018) Fast and accurate online video object segmentation via tracking parts. In: Proceedings of the IEEE conference on computer vision and pattern recognition, pp 7415–7424
35. Caelles S, Maninis KK, Pont-Tuset J, Leal-Taixé L, Cremers D, Van Gool L (2017) One-shot video object segmentation. In: Proceedings of the IEEE conference on computer vision and pattern recognition, pp 221–230
36. Perazzi F, Khoreva A, Benenson R, Schiele B, Sorkine-Hornung A (2017) Learning video object segmentation from static images. In: Proceedings of the IEEE conference on computer vision and pattern recognition, pp 2663–2672
37. Chen Y, Pont-Tuset J, Montes A, Van Gool L (2018) Blazingly fast video object segmentation with pixel-wise metric learning. In: Proceedings of the IEEE conference on computer vision and pattern recognition, pp 1189–1198
38. Song H, Wang W, Zhao S, Shen J, Lam KM (2018) Pyramid dilated deeper convlstm for video salient object detection. In: Proceedings of the European conference on computer vision (ECCV), pp 715–731
39. Tokmakov P, Alahari K, Schmid C (2017) Learning video object segmentation with visual memory. arXiv preprint arXiv:170405737
40. Jain SD, Xiong B, Grauman K (2017) Fusionseg: learning to combine motion and appearance for fully automatic segmention of generic objects in videos. arXiv preprint arXiv:170105384 2(3):6
41. Yang Z, Wang Q, Bertinetto L, Hu W, Bai S, Torr PH (2019) Anchor diffusion for unsupervised video object segmentation. In: Proceedings of the IEEE international conference on computer vision, pp 931–940
42. Wang W, Song H, Zhao S, Shen J, Zhao S, Hoi SC, Ling H (2019) Learning unsupervised video object segmentation through visual attention. In: Proceedings of the IEEE conference on computer vision and pattern recognition, pp 3064–3074
43. Kulkarni TD, Gupta A, Ionescu C, Borgeaud S, Reynolds M, Zisserman A, Mnih V (2019) Unsupervised learning of object keypoints for perception and control. In: Advances in neural information processing systems, pp 10,723–10,733
44. Minderer M, Sun C, Villegas R, Cole F, Murphy K, Lee H (2019) Unsupervised learning of object structure and dynamics from videos. NeurlPS
45. Thewlis J, Bilen H, Vedaldi A (2017) Unsupervised learning of object landmarks by factorized spatial embeddings. In: Proceedings of the IEEE international conference on computer vision, pp 5916–5925
46. Roufosse JM, Sharma A, Ovsjanikov M (2019) Unsupervised deep learning for structured shape matching. In: Proceedings of the IEEE international conference on computer vision, pp 1617–1627
47. Leordeanu M, Sukthankar R, Hebert M (2009) Unsupervised learning for graph matching. IJCV 96(1)
48. Halimi O, Litany O, Rodola E, Bronstein AM, Kimmel R (2019) Unsupervised learning of dense shape correspondence. In: The IEEE conference on computer vision and pattern recognition (CVPR)
49. Vo HV, Bach F, Cho M, Han K, LeCun Y, Perez P, Ponce J (2019) Unsupervised image matching and object discovery as optimization. In: The IEEE conference on computer vision and pattern recognition (CVPR)
50. Pei Y, Huang F, Shi F, Zha H (2011) Unsupervised image matching based on manifold alignment. IEEE Trans Pattern Anal Mach Intell 34(8):1658–1664

51. Leordeanu M, Zanfir A, Sminchisescu C (2011) Semi-supervised learning and optimization for hypergraph matching. In: ICCV
52. Rezende DJ, Eslami SA, Mohamed S, Battaglia P, Jaderberg M, Heess N (2016) Unsupervised learning of 3d structure from images. In: Advances in neural information processing systems, pp 4996–5004
53. Cha G, Lee M, Oh S (2019) Unsupervised 3d reconstruction networks. In: International conference on computer vision
54. Nunes UM, Demiris Y (2019) Online unsupervised learning of the 3d kinematic structure of arbitrary rigid bodies. In: Proceedings of the IEEE international conference on computer vision, pp 3809–3817
55. Chen Y, Schmid C, Sminchisescu C (2019) Self-supervised learning with geometric constraints in monocular video: connecting flow, depth, and camera. In: Proceedings of the IEEE international conference on computer vision, pp 7063–7072
56. Godard C, Mac Aodha O, Firman M, Brostow GJ (2019) Digging into self-supervised monocular depth estimation. In: Proceedings of the IEEE international conference on computer vision, pp 3828–3838
57. Zhou T, Brown M, Snavely N, Lowe DG (2017) Unsupervised learning of depth and ego-motion from video. In: Proceedings of the IEEE conference on computer vision and pattern recognition, pp 1851–1858
58. Ranjan A, Jampani V, Balles L, Kim K, Sun D, Wulff J, Black MJ (2019) Competitive collaboration: joint unsupervised learning of depth, camera motion, optical flow and motion segmentation. In: Proceedings of the IEEE conference on computer vision and pattern recognition, pp 12,240–12,249
59. Bian J, Li Z, Wang N, Zhan H, Shen C, Cheng MM, Reid I (2019) Unsupervised scale-consistent depth and ego-motion learning from monocular video. In: Advances in neural information processing systems, pp 35–45
60. Gordon A, Li H, Jonschkowski R, Angelova A (2019) Depth from videos in the wild: unsupervised monocular depth learning from unknown cameras. arXiv preprint arXiv:190404998
61. Yang Z, Wang P, Wang Y, Xu W, Nevatia R (2018) Lego: learning edge with geometry all at once by watching videos. In: Proceedings of the IEEE conference on computer vision and pattern recognition, pp 225–234
62. Yang Z, Wang P, Xu W, Zhao L, Nevatia R (2018) Unsupervised learning of geometry from videos with edge-aware depth-normal consistency. In: Thirty-Second AAAI conference on artificial intelligence
63. de Sa VR (1994) Unsupervised classification learning from cross-modal environmental structure. PhD thesis, University of Rochester
64. Hu D, Nie F, Li X (2019) Deep multimodal clustering for unsupervised audiovisual learning. In: The IEEE conference on computer vision and pattern recognition (CVPR)
65. Li Y, Zhu JY, Tedrake R, Torralba A (2019) Connecting touch and vision via cross-modal prediction. In: The IEEE conference on computer vision and pattern recognition (CVPR)
66. Zhang R, Isola P, Efros AA (2017) Split-brain autoencoders: unsupervised learning by cross-channel prediction. In: CVPR, vol 1, p 5
67. Pan JY, Yang HJ, Faloutsos C, Duygulu P (2004) Automatic multimedia cross-modal correlation discovery. In: Proceedings of the tenth ACM SIGKDD international conference on Knowledge discovery and data mining, ACM, pp 653–658
68. He L, Xu X, Lu H, Yang Y, Shen F, Shen HT (2017) Unsupervised cross-modal retrieval through adversarial learning. In: 2017 IEEE International conference on multimedia and expo (ICME), IEEE, pp 1153–1158
69. Zhao H, Gan C, Rouditchenko A, Vondrick C, McDermott J, Torralba A (2018) The sound of pixels. In: Proceedings of the European conference on computer vision (ECCV), pp 570–586
70. Bengio Y, Louradour J, Collobert R, Weston J (2009) Curriculum learning. In: Proceedings of the 26th annual international conference on machine learning, ACM, pp 41–48
71. Koffka K (2013) Principles of Gestalt psychology. Routledge
72. Rock I, Palmer S (1990) Gestalt psychology. Sci Am 263:84–90

73. Stretcu O, Leordeanu M (2015) Multiple frames matching for object discovery in video. In: BMVC, pp 186–1
74. Ronneberger O, Fischer P, Brox T (2015) U-net: convolutional networks for biomedical image segmentation. In: International conference on medical image computing and computer-assisted intervention. Springer, pp 234–241
75. Leordeanu M, Hebert M (2005) A spectral technique for correspondence problems using pairwise constraints. In: ICCV
76. Leordeanu M, Hebert M, Sukthankar R (2009) An integer projected fixed point method for graph matching and map inference. In: NIPS
77. Brendel W, Todorovic S (2010) Segmentation as maximum-weight independent set. In: NIPS
78. Jain A, Gupta A, Rodriguez M, Davis L (2013) Representing videos using mid-level discriminative patches. In: Computer vision and pattern recognition, pp 2571–2578
79. Semenovich D (2010) Tensor power method for efficient map inference in higher-order mrfs. In: ICPR
80. Monroy A, Bell P, Ommer B (2014) Morphological analysis for investigating artistic images. Image Visi Comput 32(6)
81. Leordeanu M, Sminchisescu C (2012) Efficient hypergraph clustering. In: International conference on artificial intelligence and statistics
82. Leordeanu M, Radu A, Baluja S, Sukthankar R (2015) Labeling the features not the samples: efficient video classification with minimal supervision. arXiv preprint arXiv:151200517
83. Haller E, Leordeanu M (2017) Unsupervised object segmentation in video by efficient selection of highly probable positive features. In: The IEEE international conference on computer vision (ICCV)
84. Haller E, Florea AM, Leordeanu M (2019) Spacetime graph optimization for video object segmentation. arXiv preprint arXiv:190703326
85. Besag J (1986) On the statistical analysis of dirty pictures. J Roy Stat Soc 48(5):259–302
86. Frank M, Wolfe P (1956) An algorithm for quadratic programming. Naval Res Logistics Q 3(1–2):95–110
87. Magnus JR, Neudecker H (1999) Matrix differential calculus with applications in statistics and econometrics. Wiley
88. Cour T, Shi J, Gogin N (2005) Learning spectral graph segmentation. In: International conference on artificial intelligence and statistics
89. Ding C, Li T, Jordan M (2008) Nonnegative matrix factorization of combinatorial optimization: spectral clustering, graph matching, and clique finding. In: IEEE international conference on data mining
90. Motzkin T, Straus E (1965) Maxima for graphs and a new proof of a theorem of turan. Canad J Math
91. Bulo S, Pellilo M (2009) A game-theoretic approach to hypergraph clustering. In: NIPS
92. Liu H, Latecki L, Yan S (2010) Robust clustering as ensembles of affinity relations. In: NIPS
93. Jia Y, Shelhamer E, Donahue J, Karayev S, Long J, Girshick R, Guadarrama S, Darrell T (2014) Caffe: convolutional architecture for fast feature embedding. In: ACM multimedia
94. Prest A, Leistner C, Civera J, Schmid C, Ferrari V (2012) Learning object class detectors from weakly annotated video. In: CVPR
95. Alexe B, Deselaers T, Ferrari V (2012) Measuring the objectness of image windows. IEEE Trans Pattern Anal Mach Intell 34(11):2189–2202
96. Meila M, Shi J (2001) A random walks view of spectral segmentation. In: AISTATS
97. Leordeanu M, Sukthankar R, Hebert M (2012) Unsupervised learning for graph matching. Int J Comput Vis 96:28–45
98. Croitoru I, Bogolin SV, Leordeanu M (2017) Unsupervised learning from video to detect foreground objects in single images. In: 2017 IEEE international conference on computer vision (ICCV), IEEE, pp 4345–4353
99. Croitoru I, Bogolin SV, Leordeanu M (2019) Unsupervised learning of foreground object segmentation. Int J Comput Vis:1–24

Chapter 2
Unsupervised Learning of Graph and Hypergraph Matching

2.1 Introduction

Graph and hypergraph matching are important problems in vision. They are applied to a wide variety of tasks that require 2D and 3D feature matching, such as image alignment, 3D reconstruction, and object or action recognition. Graph matching considers pairwise constraints that usually encode geometric and appearance relationships between local features. On the other hand, hypergraph matching incorporates higher order relations computed over sets of features, which could capture both geometric and appearance information. Therefore, using higher order constraints enables matching that is more robust (or even invariant) to changes in scale, non-rigid deformations, and outliers (Fig. 2.1).

Many objects or other entities (e.g., human activities) in the spatiotemporal domain can be represented by graphs with local information on nodes and more global information on edges or hyperedges. The problem of finding correspondences between two graph models arises in many computer vision tasks, and the types of features could vary greatly from one application to the next. We could use graph matching for registering shapes, recognizing object categories or go to more complex problems such as recognizing activities in video. Graph matching is ultimately about finding agreements between features extracted at the local level and between information computed at the higher order of edges or hyperedges (in the case of hypergraph matching). Such higher order agreements, which pass from one neighborhood to the next the local information at the nodes, ensure at the end the overall global align-

---

The material presented in this chapter is based in large part on the following papers:

Leordeanu, Marius, and Martial Hebert. "A spectral technique for correspondence problems using pairwise constraints." IEEE International Conference on Computer Vision (ICCV), 2005.

Leordeanu, Marius, Martial Hebert, and Rahul Sukthankar. "An integer projected fixed point method for graph matching and map inference." In Advances in neural information processing systems (NIPS) 2009.

Leordeanu, Marius, Rahul Sukthankar, and Martial Hebert. "Unsupervised learning for graph matching." International journal of computer vision 96, no. 1 (2012): 28–45.

Leordeanu, Marius, Andrei Zanfir, and Cristian Sminchisescu. "Semi-supervised learning and optimization for hypergraph matching." IEEE International Conference on Computer Vision, 2011.

ment between two graph models. During the matching process, which is usually iterative, messages about possible node correspondences are passed through edges or hyperedges and information is eventually spread to the whole set of potential node assignments between the two graphs and node-to-node matching convergence is eventually reached. Note that by considering information at orders beyond the pairwise edges, hypergraph matching could be more robust to changes in scale, rotations, or other transformations in geometry and appearance at the level of groups of nodes or cliques.

The amount of research focusing on graph matching has been constantly increasing in recent years [1–15], with most new approaches using or being derived from the initial IQP formulation [2, 16]. Not many papers investigate the task of learning the parameters of the graph matching model, but interest in the topic is increasing [17–21]. Graph matching with higher order constraints, known as hypergraph matching, is also receiving an increasing attention in computer vision [22–29] mainly due to a more powerful geometric modeling, often capable of similarity or even affine-invariant matching and the increased capacity to capture-invariant information at the higher level of cliques. Another very relevant research direction in graph matching is in solving the task efficiently directly in the original discrete domain, in which the final solution should be [6, 27, 30, 31]. In our original paper on the topic [6], we showed in extensive experiments that optimizing sub-optimally directly in the correct discrete domain could often be largely superior to global optimization in the relaxed, continuous domain followed by a very simplistic binarization step. Another interesting direction in graph matching is to find node correspondences among several graphs, which would enable graph matching across multiple images or object models. The multi-graph matching task has found some notable approaches in the literature [32, 33], including our initial tensor formulation [34].

The material presented in this chapter is based on our previously published works [19, 35]. It consists of a set of methods for graph and hypergraph matching and learning, generally considered to be of state-of-the-art level, in terms of accuracy, computational speed, and overall efficiency. Our approaches to unsupervised and semi-supervised learning for graph and hypergraph matching are, to the best of our knowledge, the first such algorithms proposed in the computer vision literature.

In this chapter, we discuss some of our contributions to modern graph and hypergraph matching for computer vision. We present methods for both graph and hypergraph matching, optimization, and learning, in order to convey a comprehensive, unified view of these closely related problems. The algorithms we develop for both learning and optimization are based on the same core ideas, the higher order formulation being a natural extension of the second-order graph matching case. First, we present the general QAP formulation of graph matching, which we introduced in Leordeanu and Hebert [16]. Then, we extend it to hypergraph matching, as we initially proposed it in Leordeanu et al. [36]. Our formulation follows the work of Duchenne et al. [22], which extended the power iteration solution of our initial spectral matching algorithm to the higher order case. We present two main algorithms for graph matching, that is, Spectral Matching (SM) [16] and the Integer Projected

Hyper-graph Matching Accuracy: 100%   Graph Matching Accuracy: 64%

Fig. 2.1 Hypergraph matching can be more powerful than graph matching. The geometric and appearance scale-invariant model applied by matching with higher order constraints gives superior performance than the simpler pairwise model, in cases when drastic changes in scale occur

Fixed Point Method (IPFP) for graph matching and MAP inference in conditional random fields [6]. The two methods work well in combination, the first providing an optimal solution to the relaxed problem and the second offering a powerful discretization procedure with important theoretical guarantees, such as climbing and convergence properties. Next, we present our original method for supervised and unsupervised learning for graph matching, which introduces a powerful, efficient gradient descent-based algorithm.

Our approach to learning and optimization for the graph matching problem with pairwise constraints constitutes a theoretical as well as practical basis for the development of similar methods of hypergraph matching with higher order geometric and appearance constraints. We present our extensions to the higher order case and provide experiments that show state-of-the-art performance on both synthetic and real-world images. This chapter presents a unified view of learning and optimization for efficient graph and hypergraph matching, which will then be extended to graph and hypergraph clustering in Chap. 3.

2.1.1 Relation to Principles of Unsupervised Learning

We reiterate below the unsupervised learning principle (Principle 5), introduced in the previous chapter, which is most related to graph matching and instrumental for understanding its key insights:

Principle 5
Accidental alignments are rare. When they happen they usually indicate correct alignments between a model and an image. Alignments, which could be geometric or appearance based, rare as they are, when they take place form a strong cluster of agreements that re-enforce each other in multiple ways.

We cannot stress enough how important the above principle is for inference and learning in the case of graph matching. In the context of this task, alignments often refer to agreements in geometry or appearance between nodes and edges (or hyperedges) of the two graphs. They may also refer to similarities at higher levels of semantic abstraction. Such agreements, however, do not regularly happen accidentally in nature. Accidental alignments are indeed rare and very sparse, without forming strong clusters of agreements in space and time. For example, it is possible that some parts of an object may look similar to some parts of another, unrelated object, by accident. It is also possible that by chance a random pattern in the continuously moving world may resemble some well-known object (e.g., configuration of clouds that may resemble a puppy). However, it is highly unlikely that two structures, stable in space and time, look alike at the level of whole objects in the absence of any meaningful connection between them. When such agreements, at the level of shape, appearance, and full structure, are overwhelming, they usually indicate that there is a meaningful connection between the two entities. Such relationship could be, for example, at the level of identity (e.g., it is the same object seen from different viewpoints or seen in two different moments in time), or at the level of semantic category (e.g., the objects are of the same kind, have a similar purpose or meaning).

As we will see in this chapter, the alignments at the level of nodes and edges establish strong links in a special graph of candidate assignments between the nodes of the two graphs (also discussed in Chap. 1). Through such strong agreement links, the candidate assignments that are most likely to be correct will form naturally a strong cluster. Consequently, finding the correct assignments reduces to a clustering problem that often should obey certain matching constraints, such that one node from one graph could match at most one node in the other graph. Therefore, the connection between graph matching and clustering comes naturally, which is why algorithms that were designed for one problem (e.g., spectral clustering or integer projected fixed point for graph matching), could be easily adapted for the other problem with minimal modification (usually at the level of constraints on the solution vector).

Next, we present our mathematical formulation for graph matching. Then, based on the intuitions introduced above, we introduce the spectral graph matching and integer projected fixed point algorithms. They will be further extended to the higher order matching case (hypergraph matching) and, in Chap. 3, adapted to the task of clustering with higher order constraints. It is important to remember that all methods presented in these chapters, for inference or learning, are fundamentally linked to the general concept of clustering, which is at the core of all unsupervised learning.

2.2 Graph Matching

The graph matching problem with pairwise constraints consists of solving for the indicator vector $\mathbf{x}^*$ that maximizes a quadratic score function with certain mapping constraints:

$$\mathbf{x}^* = \arg\max\ (\mathbf{x}^T \mathbf{M} \mathbf{x})\ \text{s.t.}\ \mathbf{A}\mathbf{x} = \mathbf{1},\ \mathbf{x} \in \{0, 1\}^n, \tag{2.1}$$

where $\mathbf{x}$ is an indicator vector such that $x_{ia} = 1$ if feature i from one image (or graph) is matched to feature a from the other image (or graph), and zero otherwise; $\mathbf{A}\mathbf{x} = \mathbf{1}$, $\mathbf{x} \in \{0, 1\}^n$ enforces one-to-one constraints on $\mathbf{x}$ such that one feature from one image can be matched to at most one other feature from the other. In our work, $\mathbf{M}$ is a matrix with positive elements containing the pairwise score functions, such that $M_{ia;jb}$ measures how well the pair of features (i, j) from one image agrees in terms of geometry and appearance (e.g., difference in local appearance descriptors, pairwise distances, angles, etc.) with a pair of candidate matches (a, b) from the other.

While the Spectral Matching (SM) algorithm requires $\mathbf{M}$ to have non-negative elements, the Integer Projected Fixed Point (IPFP) method does not impose this requirement. Also note that the local, unary terms of candidate assignments can be stored on the diagonal of $\mathbf{M}$; in practice, we noticed that considering such local, unary information in the pairwise scores $M_{ia;jb}$ and leaving zeros on the diagonal produce superior results; a slightly different form of the matching objective that combines both linear and quadratic terms is also possible: $\mathbf{d}\mathbf{x} + \mathbf{x}^T \mathbf{M} \mathbf{x}$. Here, the elements of vector $\mathbf{d}$ represent the unary scores, and $\mathbf{M}$ contains the pairwise constraints, as before. Note that this formulation is not applicable to spectral graph matching (Sect. 2.4.1), but it is valid for IPFP (Sect. 2.4.2) and similar to quadratic formulations of the MAP (Maximum A Posteriori) inference problem in Pairwise Markov Networks [37–39]. The main difference is that for the MAP inference task, the constraints $\mathbf{A}\mathbf{x} = \mathbf{1}$, $\mathbf{x} \in \{0, 1\}^n$ on the solution are usually many-to-one: many nodes from the graph can have the same label. Many algorithms, including our IPFP, can handle both problems, with slight modifications. In this chapter, we present similar learning algorithms applicable to both problems, graph matching and MAP Inference, in order to emphasize once more that the two tasks are closely related, and give a general, encompassing view of our optimization and learning approaches.

Coming back to the main QAP graph matching formulation, we mention that $M_{ia;jb}$ is essentially a function that is defined by a certain parameter vector $\mathbf{w}$. The type of pairwise scores $M_{ia;jb}$ that we use, also closely related to the type of relationships we apply to hypergraph matching, is

$$M_{ia;jb} = \exp(-\mathbf{w}^T \mathbf{g}_{ia;jb}), \tag{2.2}$$

where $\mathbf{w}$ is a vector of learned parameter weights, and $\mathbf{g}_{ia;jb}$ is a vector that usually contains non-negative errors and deformations, to describe the changes in geometry and appearance when we match the pair of features (i, j) to the pair (a, b).

2.3   Hypergraph Matching

Hypergraph matching follows naturally from graph matching, by extending Eq. 2.1 to a tensor formulation that is able to include higher order constraints. The first hypergraph matching approaches in computer vision are the probabilistic method [24] and the tensor-based higher order power method [22]. The formulation we present applies to hypergraph models of any order. However, in practice, third-order constraints offer a good compromise between efficiency and the ability to capture the appearance and geometry of objects. The classical graph matching approach deals with pairwise relationships between features that are not invariant under similarity, affine, or projective transformations. Invariance to such transformations is possible if third-order constraints are used [22]. Third-order relationships could still be handled efficiently, while being able to model both higher order geometry and appearance. Objects could be described by triangulated meshes, with features located at corners and appearance computed over the triangles' interiors. In the rest of the chapter, we discuss only third-order hypergraph matching, while keeping in mind that the same formulation can be extended, in principle, to matching using relations of any order.

Given two sets of features, one from a model image I_m and the other from a test image I_t, third-order (hypergraph) matching consists of finding correspondences between the two sets, such that a matching score, as a sum over triplets of candidate matches, is maximized. The score could consider geometric and appearance information over the triplets of potential assignments, extending the pairwise scores used for graph matching.

Having a list of candidate assignments $L_{ia} = (i, a)$, with a unique index ia for each potential correspondence (i, a) (with feature i from I_m and feature a from I_t), we express a possible solution by an indicator vector $\mathbf{x}$ (of the same length as L), such that $x_{ia} = 1$ if feature i from the model is matched to feature a from the test image, and $x_{ia} = 0$ otherwise. One-to-one mapping constraints are usually imposed, such that one feature from one image can match a single feature from the other image, and vice versa: $\sum_i x_{ia} \leq 1$, $\sum_a x_{ia} \leq 1$ and $\mathbf{x} \in \{0, 1\}^n$. The mapping constraints are written in matrix form $\mathbf{Ax} \leq \mathbf{1}$, $\mathbf{x} \in \{0, 1\}^n$, with $\mathbf{A}$ a binary matrix. The third-order matching score is

$$S(\mathbf{x}) = \sum_{ia;jb;kc} H_{ia;jb;kc} x_{ia} x_{jb} x_{kc}. \tag{2.3}$$

Here, $\mathbf{H}$ is a super-symmetric tensor with non-negative elements. They are increasing with the quality of the match between tuples of features: $H_{ia;jb;kc}$ indicates how well features (i, j, k) from the model image match, in terms of appearance and

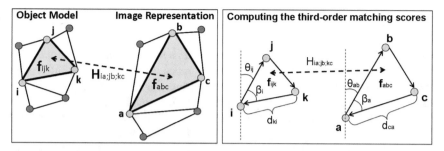

Fig. 2.2 Left: matching the object model to the image by using third-order representations (matching scores computed over triangles, considering both geometry and appearance). Right: the third-order scores $H_{ia;jb;kc}$ can be functions of both geometry and appearance. Geometric information could be represented by rotation-invariant angles such as β_i, rotation-dependent angles such as θ_{ij} and pairwise distances d_{ki}, which could be scale-invariant if divided by the perimeter. Appearance could be represented by a feature vector $\mathbf{f}_{ijk}$, encoding gradient, color, or texture information. The relative importance of each cue is application-dependent and should be automatically learned during training

geometry, features (a, b, c) from the test image. Solving hypergraph matching means finding the solution that maximizes $S(\mathbf{x})$, under one-to-one mapping constraints:

$$\mathbf{x}^* = \arg\max_{\mathbf{x}} \sum_{ia;jb;kc} H_{ia;jb;kc} x_{ia} x_{jb} x_{kc}. \tag{2.4}$$

In practice, one can use any type of information to design the scores $H_{ia;jb;kc}$. Most authors focus on simple geometric relationships [23, 24], such as differences in the angles β_i of the triangles formed by triplets of features (Fig. 2.2). In this work, we show that by learning powerful scores that include both geometric and appearance information we can significantly improve matching performance. The scores we use are

$$H_{ia;jb;kc} = \exp(-\mathbf{w}^T \mathbf{g}_{ia;jb;kc}), \tag{2.5}$$

where $\mathbf{w}$ is a vector of weights/parameters to be learned and $\mathbf{g}_{ia;jb;kc}$ is a vector containing non-negative errors/deformations, modeling the change in geometry and appearance when matching the triangle defined by features (i, j, k) in the model image I_m to the triangle (a, b, c) in the test image I_t. See Fig. 2.2 for details.

2.4 Solving Graph Matching

Graph and hypergraph matching are NP-hard problems, so we must look for efficient approximate methods. Here, we briefly review our main methods for graph matching, spectral matching [16], and Integer Projected Fixed Point (IPFP) [6], which relate to our later work on hypergraphs, which will be discussed during the later part of the chapter.

2.4.1 Spectral Matching

Since its publication, our approach to spectral graph matching has been applied successfully in a wide range of computer vision applications, significantly influencing both research works and commercial patents. Here, we enumerate some representative research directions that used spectral graph matching in various scientific domains: action recognition and modeling of object interactions in video [40–44], discovering and modeling of object categories [45–50], capturing 3D human performance [51], 3D scene acquisition [52], matching objects in 2D [53], and symmetry detection and analysis [54, 55], automatic annotation in 3D shape collections [56], detecting sulcal cerebral cortex patterns in medical images [57–59], medical image registration [60], the alignment of multiple protein interaction networks [61], Wi-Fi-based indoor localization [62], person re-identification [63], surveillance video summarization [64], and detecting and tracking skin lesions [65].

Spectral matching was also the starting point in the development of other matching algorithms, such as spectral matching with affine constraints (SMAC) [3], integer projected fixed point (IPFP) [6], tensor higher order matching [22], probabilistic and approximate spectral graph and hypergraph matching [8, 24, 66–68], and algorithms for MAP inference based on spectral relaxations [37, 38].

Spectral matching optimally solves the following relaxed variant of Problem 2.1:

$$\mathbf{x}^* = \arg \max \left(\mathbf{x}^T \mathbf{M} \mathbf{x} \right) \quad \text{s.t. } \mathbf{x}^T \mathbf{x} = \mathbf{1}. \tag{2.6}$$

The solution to this problem is given by the first eigenvector of $\mathbf{M}$. Since $\mathbf{M}$ has only positive elements, by the Perron–Frobenius theorem, the eigenvector elements are also positive, which makes the post-processing discretization of the eigenvector easier. This eigenvector also has an intuitive interpretation due to the statistical properties of $\mathbf{M}$. We observe that $\mathbf{M}$ can be interpreted as the adjacency matrix of a graph whose nodes represent candidate assignments and edges $M_{ia;jb}$ represent agreements between these possible assignments. This graph has a particular structure, which helps us understand why using the first eigenvector to find an approximate solution to Problem 2.1 is a good idea. It contains the following:

1. Strongly connected cluster formed mainly by the correct assignments that tend to establish agreement links (strong edges) among each other. These agreement links are formed when pairs of assignments agree at the level of pairwise relationships (e.g., geometry) between the features they are putting in correspondence (see Fig. 2.3).
2. A lot of incorrect assignments, mostly outside of that cluster or weakly connected to it (through weak edges), which do not form strongly connected clusters due to their small probability of establishing agreement links and random, unstructured way in which they form these links.

These statistical properties motivate the spectral approach to the problem. The eigenvector value corresponding to a given assignment indicates the level of *associ-*

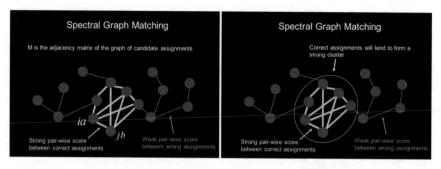

Fig. 2.3 **Left**: each node in the graph represents a potential assignment, such as ia, where node i from one graph could potentially matching node a from the other graph. Matrix **M**, which is used to define graph matching as an Integer Quadratic Program (Eq. 2.1) becomes the adjacency matrix of the graph of candidate assignments. **Right**: the correct assignments are more likely to establish strong agreement links (thick blue lines in the figure) and then incorrect ones. Thus, a natural cluster is formed by the correct candidate assignments in the graph—which could be found rapidly by our spectral approach, as they correspond to high values in the principal eigenvector of **M**. In this manner, a notoriously hard problem becomes easy to solve by exploiting the natural statistics in the data, as they are reflected in **M**

ation of that assignment with the main cluster. We can employ a variety of discretization procedures in order to find an approximate solution. One idea is to apply the Hungarian method, which efficiently finds the binary solution that obeys the one-to-one mapping constraints and maximizes the dot product with the eigenvector of **M**. Another idea is to use the greedy discretization algorithm that we originally proposed in Leordeanu and Hebert [16]: we interpret each element of the principal eigenvector **v** of **M** as the confidence that the corresponding assignment is correct. We start by choosing the element of maximum confidence as correct, then we remove (zero out in **v**) all the assignments in conflict (w.r.t. the one-to-one mapping constraints) with the assignment chosen as correct, and then we repeat this procedure until all assignments are labeled as either correct or incorrect. The eigenvector relaxation of the graph matching problem, combined with the greedy discretization procedure, makes spectral matching one of the most efficient algorithms for matching using pairwise constraints.

2.4.2 Integer Projected Fixed Point Algorithm

In a more recent paper Leordeanu et al. [6], we presented another efficient graph matching algorithm (IPFP) that outperforms many state-of-the-art methods. Since its publication, modified versions of IPFP have been applied to tasks such as segmentation [69], higher level video representation [70], hyper-order MRFs [71], analysis of artistic images [72], generalizing IPFP algorithm for outlier detection [73], and fast IPFP for large-scale graph matching [74], among others. An algorithm that shares many of IPFP's properties in a different formulation was independently developed by Zaslavskiy et al. [75] and Zaslavskiy [76].

IPFP can be used as a standalone algorithm, or as a discretization procedure for other graph matching algorithms, such as spectral matching. Moreover, IPFP can also be used for MAP inference in MRF's and CRF's, formulated as quadratic integer programming problems. It solves efficiently (though not necessarily optimally) a tighter relaxation of Problem 2.1 that is also known to be NP-hard:

$$\mathbf{x}^* = \arg \max (\mathbf{x}^T \mathbf{M} \mathbf{x}) \quad \text{s.t. } \mathbf{A} \mathbf{x} = \mathbf{1}, \ \mathbf{x} \geq \mathbf{0}. \tag{2.7}$$

The only difference between Problems 2.1 and 2.7 is that in the latter the solution is allowed to take continuous values. Integer Projected Fixed Point (IPFP) takes as input any initial solution, continuous or discrete, and quickly finds a solution obeying the initial discrete constraints of Problem 2.1 with a better score (most often significantly better) than the initial one:

Algorithm 2.1 Sequential second-order expansion for higher order matching

---

Initialize $\mathbf{x}^* = \mathbf{x}_0$, $S^* = \mathbf{x}_0^T \mathbf{M} \mathbf{x}_0$, $k = 0$, where $x_i \geq 0$ and $\mathbf{x} \neq \mathbf{0}$;
repeat
   1) Let $\mathbf{b}_{k+1} = P_d(\mathbf{M} \mathbf{x}_k)$, $C = \mathbf{x}_k^T \mathbf{M}(\mathbf{b}_{k+1} - \mathbf{x}_k)$, $D = (\mathbf{b}_{k+1} - \mathbf{x}_k)^T \mathbf{M}(\mathbf{b}_{k+1} - \mathbf{x}_k)$;
   2) If $D \geq 0$, set $\mathbf{x}_{k+1} = \mathbf{b}_{k+1}$. Else let $r = \min(-C/D, 1)$ and set $\mathbf{x}_{k+1} = \mathbf{x}_k + r(\mathbf{b}_{k+1} - \mathbf{x}_k)$;
   3) If $\mathbf{b}_{k+1}^T \mathbf{M} \mathbf{b}_{k+1} \geq S^*$ then set $S^* = \mathbf{b}_{k+1}^T \mathbf{M} \mathbf{b}_{k+1}$ and $\mathbf{x}^* = \mathbf{b}_{k+1}$;
   4) Set $k = k + 1$
until $\mathbf{x}_{k-1} = \mathbf{x}_k$
return $\mathbf{x}^*$

Here, the $P_d(.)$ in step 2 denotes a projection on to the discrete domain, discussed below.

Relation to the Power Iteration Algorithm: This algorithm is loosely related to the power iteration method for eigenvectors, also used by spectral matching: at step 2, it replaces the fixed point iteration of the power method $\mathbf{v}_{k+1} = P(\mathbf{M} \mathbf{v}_k)$, where $P(.)$ denotes the projection on the unit sphere, with the analogous update $\mathbf{b}_{k+1} = P_d(\mathbf{M} \mathbf{x}_k)$, in which $P_d(.)$ denotes projection on the one-to-one (for graph matching) or many-to-one (for MAP inference) discrete constraints. Since all possible discrete solutions have the same norm, $P_d(.)$ boils down to finding the discrete vector $\mathbf{b}_{k+1} = \arg \max_{\mathbf{b}}(\mathbf{b}^T \mathbf{M} \mathbf{x}_k)$. For one-to-one constraints, this can be efficiently accomplished using the Hungarian method; for many-to-one constraints, the projection can easily be achieved in linear time.

Relation to Taylor Approximation: The intuition behind this algorithm is the following: at every iteration, the quadratic score $\mathbf{x}^T \mathbf{M} \mathbf{x}$ can be approximated by the first-order Taylor expansion around the current solution $\mathbf{x}_k$: $\mathbf{x}^T \mathbf{M} \mathbf{x} \approx \mathbf{x}_k^T \mathbf{M} \mathbf{x}_k + $

$2\mathbf{x}_k^T\mathbf{M}(\mathbf{x} - \mathbf{x}_k)$. This approximation is maximized within the discrete domain of Problem 2.1, in step 2, where $\mathbf{b}_{k+1}$ is found.

From Leordeanu et al. [6], Proposition 2.1 we know that the same discrete $\mathbf{b}_{k+1}$ also maximizes the linear approximation in the continuous domain of Problem 2.7. The role of $\mathbf{b}_{k+1}$ is to provide a direction of largest possible increase (or ascent) in the first-order approximation, simultaneously within both the continuous and discrete domains. Along this direction, the original quadratic score can be further maximized in the continuous domain of Problem 2.7 (as long as $\mathbf{b}_{k+1} \neq \mathbf{x}_k$). At step 3, we find the optimal point along this direction, also inside the continuous domain of Problem 2.7. The hope, also confirmed in practice, is that the algorithm will tend to converge towards discrete solutions that are, or are close to, maxima of Problem 2.7. For MAP inference problems, as shown in Leordeanu et al. [6], IPFP always converges to discrete solutions, while for graph matching we observe that it typically converges to discrete solutions, but not always.

Relation to Iterative Conditional Modes Algorithm: IPFP can also be seen as an extension to the popular Iterated Conditional Modes (ICM) algorithm [77], having the advantage of updating the solution for all nodes in parallel, while retaining the optimality and convergence properties.

Relation to Frank–Wolfe Algorithm: IPFP is related to the Frank–Wolfe method (FW) [78], a classical optimization algorithm from 1956 most often used in operations research [79]. The Frank–Wolfe method is applied to convex programming problems with linear constraints. A well-known fact about FW is that it has slow convergence rate around the optimum due to the convexity of the objective function, which is why in practice it is stopped earlier for obtaining an approximate solution. In contrast, in the case of graph matching IPFP (generally applied to non-convex problems) the local optimum is most often discrete (for MAP, it is always discrete). When the solution is discrete the optimum is actually found during the optimization of the linear approximation, when the discrete point is found, so the convergence is immediate. This insight is also demonstrated in our experiments, where IPFP most often converges quickly. The problem domain of IPFP, with its particular one-to-one or many-to-one constraints, gives IPFP a nice property over the general case FW, that is, the discrete solution to the projection function (Step 2) with a simple optimization algorithm, applied directly in the discrete domain. Also, the quadratic function in one parameter that defines the line search step has a closed-form, optimal solution. This is another important property of IPFP for efficient graph matching.

2.5 Theoretical Analysis

Proposition 2.1 *For any vector* $\mathbf{x} \in R^n$, *there exists a global optimum* $\mathbf{y}^*$ *of* $\mathbf{x}^T\mathbf{M}\mathbf{y}$ *in the domain of Problem 2.2 that has binary elements (thus it is also in the domain of Problem 2.1).*

Proof Maximizing $\mathbf{x}^T\mathbf{M}\mathbf{y}$ with respect to $\mathbf{y}$, subject to $\mathbf{A}\mathbf{y} = \mathbf{1}$ and $\mathbf{y} > 0$ is linear program for which an integer optimal solution exists because the constraints matrix $\mathbf{A}$ is totally unimodular [9]. This is true for both one-to-one and many-to-one constraints.

It follows that the maximization from step 2 $\mathbf{b}_{k+1} = \arg\max \mathbf{b}^T\mathbf{M}\mathbf{x}_k$ in the original discrete domain, also maximizes the same dot product in the continuous domain of Problem 2.2, of relaxed constraints $\mathbf{A}\mathbf{x} = \mathbf{1}$ and $\mathbf{x} > 0$. This ensures that the algorithm will always move towards some discrete solution that also maximizes the linear approximation of the quadratic function in the domain of Problem 2.2. Most often in practice, that discrete solution also maximizes the quadratic score, along the same direction and within the continuous domain. Therefore, $\mathbf{x}_k$ is likely to be discrete at every step.   □

Proposition 2.2 *The quadratic score* $\mathbf{x}_k^T\mathbf{M}\mathbf{x}_k$ *increases at every step k and the sequence of* $\mathbf{x}_k$ *converges*

Proof For a given step k, if $\mathbf{b}_{k+1} = \mathbf{x}_k$ we have convergence. If $\mathbf{b}_{k+1} \neq \mathbf{x}_k$, let $\mathbf{x}$ be a point on the line between $\mathbf{x}_k$ and $\mathbf{b}_{k+1}$, $\mathbf{x} = \mathbf{x}_k + t(\mathbf{b}_{k+1} - \mathbf{x}_k)$. For any $0 \leq t \leq 1$, $\mathbf{x}$ is in the feasible domain of Problem 2.2. Let $S_k = \mathbf{x}_k^T\mathbf{M}\mathbf{x}_k$. Let us define the quadratic function $f(t) = \mathbf{x}^T\mathbf{M}\mathbf{x} = S_k + 2tC + t^2D$, which is the original function in the domain of Problem 2.2 on the line between $\mathbf{x}_k$ and $\mathbf{b}_{k+1}$. Since $\mathbf{b}_{k+1}$ maximizes the dot product with $\mathbf{x}_k^T\mathbf{M}$ in the discrete (and the continuous) domain, it follows that $C \geq 0$. We have two cases: $D \geq 0$, when $\mathbf{x}_{k+1} = \mathbf{b}_{k+1}$ (step 3) and $S_{k+1} = \mathbf{x}_{k+1}^T\mathbf{M}\mathbf{x}_{k+1} = f_q(1) \geq S_k = \mathbf{x}_k^T\mathbf{M}\mathbf{x}_k$; and $D < 0$, when the quadratic function $f_q(t)$ is convex with the maximum in the domain of Problem 2.2 attained at point $\mathbf{x}_{k+1} = \mathbf{x}_k + r(\mathbf{b}_{k+1} - \mathbf{x}_k)$. Again, it also follows that $S_{k+1} = \mathbf{x}_{k+1}^T\mathbf{M}\mathbf{x}_{k+1} = f_q(r) \geq S_k = \mathbf{x}_k^T\mathbf{M}\mathbf{x}_k$. Therefore, the algorithm is guaranteed to increase the score at every step. Since the score function is bounded above on the feasible domain, it has to converge, which happens when $C = 0$.   □

By always improving the quadratic score in the continuous domain, at each step the next solution moves towards discrete solutions that are better suited for solving the original Problem 2.1.

For many-to-one constraints (MAP inference), it basically follows that the algorithm will converge to a discrete solution, since maxima of Problem 2.2 are in the discrete domain [11, 12, 13]. This is another similarity with ICM, which also converges to a maximum. Therefore, combining ours with ICM cannot improve the performance of ICM, and vice versa.

Proposition 2.3 *If* $\mathbf{M}$ *is positive semi-definite with positive elements, then the algorithm converges in a finite number of iterations to a discrete solution, which is a maximum of Problem 2.2.*

Proof Since $\mathbf{M}$ is positive semi-definite, we always have $D \geq 0$, and thus $\mathbf{x}_k$ is always discrete for any k. Since the number of discrete solutions is finite, the algorithm must converge in a finite number of steps to a local (or global) maximum, which must be discrete. This result is obviously true for both one-to-one and many-to-one constraints. □

When $\mathbf{M}$ is positive semi-definite, Problem 2.2 is a concave minimization problem for which it is well known that the global optimum has integer elements, so it is also a global optimum of the original Problem 2.1. In this case, our algorithm is only guaranteed to find a local optimum in a finite number of iterations. Global optimality of concave minimization problems is a notoriously difficult task since the problem can have an exponential number of local optima. In fact, if a large enough constant is added to the diagonal elements of $\mathbf{M}$, every point in the original domain of possible solutions becomes a local optimum for one-to-one problems. Therefore, adding a large constant to make the problem concave is not a good idea, even if the global optimum does not change. In practice, $\mathbf{M}$ is rarely positive semi-definite, but it can be close to being one if the first eigenvalue is much larger than the others, which is the assumption made by our spectral matching algorithm, for example.

Proposition 2.4 *If $\mathbf{M}$ has non-negative elements and is rank-1, then the algorithm will converge and return the global optimum of the original problem after the first iteration.*

Proof Let $\mathbf{v}, \lambda$ be the leading eigenpair of $\mathbf{M}$. Then, since $\mathbf{M}$ has non-negative elements both $\mathbf{v}$, and λ are positive. Since $\mathbf{M}$ is also rank one, we have $\mathbf{M}\mathbf{x}_0 = \lambda(\mathbf{v}^T\mathbf{x}_0)\mathbf{v}$. Since both $\mathbf{x}_0$ and $\mathbf{v}$ have positive elements, it immediately follows that $\mathbf{x}_1$ after the first iteration is the indicator solution vector that maximizes the dot product with the leading eigenvector ($\mathbf{v}^T\mathbf{x}_0 = 0$ is a very unlikely case that never happens in practice). It is clear that this vector is the global optimum, since in the rank-1 case we have $\mathbf{x}^T\mathbf{M}\mathbf{x} = \lambda_1(\mathbf{v}^T\mathbf{x})^2$, for any $\mathbf{x}$. □

The assumption that $\mathbf{M}$ is close to being rank-1 is used by two recent algorithms [16, 24]. Spectral matching [16] also returns the optimal solution in this case, and it assumes that the rank-1 assumption is the ideal matrix to which a small amount of noise is added. Probabilistic graph matching [24] makes the rank-1 approximation by assuming that each second-order element of $M_{ia;jb}$ is the product of the probability of feature i being matched to a and feature j being matched to b, independently. However, instead of maximizing the quadratic score function, they use this probabilistic interpretation of the pairwise terms and find the solution by looking for the closest rank-1 matrix to $\mathbf{M}$ in terms of the KL-divergence. If the assumptions in Zass and Shashua [24] were perfectly met, then spectral matching, probabilistic graph matching, and our algorithm would all return the same solution. For a comparison of all these algorithms on real-world experiments, please see the experiments in Sect. 2.9.

2.6 Solving Hypergraph Matching

Hypergraph matching, as well as graph matching, are NP-hard, so one is usually after efficient algorithms that can find good approximate solutions. The complexity increases dramatically in the transition from second- to higher order matching. Therefore, it becomes harder to develop fast and accurate algorithms. Some current methods [22–24] take advantage of the Rank-One Approximation (ROA) in order to reduce complexity. This leads to either relaxing the one-to-one matching constraints of the original problem [22], or modifying the formulation based on probabilistic interpretations [23, 24].

There are only a few papers published on higher order matching for real-world computer vision applications. One example is on finding skin lesions in medical images [80]. Despite the theoretical advantages of higher order potentials, hypergraph matching could be extremely expensive, which could explain the limited number of published works on the problem. Even in the case of graph matching, the matrix **M** could be huge. Only a carefully designed, efficient method that takes advantage of the inner statistics of the matching problem could handle the task, by constructing and working only with sparse matrices or algorithmically reducing the complexity of the problem. This could be achieved by factorizing the matrix [9] or by algebraic manipulations that take advantage of certain local geometric transformation constraints [81]. When approaching the NP-hard hypergraph matching problem, one must keep in mind that the task is extremely expensive, both in terms of storage and computation.

In this section, we present a novel method for hypergraph matching that can be used either standalone, starting from a uniform initial solution, or as a refinement procedure started at the solution returned by another algorithm. As opposed to other methods, ours aim is to maximize the original objective score, while preserving to the best possible the original discrete one-to-one matching constraints. We experimentally show that local optimization in the original domain turns out to be more effective than global optimization in a relaxed domain, a fact that was also observed by Leordeanu et al. [6] in the case of second-order matching.

Our proposed method (Algorithm 2.2) is an iterative procedure that, at each step k, approximates the higher order score $S(\mathbf{x})$ by its second-order Taylor expansion around the current solution $\mathbf{x}_k$. This transforms the hypergraph matching problem into an Integer Quadratic Program (IQP), defined around the current solution $\mathbf{x}_k$. In turn, the second-order approximation can be optimized locally, quite efficiently in the continuous domain $\mathbf{Ax} = \mathbf{1}$, $\mathbf{x} \in [0, 1]^n$.

Given a possible solution $\mathbf{x}$ in the continuous domain $\mathbf{Ax} = \mathbf{1}$, $\mathbf{x} \in [0, 1]^n$, let matrix $\mathbf{M}(\mathbf{x})$ be a function of $\mathbf{x}$, obtained by marginalizing the tensor $\mathbf{H}$ as follows:

$$M(\mathbf{x})_{ia;jb} = \sum_{kc} H_{ia;jb;kc} x_{kc}. \tag{2.8}$$

Similarly, we define the column vector $\mathbf{d}$:

$$d(\mathbf{x})_{ia} = \sum_{jb;kc} H_{ia;jb;kc} x_{jb} x_{kc}. \tag{2.9}$$

Having defined **M** and **d**, we can now write the second-order Taylor approximation of the third-order matching score (Formula 2.5) as follows:

$$S(\mathbf{x}) \approx S(\mathbf{x}_0) + 3(\mathbf{x}^T \mathbf{M}(\mathbf{x}_0)\mathbf{x} - \mathbf{d}(\mathbf{x}_0)^T \mathbf{x}). \tag{2.10}$$

We can locally improve the score around $\mathbf{x}_0$ by optimizing the quadratic assignment cost $\mathbf{x}^T \mathbf{M}(\mathbf{x}_0)\mathbf{x} - \mathbf{d}(\mathbf{x}_0)^T \mathbf{x}$ in the continuous domain $\mathbf{Ax} = \mathbf{1}$, $\mathbf{x} \in [0, 1]^n$, since the first term $S(\mathbf{x}_0)$ is constant. Note that in practice matrix **M** is often close to semi-positive definite. In this case, the quadratic approximation is a concave minimization problem, which in general is notoriously difficult to solve globally (as opposed to convex minimization). Since, at each iteration, local maximization suffices, we use the efficient method of Leordeanu et al. [6] (the inner loop of Algorithm 2.2).

Algorithm 2.2 Sequential second-order expansion for higher order matching

---
$\mathbf{x} \leftarrow \mathbf{x}_0$
$\mathbf{x}^* \leftarrow \mathbf{x}_0$
$S^* \leftarrow \sum_{ia;jb;kc} H_{ia;jb;kc} x_{ia} x_{jb} x_{kc}$
repeat
  $M_{ia;jb} \leftarrow \sum_{kc} H_{ia;jb;kc} x_{kc}$
  $d_{ia} \leftarrow \sum_{jb;kc} H_{ia;jb;kc} x_{jb} x_{kc}$
  repeat
    $\mathbf{b} \leftarrow \pi(2\mathbf{Mx} - \mathbf{d}^T\mathbf{x})$
    $S \leftarrow \sum_{ia;jb;kc} H_{ia;jb;kc} b_{ia} b_{jb} b_{kc}$
    if $S > S^*$ **then**
      $S^* \leftarrow S$
      $\mathbf{x}^* \leftarrow \mathbf{b}$
    end if
    $t^* = \arg \max \mathbf{x}(t)^T \mathbf{M}\mathbf{x}(t) + \mathbf{d}^T \mathbf{x}(t), \quad t \in [0, 1]$
    where $\mathbf{x(t)} = \mathbf{x} + t(\mathbf{b} - \mathbf{x})$
    $\mathbf{x} \leftarrow \mathbf{x} + t^*(\mathbf{b} - \mathbf{x})$
  until convergence
until S^* does not improve
return $\mathbf{x}^*$

The algorithm is initialized at $\mathbf{x}_0$, which can be either a uniform vector or the solution given by another hypergraph matching method, such as HOPM [22]. The inner loop (method of Leordeanu et al. [6]) locally optimizes the quadratic approximation of the hypergraph matching model around the current solution. This inner loop can be viewed as a graph matching problem that locally approximates the original hypergraph matching problem. This relates our approach to the method recently proposed in Chertok and Keller [23], which also boils down to a graph matching problem by marginalizing the tensor down to a matrix. Different from Chertok and Keller [23], our algorithm is based on a second-order Taylor approximation, not the ROA of the tensor. This novel approach gives better objective scores in our experiments. Our method is also different from the probabilistic algorithm [24], which marginalizes the tensor down to a vector, thus transforming the higher order problem into a linear one. This may explain why the probabilistic method

[24] performs less well than both of our methods [23]. The inner loop consists of $\mathbf{b} \leftarrow \pi(2\mathbf{Mx} - \mathbf{d}^T\mathbf{x})$, which is a projection on the discrete domain followed by line maximization $t^* = \arg\max \mathbf{x}(t)^T\mathbf{Mx}(t) + \mathbf{Dx}(t), \quad t \in [0, 1]$, with closed-form solution (see Leordeanu et al. [6] for more details). Similar ideas have been recently implemented also for image segmentation [69] and MAP inference in higher order MRFs [82].

By tracking the best discrete solution $\mathbf{x}^* \leftarrow \mathbf{b}$, our method always returns a solution at least as good as the starting one. Since the score S is bounded above and the number of discrete solutions is finite, the algorithm is guaranteed to improve the matching score and stop after a finite number of iterations. Note that the algorithm stops if the previous iteration of the inner loop did not improve. In our experiments, convergence requires on average less than five outer-loop iterations.

2.7 Learning Graph Matching

In this section, we present supervised, semi-supervised, and unsupervised learning variants of our approach to graph matching, and detail parameter learning for conditional random fields.

2.7.1 Theoretical Analysis

Our proposed learning algorithm is motivated by the statistical properties of the matrix $\mathbf{M}$ and of its principal eigenvector $\mathbf{v}$, which is the continuous solution given by the spectral graph matching algorithm [16]. In order to analyze the properties of $\mathbf{M}$ theoretically, we need a few assumptions and approximations. The assumptions we make are intuitive and not necessarily rigorous. They are validated by our numerous experiments. Each instance of the matching problem is unique so nothing can be said with absolute certainty about $\mathbf{M}$ and its eigenvector $\mathbf{v}$, nor the quality of the solution returned. Therefore, we must be concerned with the average (or expected) properties of $\mathbf{M}$ rather than the infinitely many particular cases. We propose a model for $\mathbf{M}$ (Fig. 2.4) that we validate through experiments.

For a given matching experiment with its corresponding matrix $\mathbf{M}$, let $p_1 > 0$ be the average value of the second-order scores between correct assignments $E(M_{ia;jb})$ for any pair (ia, jb) of correct assignments. Similarly, let $p_0 = E(M_{ia;jb}) \geq 0$ if at least one of the assignments ia and jb is wrong. p_1 should be higher than p_0, since the pairs of correct assignments are expected to agree both in appearance and geometry and have strong second-order scores, while the wrong assignments have such high pairwise scores only accidentally. Intuitively, we expect that the higher p_1 and the lower p_0, the higher the matching rate. We also expect that the performance depends on their ratio $p_r = p_0/p_1$ rather than on their absolute values, since multiplying $\mathbf{M}$ by a constant does not change the leading eigenvector. Similarly, we define the

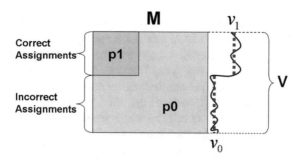

Fig. 2.4 Pairwise scores (elements of the matching matrix **M**) between correct assignments have a higher expected value p_1 than elements with at least one wrong assignment, with average value p_0. This will be reflected in the eigenvector **v** that will have higher average value v_1 for correct assignments than v_0 for wrong ones

average eigenvector value $v_1 = E(v_{ia})$ over correct assignments ia, and $v_0 = E(v_{jb})$, over wrong assignments jb. The spectral matching algorithm assumes that correct assignments will correspond to large elements of the eigenvector **v** and the wrong assignments to low values in **v**, so the higher v_1 and the lower v_0 the better the matching rate. As in the case of p_r, if we could minimize during learning the average ratio $v_r = v_0/v_1$ (since the norm of the eigenvector is irrelevant) over all image pairs in a training sequence, then we would expect to optimize the overall training matching rate. This model assumes fully connected graphs, but it can be verified that the results we obtain next are also valid for weakly connected graphs, as also shown in our experiments.

It is useful to investigate the relationship between v_r and p_r for a given image pair. We know that $\lambda v_{ia} = \sum_{jb} M_{ia;jb} v_{jb}$. For clarity of presentation, we assume that for each of the n features in the left image there are k candidate correspondences in the right image. We make the following approximation $E(\sum_{jb} M_{ia;jb} v_{jb}) \approx \sum_{jb} E(M_{ia;jb}) E(v_{jb})$, by considering that any v_{jb} is *almost* independent of any particular $M_{ia;jb}$, since **M** is large. The approximation is actually a "$\geq$" inequality, since the correlation is expected to be positive (but very small). For our given matrix **M**, let us call its first eigenvalue λ. It follows that for a correct correspondence ia, $\lambda E(v_{ia}) = \lambda v_1 \approx n p_1 v_1 + n(k-1) p_0 v_0$. Similarly, if ia is a wrong correspondence then $\lambda E(v_{ia}) = \lambda v_0 \approx n p_0 v_1 + n(k-1) p_0 v_0$. Dividing both equations by $p_1 v_1$ and taking the ratio of the two, we obtain

$$v_r \approx \frac{p_r + (k-1) p_r v_r}{1 + (k-1) p_r v_r}. \tag{2.11}$$

Solving this quadratic equation for v_r, we get

$$v_r \approx \frac{(k-1)p_r - 1 + \sqrt{(1 - (k-1)p_r)^2 + 4(k-1)p_r^2}}{2(k-1)p_r}. \tag{2.12}$$

Using Eqs. 2.11 and 2.12, it can be verified that v_r is a monotonically increasing function of p_r, for $k > 1$. This is in fact not surprising since we expect that the smaller $p_r = p_0/p_1$, the smaller $v_r = v_0/v_1$ and the more binary the eigenvector $\mathbf{v}$ would be (and closer to the binary ground truth $\mathbf{t}$), with the elements of the wrong assignments approaching 0. This approximation turns out to be very accurate in practice, as shown by our experiments in Figs. 2.6, 2.7, and 2.8. Also, the smaller v_r, the higher the expected matching rate, by which we mean the number of correctly matched features divided by the total number of features. For the sake of clarity, during this analysis, we assume an equal number of features in both images. We also assume that for each feature in the left image there is one correct match in the right image. However, as we show in the experimental section, our algorithm is robust to the presence of outliers.

One way to minimize v_r is to maximize the correlation between $\mathbf{v}$ and the ground truth indicator vector $\mathbf{t}$, while making sure that one feature from the left images matches one feature in the right image. However, in this chapter, we want to minimize v_r in an unsupervised fashion that is without knowing $\mathbf{t}$ during training. Our proposed solution is to maximize instead the correlation between $\mathbf{v}$ and its binary version (that is, the binary solution returned by the matching algorithm). How do we know that this procedure will ultimately give a binary version of $\mathbf{v}$ that is close to the real ground truth? We will investigate this question next.

Let $\mathbf{b}(\mathbf{v})$ be the binary solution obtained from $\mathbf{v}$, respecting the one-to-one mapping constraints, as returned by spectral matching for a given pair of images. Let us assume for now that we know how to maximize the correlation $\mathbf{v}^T \mathbf{b}(\mathbf{v})$. We expect that this will lead to minimizing the ratio $v_r^* = E(v_{ia}|b_{ia}(\mathbf{v}) = 0)/E(v_{ia}|b_{ia}(\mathbf{v}) = 1)$. If we let n_m be the number of misclassified assignments, n the number of true correct assignments (same as the number of features, equal in both images), and k the number of candidate assignments for each feature, we can obtain the next two equations: $E(v_{ia}|b_{ia}(\mathbf{v}) = 0) = \frac{n_m v_1 + (n(k-1) - n_m) v_0}{n(k-1)}$ and $E(v_{ia}|b_{ia}(\mathbf{v}) = 1) = \frac{n_m v_0 + (n - n_m) v_1}{n}$.

Dividing both by v_1 and taking the ratio of the two, we finally obtain

$$v_r^* = \frac{m/(k-1) + (1 - m/(k-1))v_r}{1 - m + mv_r}, \tag{2.13}$$

where m is the matching error rate $m = n_m/n$. If we reasonably assume that $v_r < 1$ (eigenvector values higher on average for correct assignments than for wrong ones) and $m < (k-1)/k$ (error rate lower than random), this function of m and v_r has both partial derivatives strictly positive. Since m also increases with v_r, by maximizing $\mathbf{v}^T \mathbf{b}(\mathbf{v})$, we minimize v_r^*, which minimizes both v_r and the true error rate m, so the unsupervised algorithm can be expected to do the right thing. In all of our experiments, we obtained values for all p_r, v_r, v_r^*, and m that were very close to zero, which is sufficient in practice, even if our gradient-based method might not necessarily have found the global minimum.

The model for $\mathbf{M}$ and the equations we obtained in this section are validated experimentally in Sect. 2.9. By maximizing the correlation between $\mathbf{v}$ and $\mathbf{b}(\mathbf{v})$ over the training sequence, we do indeed lower the true misclassification rate m, maximize $\mathbf{v}^T\mathbf{t}$, and also lower p_r, v_r, and v_r^*.

2.7.2 Supervised Learning for Graph Matching

We want to find the geometric and appearance parameters $\mathbf{w}$ that maximize (in the supervised case) the expected correlation between the principal eigenvector of $\mathbf{M}$ and the ground truth $\mathbf{t}$, which is empirically proportional to the following sum over all training image pairs:

$$J(\mathbf{w}) = \sum_{i=1}^{N} \mathbf{v}^{(i)}(\mathbf{w})^T \mathbf{t}^{(i)}, \qquad (2.14)$$

where $\mathbf{t}^{(i)}$ is the ground truth indicator vector for the ith training image pair. We maximize $J(\mathbf{w})$ by coordinate gradient ascent:

$$w_j^{(k+1)} = w_j^{(k)} + \eta \sum_{i=1}^{N} \mathbf{t}_i^T \frac{\partial \mathbf{v}_i^{(k)}(\mathbf{w})}{\partial w_j}. \qquad (2.15)$$

To simplify notations throughout the rest of this chapter, we use $\mathbf{F}'$ to denote the vector or matrix of derivatives of any vector or matrix $\mathbf{F}$ with respect to some element of $\mathbf{w}$. One possible way of taking partial derivatives of an eigenvector of a symmetric matrix (when λ has order 1) is given in Sect. 8.8 of Magnus and Neudecker [83] and also in Cour et al. [84] in the context of spectral clustering:

$$\mathbf{v}' = (\lambda\mathbf{I} - \mathbf{M})^\dagger (\lambda'\mathbf{I} - \mathbf{M}')\mathbf{v}, \qquad (2.16)$$

where $\mathbf{A}^\dagger$ denotes the pseudo-inverse of $\mathbf{A}$ and

$$\lambda' = \frac{\mathbf{v}^T\mathbf{M}'\mathbf{v}}{\mathbf{v}^T\mathbf{v}}. \qquad (2.17)$$

These equations are obtained by using the fact that $\mathbf{M}$ is symmetric and the equalities $\mathbf{v}^T\mathbf{v}' = 0$ and $\mathbf{M}\mathbf{v} = \lambda\mathbf{v}$. However, this method is general and therefore does not take full advantage of the fact that in this case $\mathbf{v}$ is the principal eigenvector of a matrix with large eigengap. $\mathbf{M} - \lambda\mathbf{I}$ is large and also rank deficient so computing its pseudo-inverse is not efficient in practice. Instead, we use the power method to compute the partial derivatives of the approximate principal eigenvector: $\mathbf{v} = \frac{\mathbf{M}^n\mathbf{1}}{\sqrt{(\mathbf{M}^n\mathbf{1})^T(\mathbf{M}^n\mathbf{1})}}$. This is related to Bach and Jordan [85], but in Bach and Jordan [85] the method is used for segmentation and as also pointed out by Cour et al. [84], it could be very unstable in that case, because in segmentation and typical clustering problems the eigengap between the first two eigenvalues is not large.

Fig. 2.5 Experiments on the House sequence. The plots show the normalized correlation between the eigenvector and the ground truth solution for different numbers of recursive iterations n used to compute the approximative derivative of the eigenvector (averages over 70 experiments). Even for n as small as 5 the learning method converges in the same way, returning the same result

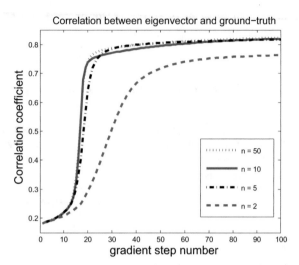

Here, $\mathbf{M}^n\mathbf{1}$ is computed recursively by $\mathbf{M}^{k+1}\mathbf{1} = \mathbf{M}(\mathbf{M}^k\mathbf{1})$. Since the power method is the preferred choice for computing the leading eigenvector, it is justified to use the same approximation for learning. Thus, the estimated derivatives are not an approximation, but actually the exact ones, given that $\mathbf{v}$ is itself an approximation based on the power method. Thus, the resulting partial derivatives of $\mathbf{v}$ are computed as follows:

$$\mathbf{v}' = \frac{\|\mathbf{M}^n\mathbf{1}\|^2(\mathbf{M}^n\mathbf{1})' - ((\mathbf{M}^n\mathbf{1})^T(\mathbf{M}^n\mathbf{1})')(\mathbf{M}^n\mathbf{1})}{\|\mathbf{M}^n\mathbf{1}\|^3}. \tag{2.18}$$

In order to obtain the derivative of $\mathbf{v}$, we first need to compute the derivative of $\mathbf{M}^n\mathbf{1}$, which can be obtained recursively:

$$(\mathbf{M}^n\mathbf{1})' = \mathbf{M}'(\mathbf{M}^{n-1}\mathbf{1}) + \mathbf{M}(\mathbf{M}^{n-1}\mathbf{1})'. \tag{2.19}$$

Since $\mathbf{M}$ has a large eigengap, as shown in Leordeanu and Hebert [16], this method is stable and efficient. Figure 2.5 demonstrates this point empirically. The method is linear in the number of iterations n, but qualitatively insensitive to n, as it works equally well with n as low as 5. These results are averaged over 70 experiments (described later) on 900×900 matrices.

To get a better feeling of the efficiency of our method as compared to Eq. 2.16, computing Eq. 2.16 takes 1500 times longer in Matlab (using the function *pinv*) than our method for $n = 10$ on the 900×900 matrices used in our experiments on the House and Hotel datasets. In practice, we manually selected the gradient step size once and used this value in all our experiments.

2.7.3 Unsupervised and Semi-supervised Learning for Graph Matching

The idea for unsupervised learning (introduced in Sect. 2.7.1) is to maximize v_r^* instead of v_r, which could be achieved through the maximization of the dot product between the eigenvector and the binary solution obtained from the eigenvector. Thus, during unsupervised training, we maximize the following function:

$$J(\mathbf{w}) = \sum_{i=1}^{N} \mathbf{v}^{(i)}(\mathbf{w})^T \mathbf{b}(\mathbf{v}^{(i)}(\mathbf{w})). \tag{2.20}$$

The difficulty here is that $\mathbf{b}(\mathbf{v}^{(i)}(\mathbf{w}))$ is not a continuous function and also it may be impossible to express in closed-form, in terms of $\mathbf{w}$, since $\mathbf{b}(\mathbf{v}^{(i)}(\mathbf{w}))$ is the result of an iterative discretization procedure. However, it is important that $\mathbf{b}(\mathbf{v}^{(i)}(\mathbf{w}))$ is piecewise constant and has zero derivatives everywhere except for a finite set of discontinuity points. We can therefore expect that we will evaluate the gradient only at points where $\mathbf{b}$ is constant and has zero derivatives. Also, at those points, the gradient steps will lower v_r (Eq. 2.13) because changes in $\mathbf{b}$ (when the gradient updates pass through discontinuity points in $\mathbf{b}$) do not affect v_r. Lowering v_r will increase $\mathbf{v}^T \mathbf{t}$ and also decrease m, so the desired goal will be achieved without having to worry about the discontinuity points of $\mathbf{b}$. This has been verified every time in our experiments. Then, the learning step function becomes

$$w_j^{(k+1)} = w_j^{(k)} + \eta \sum_{i=1}^{N} \mathbf{b}(\mathbf{v}_i^{(k)}(\mathbf{w}))^T \frac{\partial \mathbf{v}_i^{(k)}(\mathbf{w})}{\partial w_j}. \tag{2.21}$$

In most practical applications, the user has knowledge of some correct assignments, in which case a semi-supervised approach becomes more appropriate. Our algorithm can easily accommodate such a semi-supervised scenario by naturally combining the supervised and unsupervised learning steps: the discrete solution $\mathbf{b}$ from each step has fixed values for assignments for which the ground truth information is available, while for the rest of unlabeled assignments we use, as in the unsupervised case, the solution returned by the graph matching algorithm. The ability of easily combining the supervised case with the unsupervised one in a principled manner is another advantage of our proposed method.

2.7.4 Learning Pairwise Conditional Random Fields

The spectral inference method presented in Leordeanu and Hebert [37] is based on a fixed point iteration similar to the power method for eigenvectors, which maximizes the quadratic score under the L2 norm constraints $\sum_{b=1}^{L} v_{ib}^{*\,2} = 1$. These constraints require that the sub-vectors corresponding to the candidate labels for each site i have norm 1:

$$v_{ia}^* = \frac{\sum_{jb} M_{ia;jb} v_{jb}^*}{\sqrt{\sum_{b=1}^{L} v_{ib}^{*\,2}}}. \tag{2.22}$$

This equation looks similar to the eigenvector equation $\mathbf{Mv} = \lambda\mathbf{v}$. Here, instead of the global L2 normalization applied to eigenvectors, we have site-wise L2 normalization. Starting from a vector with positive elements, the fixed point $\mathbf{v}^*$ of the above equation has positive elements is unique and it is a global maximum of the quadratic score under the constraints $\sum_a v_{ia}^2 = 1$, due to the fact that $\mathbf{M}$ has non-negative elements (Theorem 5 in Baratchart et al. [86]).

The learning method for the MAP problem, which we propose here, is based on gradient ascent, similar to the one for graph matching, and requires taking the derivatives of $\mathbf{v}$ with respect to the parameters $\mathbf{w}$.

Let $\mathbf{M}_i$ be the non-square submatrix of $\mathbf{M}$ of size nLabels $\times$ nLabels $*$ nSites, corresponding to a particular site i. Also let $\mathbf{v}_i$ be the corresponding sub-vector of $\mathbf{v}$, which is computed by the following iterative procedure. Let n be a particular iteration number:

$$\mathbf{v}_i^{(n+1)} = \frac{\mathbf{M}_i \mathbf{v}_i^{(n)}}{\sqrt{(\mathbf{M}_i \mathbf{v}_i^{(n)})^T (\mathbf{M}_i \mathbf{v}_i^{(n)})}}. \tag{2.23}$$

Let $\mathbf{h}_i$ be the corresponding sub-vector of an auxiliary vector $\mathbf{h}$ defined at each iteration as follows:

$$\mathbf{h}_i = \frac{\mathbf{M}_i' \mathbf{v}_i^{(n)} + \mathbf{M}_i' (\mathbf{v}_i^{(n)})'}{\sqrt{(\mathbf{M}_i \mathbf{v}_i^{(n)})^T (\mathbf{M}_i \mathbf{v}_i^{(n)})}}. \tag{2.24}$$

Then, the derivatives of $\mathbf{v}_i^{(n+1)}$ with respect to some element of $\mathbf{w}$, at step $n + 1$, can be obtained recursively as a function of the derivatives of $\mathbf{v}_i^{(n)}$ at step n, by iterating the following update rule:

$$(\mathbf{v}^{(n+1)})' = \mathbf{h} - (\mathbf{h}^T \mathbf{v}^{(n+1)}) \mathbf{v}^{(n+1)}. \tag{2.25}$$

This update rule, which can be easily verified, is similar to the one used for computing the derivatives of eigenvectors.

The partial derivatives of the individual elements of $\mathbf{M}$ with respect to the individual elements of $\mathbf{w}$ are computed from the equations that define these pairwise potentials, given in Eq. 2.31. Of course, other differential functions can also be used to define these potentials.

In both supervised and unsupervised cases, the learning update step is similar to the one used for learning graph matching. Here, we present the supervised case. In the case of MAP problems, we have noticed that unsupervised learning can be successfully applied only to simpler problems, as shown in the experiments in Sect. 2.9.3.

This is due to the fact that in MAP problems it is easily possible to find parameters that will strongly favor one label and make the solution of the relaxed problem almost perfectly discrete. The supervised learning rule for MAP is

$$w_j^{k+1} = w_j^k + \eta \sum_{i=1}^{N} \mathbf{t}_i^T \frac{\partial \mathbf{v}_i^{(k)}(\mathbf{w})}{\partial w_j}, \tag{2.26}$$

where $\mathbf{t}_i$ is the ground truth labeling for the ith training image.

2.8 Learning Hypergraph Matching

Recent hypergraph matching methods [22–24] are based on a Rank-One Approximation (ROA) of the tensor $\mathbf{H}$. The ROA of $\mathbf{H}$ is based on the assumption that the assignments are probabilistically independent. Let us define the probability assignment vector $\mathbf{p}$, of the same length as the list L of candidate assignments, where $p_{ia} \in [0, 1]$ is the probability that feature i from I_m is matched to feature a from I_t. Under a statistical independence assumption, tensor $\mathbf{H}$ can be approximated by the following decomposition using outer products [23, 24]:

$$\mathbf{H} \approx \prod_{i=1\ldots3} \otimes \mathbf{p}. \tag{2.27}$$

This approximation is also related to the symmetric higher order power method (S-HOPM, Algorithm 2.3) used by Duchenne et al. [22], whose stationary point $\mathbf{v}$ is an approximation of the assignment probability vector and gives a rank-one approximation of $\mathbf{H}$. The symmetric higher order power method [87] applies to super-symmetric tensors and is guaranteed to converge to a stationary point of the third-order score (Eq. 2.5), under unit norm constraints $\|x\| = 1$. In our experiments, S-HOPM was convergent and numerically stable. Since the super-symmetric tensor $\mathbf{H}$ has non-negative elements, the stationary point $\mathbf{v}$ of S-HOPM also has non-negative elements. The connection between the stationary point $\mathbf{v}$ of the S-HOPM and the probability of assignments $\mathbf{p}$ further justifies the discretization of $\mathbf{v}$ [22] in order to obtain an integer solution satisfying the initial constraints.

Algorithm 2.3 Symmetric higher order power method (S-HOPM)

$\mathbf{v} \leftarrow 1$
repeat
  $v_{ia} \leftarrow \sum_{ia,jb,kc} H_{ia;jb;kc} v_{jb} v_{kc}$
  $\mathbf{v} \leftarrow \mathbf{v}/\|\mathbf{v}\|$
until convergence
return $\mathbf{v}$

The approximation $\mathbf{v}$ of $\mathbf{p}$ only works if the third-order scores $H_{ia;jb;kc}$ correctly reflect the structure and similarities in the underlying problem. We propose to learn the parameters $\mathbf{w}$ (Eq. 2.5) that maximize the dot product between $\mathbf{v}$ and a binary solution vector $\mathbf{b}$. The elements of $\mathbf{b}$ are equal to the ground truth for the known assignments and the assignments produced by the hypergraph matching algorithm for the unknown ones. This idea extends the approach to learning for graph matching presented in the previous sections and in Leordeanu et al. [19]. The objective function maximized during learning is the sum of the dot products between each $\mathbf{v}$ and the corresponding $\mathbf{b}$ over the training set of image pairs:

$$\mathbf{w}^* = \arg\max_{\mathbf{w}} \sum_i \mathbf{b}_i^T \mathbf{v}_i(\mathbf{w}). \tag{2.28}$$

Note that $\mathbf{v}(\mathbf{w})$ is a function of the parameters $\mathbf{w}$, since it depends on the elements of the tensor $\mathbf{H}$, which are all functions of $\mathbf{w}$. We optimize the objective by a gradient ascent update:

$$w_j \leftarrow w_j + \eta \sum_{i=1}^{N} \mathbf{b}_i(\mathbf{w})^T \frac{\partial \mathbf{v}_i(\mathbf{w})}{\partial w_j}. \tag{2.29}$$

Each iteration of S-HOPM computes the update $\mathbf{v} \leftarrow \mathbf{v}/\|\mathbf{v}\|$. It follows that the gradient of the estimate of $\mathbf{v}$ at iteration $k+1$ can be expressed as a function of the gradient at the previous step. Hence, the gradient can be computed recursively, jointly with the computation of $\mathbf{v}$. A recursive algorithm based on these ideas (Algorithm 2.4) can be derived based on these computational ideas.

Algorithm 2.4 Compute the gradient $\mathbf{dv}$ of $\mathbf{v}$

$\mathbf{v} \leftarrow \mathbf{1}$
$\mathbf{dv} \leftarrow \mathbf{0}$
$\mathbf{dH} \leftarrow \frac{\partial \mathbf{H}}{\partial w_j}$
repeat
    $h_{ia} \leftarrow \sum_{jb;kc} dH_{ia;jb;kc} v_{jb} v_{kc} + 2 H_{ia;jb;kc} v_{jb} dv_{kc}$
    $v_{ia} \leftarrow \sum_{jb;kc} H_{ia;jb;kc} v_{jb} v_{kc}$
    $\mathbf{h} \leftarrow \frac{\mathbf{h}}{\|\mathbf{v}\|}$
    $\mathbf{v} \leftarrow \frac{\mathbf{v}}{\|\mathbf{v}\|}$
    $\mathbf{dv} \leftarrow \mathbf{h} - (\mathbf{v}^T\mathbf{h})\mathbf{v}$
until convergence
return $\mathbf{dv}$

2.9 Experiments on Graph Matching

In the case of graph matching, we focus on two objectives. The first one is to validate the theoretical results from Sect. 2.7.1, especially Eq. 2.12, which establishes a relationship between p_r and v_r, and Eq. 2.13, which connects v_r^* to v_r and the error rate m. Each p_r is empirically estimated from each individual matrix $\mathbf{M}$ over the training sequence, and similarly each v_r^* and v_r from each individual eigenvector. Equation 2.12 is important because it shows that the more likely the pairwise agreements between correct assignments as compared to pairwise agreements between incorrect ones (as reflected by p_r), the closer the eigenvector $\mathbf{v}$ is to the binary ground truth $\mathbf{t}$ (as reflected by v_r), and, as a direct consequence, the better the matching performance. This equation also validates our model for the matching matrix $\mathbf{M}$, which is defined by two average values, p_0 and p_1, respectively. Equation 2.13 is important because it explains why by maximizing the correlation $\mathbf{v}^T\mathbf{b}(\mathbf{v})$ (and implicitly minimizing v_r^*) we in fact minimize v_r and the matching error m. Equation 2.13 basically shows why the unsupervised algorithm will indeed maximize the performance with respect to the ground truth. We mention that by matching rate/performance we mean the ratio of features that are correctly matched (out of the total number of matched features), while the error rate is 1 minus the matching rate.

The results that validate our theoretical claims are shown in Figs. 2.6, 2.7, 2.8 and 2.8 on the House, Hotel, Faces, Cars, and Motorbikes experiments, respectively. The details of these experiments are given below.

There are a few relevant results to consider. On all four different experiments, the correlation between $\mathbf{v}$ and the ground truth $\mathbf{t}$ increases at every gradient step even though the ground truth is unknown to the learning algorithm. The matching rate improves at the same time and at a similar rate with the correlation, showing that maximizing this correlation also maximizes the final performance. In Fig. 2.7, we display a representative example of the eigenvector for one pair of faces, as it becomes more and more binary during training. If after the first iteration the eigenvector is almost flat, at the last iteration it is very close to the binary ground truth, with all the correct assignments having larger confidences than any of the wrong ones. Also, on all individual experiments both approximations from Eqs. 2.12 and 2.13 become increasingly accurate with each gradient step, from less than 10% accuracy at the first iteration to less than 0.5% error at the last. In all our learning experiments, we started from a set of parameters $\mathbf{w}$ that does not favor any assignment ($\mathbf{w} = 0$, which means that before the very first iteration all non-zero scores in $\mathbf{M}$ are equal to 1). These results motivate both the models proposed for $\mathbf{M}$ (Eq. 2.12), but also the results (Eq. 2.13) that support the unsupervised learning scheme.

The second objective of our experiments is to evaluate the matching performance, before and after learning, on new test image pairs. The goal is to show that, at testing time, the matching performance after learning is significantly better than if no learning was done.

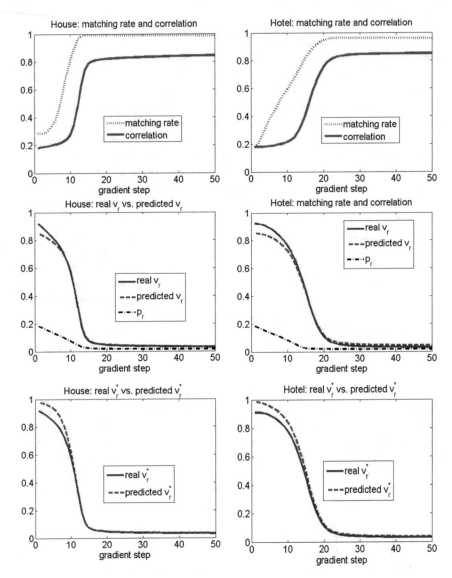

Fig. 2.6 Unsupervised learning stage. First row: matching rate and correlation of eigenvector with the ground truth during training per gradient step. The remaining plots show how the left-hand side of Eqs. 2.12 and 2.13, that is, v_r and v_r^*, estimated empirically from the eigenvectors obtained for each image pair, agree with their predicted values (right-hand side of Eqs. 2.12 and 2.13). Results are averages over 70 different experiments

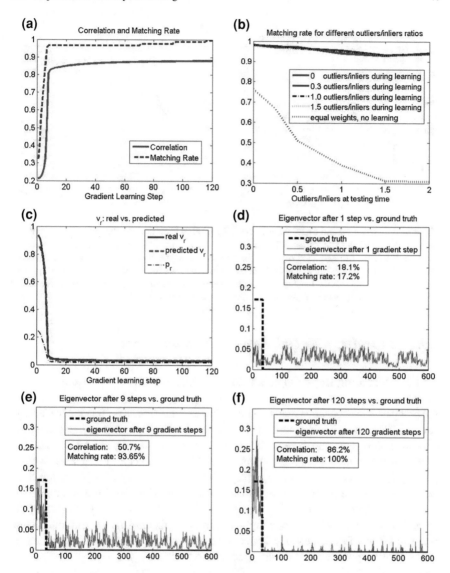

Fig. 2.7 Results on faces: correlation between eigenvectors and ground truth, and matching rate during training (top left), matching rate at testing time, for different outliers/inliers ratios at both learning and test time (top right), verifying Eq. 2.12 (middle left), example eigenvector for different learning steps. Results in the first three plots are averages over 30 different experiments

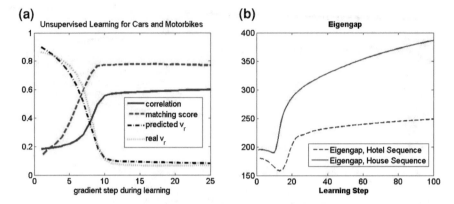

Fig. 2.8 a Correlation and matching rate w.r.t. the ground truth during unsupervised learning for Cars and Motorbikes from Pascal 2007 challenge. Real and predicted v_r decrease as predicted by the model. Results are averaged over 30 different experiments. **b** During unsupervised learning, the normalized eigengap (eigengap divided by the mean value in **M**) starts increasing after a few iterations, indicating that the leading eigenvector becomes more and more stable. Results are on the House and Hotel datasets averaged over 70 random experiments

2.9.1 Learning with Unlabeled Correspondences

2.9.1.1 Matching Rigid Objects Under Perspective Transformations

We first perform experiments on two tasks that are the same as performed in Caetano et al. [17] and our previous work [18]. We use exactly the same image sequences (House: 110 images and Hotel: 100 images) both for training and testing and the same features, which were manually selected by Caetano et al. For testing, we use all the pairs between the remaining images. The pairwise scores $M_{ia;jb}$ are the same as the ones that we previously used in Leordeanu and Hebert [18], using the shape context descriptor [88] for local appearance, and pairwise distances and angles for the second-order relationships. They measure how well features (i, j) from one image agree in terms of geometry and appearance with their candidate correspondences (a, b). More explicitly, the pairwise scores have the form $M_{ia;jb} = \exp(-\mathbf{w}^T \mathbf{g}_{ia;jb})$, where $\mathbf{g}_{ia;jb} = [|s_i - s_a|, |s_j - s_b|, \frac{|d_{ij}-d_{ab}|}{|d_{ij}+d_{ab}|}, |\alpha_{ij} - \alpha_{ab}|]$. Here, s_a denotes the shape context of features a; d_{ij} is the distance between features (i, j); and α_{ij} is the angle between the horizontal axis and the vector $\overrightarrow{ij}$. Learning consists of finding the vector of parameters **w** that maximizes the matching performance on the training sequence.

As in both Caetano et al. [17] and Leordeanu and Hebert [18], we first obtain a Delaunay triangulation and allow non-zero pairwise scores $M_{ia;jb}$ if and only if both (i, j) and (a, b) are connected in their corresponding triangulation. Our previous method [18] is supervised and based on a global optimization scheme that is more likely to find the true global optimum than the unsupervised gradient-based method proposed in this chapter. Therefore, it is significant to note that the proposed

Table 2.1 Matching performance on the hotel and house datasets at testing time. In our experiments, we used only 5 training images from the "House" sequence, while for the method from Caetano et al. [17], we report upper bounds of their published results using both 5 and 106 training images. Notation: "S" and "U" denote "supervised", and "unsupervised", respectively

Dataset	Ours: S(5) (%)	Ours: U(5) (%)	Caetano et al. [17]: S(5) (%)	Caetano et al. [17]: S(106) (%)
House	99.8	99.8	<84	<96
Hotel	94.8	94.8	<87	<90

unsupervised learning method matches our previous results, while significantly outperforming the method in Caetano et al. [17] (Table 2.1).

We point out that the main reason for the significant difference in performance between ours and Caetano et al. [17] has to do with the fact that [17] puts less emphasis on the second-order geometry in the pairwise scores $M_{ia;jb}$, by using only information from the 0–1 Delaunay triangulation and no information about pairwise distances and angles. On the contrary, we emphasize the importance of second-order relationships, since in our experiments, even when we leave out completely the shape context descriptors and use only the pairwise geometric information, the performance of our method does not degrade. Of course, it is also important to stress that, while the method by Caetano et al. [17] learns the parameters in a supervised way, ours is the first to do so in an unsupervised fashion.

Next, we investigate the performance at learning and testing stages of the unsupervised learning method versus its supervised variant (when the ground truth assignments are known). We perform 70 different experiments using both datasets, by randomly choosing 10 training images (and using all image pairs from the training set) and leaving all pairs of the rest of images for testing. As expected, we observe that the unsupervised method learns somewhat slower on average than the supervised one, but the parameters they learn are almost identical. In Fig. 2.9, we plot the average correlation (between the eigenvectors and ground truth) and matching rate at each gradient step for all training pairs and all experiments versus each gradient step, for both the supervised and unsupervised cases. It is interesting that while the unsupervised version tends to converge slower, after several iterations their performances (and also parameters) converge to the same values. During testing, the two methods performed identically in terms of matching performance (average percentage of correctly matched features over all 70 experiments). As compared to the same matching algorithm without learned parameters, the two algorithms performed clearly better (Table 2.2). Without learning, the default parameters (elements of $\mathbf{w}$) were chosen to be all equal.

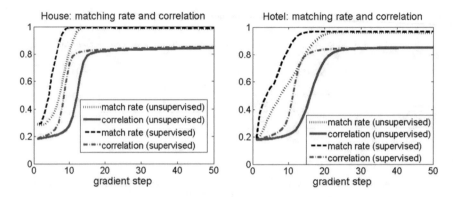

Fig. 2.9 Supervised versus unsupervised learning: Average match rate and correlation between the eigenvector and the ground truth over all training image pairs, over 70 different experiments using 10 randomly chosen training images from the House and Hotel sequences, respectively. Notice how unsupervised learning converges to the same correlation and matching rate as the supervised one

Table 2.2 Comparison of average matching performance at testing time on the house and hotel datasets for 70 different experiments (10 training images, the rest used for testing). We compare the case of unsupervised learning versus no learning. First column: unsupervised learning; Second: no learning, equal default weights **w**

Datasets	Unsupervised learning	No learning
House + Hotel (%)	99.14	93.24

2.9.1.2   Matching Deformable 2D Shapes with Outliers

The third dataset used for evaluation consists of 30 random image pairs selected from Caltech-4 Faces dataset.[1] The experiments on this dataset are different from the previous ones for two reasons: the images contain not only faces but also background clutter. Also, the faces belong to different people, both women and men, with different facial expressions, so there are significant non-rigid deformations between the faces that have to be matched. The features we used are oriented points sampled along contours extracted in the image in a similar fashion as in our previous work [50] (Fig. 2.10). The orientation considered for each point is the normal vector of the contour, at that particular point. The points on the faces that have to be matched (the inliers) were selected manually, while the outliers (features in the background) were selected randomly, while making sure that each outlier is not too close (15 pixels) to any other point. For each pair of faces, we manually selected the ground truth (the correct matches) for the inliers only. The pairwise scores contain only geometric information about pairwise distances and angles:

$$M_{ia;jb} = e^{-\mathbf{w}^T \mathbf{g}_{ia;jb}}, \qquad (2.30)$$

---

[1]http://www.vision.caltech.edu/archive.html

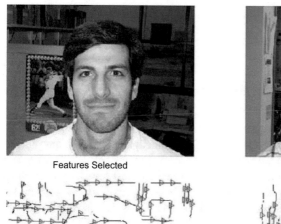

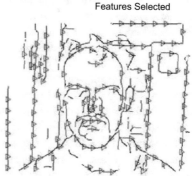

Features Selected

Features Selected

Fig. 2.10 Top row: a pair of faces from Caltech-4 dataset used in our experiments. Bottom row: the contours extracted and the points selected as features

where **w** is a vector of seven parameters (that have to be learned) and $\mathbf{g}_{ia;jb} = [|d_{ij} - d_{ab}|/d_{ij}, |\theta_i - \theta_a|, |\theta_j - \theta_b|, |\sigma_{ij} - \sigma_{ab}|, |\sigma_{ji} - \sigma_{ba}|, |\alpha_{ij} - \alpha_{ab}|, |\beta_{ij} - \beta_{ab}|]$. Here, d_{ij} is the distance between the features (i, j), θ_i is the angle between the normal of feature i and the horizontal axis, σ_{ij} is the angle between the normal at point i and the vector $\overrightarrow{ij}$, α_{ij} is the angle between $\overrightarrow{ij}$ and the horizontal axis, and β_{ij} is the angle between the normals of i and j.

We performed 30 experiments by randomly picking 10 pairs for training and leaving the rest 20 for testing. The results shown in Fig. 2.7 are averages over the 30 experiments. The top-left plot shows how, as in the previous experiments, both the correlation $\mathbf{v}^T\mathbf{t}$ and the matching performance during training improve with every learning step. During training and testing, we used different percentages of outliers to evaluate the robustness of the method (top-right plot). The learning method is robust to outliers, since the matching performance during testing does not depend on the percentage of outliers introduced during training (the percentage of outliers is always the same in the left and the right images), but only on the percentage of outliers present at testing time. Without learning (the dotted black plot), when the default parameters chosen are all equal, the performance is much worse and degrades faster as the percentage of outliers at testing time increases. This suggests that learning not only increases the matching rate, but it also makes it more robust to the presence of outliers.

2.9.2   Learning for Different Graph Matching Algorithms

We perform unsupervised learning for five state-of-the-art graph matching algorithms on the Cars and Motorbikes data from Pascal'07 challenge (Fig. 2.11), the same as that used in [19]. We use the same features (oriented points selected from pieces of contours) as described before. The algorithms are Spectral Matching [16] (SM), Spectral Matching with Affine Constraints [3] (SMAC), Graduated Assignment [1] (GA), Probabilistic graph Matching [24] (PM), and our Integer Projected Fixed Point [6] (IPFP) algorithm.

For each algorithm, we perform 30 different learning and testing experiments for each class and we average the results. For each experiment, we randomly pick 10 pairs of images for learning (with outliers) and leave the remaining 20 for testing (with and without outliers). During training, we add outliers to one image in every pair, such that the ratio of outliers to inliers is 0.5. The other image from the pair contains no outliers. We introduce this moderate amount of outliers during training in order to test the robustness of the unsupervised learning method in real-world experiments, where, especially in the unsupervised case, it is time-consuming to enforce an equal number of features in both images in every pair. During testing, we have two cases: we had no outliers in both images in the first case, and allowed all outliers possible in only one image in the second case. The number of outliers introduced was significant, the ratio of outliers to inliers ranging from 1.4 to 8.2 for the Cars class (average of 3.7), and from 1.8 to 10.5 for the Motorbikes class (average of 5.3). As in all our other tests, by an inlier, we mean a feature for which there exists a correspondence in the other image, according to the ground truth, whereas an outlier has no such correct correspondence. The inliers were manually picked, in the same manner as the ones used in the previous experiments, whereas outliers were chosen randomly on pieces of contours such that no outlier is closer than 15 pixels to any other feature selected.

In Fig. 2.12, we display the behavior of each algorithm during learning: average matching rate and average correlation of the eigenvector with the ground truth at each learning step. There are several important aspects to notice: the correlation between the eigenvector and the ground truth increases with every gradient step for all algorithms, SMAC converging much faster than the others. This is reflected also in the matching rate that increases much faster for SMAC. All algorithms benefit from learning, as all matching rates improve significantly after several iterations. The vector of parameters **w** was initialized to zero, and the final **w**'s learned are similar for all the algorithms. GA and SMAC have a rapid improvement in the first 10 steps, followed by a small decline and a plateau for the remaining iterations. This might suggest that for GA and SMAC learning should be performed only for a few

Fig. 2.11 Matching results on image pairs from Pascal 2007 challenge. Notice the significant differences in shape, viewpoint, and scale. Best viewed in color

iterations. For the other three algorithms, learning constantly improves the matching rate during training.

In Table 2.3, we show the test results of all algorithms with and without learning, for both datasets, when outliers are introduced. Without learning all algorithms, use a parameter vector $\mathbf{w} = [0.2, 0.2, 0.2, 0.2, 0.2]$ on both datasets. In our experiments, the more outliers we introduce during testing the more beneficial learning becomes. This is also in agreement with our experiments on faces (Fig. 2.7). Table 2.3 shows the results with and without learning in the presence of a significant number of outliers

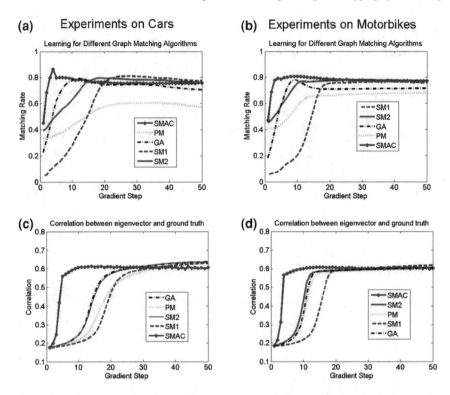

Fig. 2.12 First row: the average matching rate with respect to the ground truth, during training for each algorithm at each gradient step. Second row: average correlation between the principal eigenvector and the ground truth during learning for each algorithm at each gradient step. SMAC converges faster than the other algorithms. The final parameters learned by all algorithms are similar

(no outliers in one image and all possible outliers in the other image, as explained previously). It is evident that learning significantly improves the performance of all algorithms. The results shown in this section strongly suggest that our unsupervised learning scheme can significantly improve the performance of other algorithms on difficult data (such as the Cars and Motorbikes from Pascal'07) in the presence of a large number of outliers.

2.9.3 Experiments on Conditional Random Fields

In order to compare our method to previous work on CRFs, we have followed exactly the experiments of Kumar on image denoising, following the implementation details

Table 2.3 Comparison of matching rates at testing time for different graph matching algorithms before (NL) and after (WL) unsupervised learning on Cars and Motorbikes from Pascal'07 database, with outliers: all outliers allowed in the right image, no outliers in the left image. The algorithm used during testing was the same as the one used for learning. Results are averages over 30 different experiments. The same parameters were used by all algorithms for the case of no learning with and without outliers: all elements of **w** being equal

Dataset	IPFP (%)	SM (%)	SMAC (%)	GA (%)	PM (%)
Cars (NL)	**50.9**	26.3	39.1	31.9	20.9
Cars (WL)	**73.1**	61.6	64.8	46.6	33.6
Motorbikes (NL)	32.9	29.7	**39.2**	34.2	26.1
Motorbikes (WL)	**55.7**	54.8	52.4	46.8	42.2

and test data provided in Kumar [89]. The task is to obtain denoised images from corrupted binary 64×64 images. We used the same four images and the same noise models. For the easier task, the noise model is Gaussian with mean $\mu = 0$ and standard deviation $\sigma = 0.3$ added to the 0–1 binary images. For the more difficult task, we used as in Kumar [89], for each class, a different mixture of two Gaussians with equal mixing weights yielding a bimodal noise. The model parameters (mean, std) for the two Gaussians were [(0.08, 0.03), (0.46, 0.03)] for the foreground class and [(0.55, 0.02), (0.42, 0.10)] for the background class. The original images together with examples of their noisy versions are shown in Fig. 2.13.

Unlike Kumar's approach [89], which uses 50 randomly generated noisy versions of the first image for training, we used only 5 such images. For the simpler task, we also performed completely unsupervised learning getting almost identical results (Table 2.4). Our results were significantly better for the simpler noise model, while matching the results from Kumar [89] for the more difficult noise model. Also, note that our learning method is easier to implement and improves the performance of IPFP (Fig. 2.14), not just L2QP (our spectral MAP inference algorithm from Leordeanu and Hebert [37] for which it was originally designed). The pairwise potentials we used are

$$M_{ia;jb} = \sigma(\mathbf{w}^T[t_a; \ t_a I_i; \ t_b; \ t_b I_j; \ t_a t_b |I_i - I_j|]), \qquad (2.31)$$

where I_i is the value of pixel i in the image and $|I_i - I_j|$ is the absolute difference in image pixel values between connected sites i and j. Following Kumar [89], we used four-connected lattices. We also experimented with eight-connected neighborhoods with no significant difference in performance.

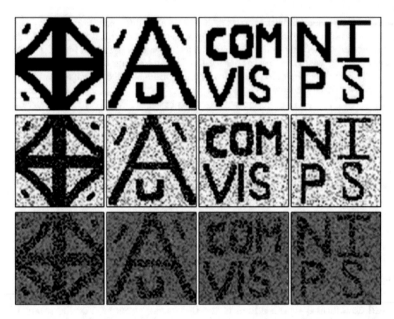

Fig. 2.13 First row: original binary images (left one used for training, next three for testing). Second row: images corrupted with unimodal noise. Third row: images corrupted with bimodal noise

Table 2.4 Comparisons with Kumar [89] on the same experiments. In Kumar [89], 50 noisy versions of the first image are used for training. We used only five noisy versions of the first image are used for training. For testing both approaches, use 50 noisy versions of the remaining three images. Note that the unsupervised learning matches the performance of the supervised one. The inference method used in Kumar [89] is graph cuts, and the learning methods are maximum pseudo-likelihood (PL) and maximum penalized pseudo-likelihood (PPL)

Algorithm	L2QP (%)	IPFP (%)	Kumar [89]: PPL (%)	Kumar [89]: PL (%)
Unimodal (sup.)	0.75	**0.73**	2.3	3.82
Unimodal (unsup.)	0.85	**0.69**	NA	NA
Bimodal (sup.)	7.15	15.94	**6.21**	17.69

The parameter values for all of our learning experiments were

- Initial,
  $\mathbf{w} = [0.5; -1; 0.5; -1; -0.5; 1]$;
- Unsupervised, unimodal noise,
  $\mathbf{w} = [1.27; -2.55; 1.27; -2.55; -2.50; 0.47]$;
- Supervised, unimodal noise,
  $\mathbf{w} = [1.27; -2.55; 1.26; -2.55; -2.63; 0.26]$;
- Supervised, bimodal noise,
  $\mathbf{w} = [1.98; -5.24; 1.98; -5.24; -2.99; 0.22]$.

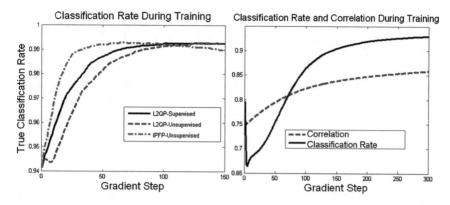

Fig. 2.14 Left plot: supervised and unsupervised learning when unimodal noise is used. Right plot: supervised learning when bimodal noise is used. Results are averages over five training images for each learning gradient step

Our learning method avoids the computational bottlenecks of most probabilistic approaches such as maximum likelihood and pseudo-likelihood, which need the estimation of the normalization function Z. The main reason for unsupervised learning to not do as well for MAP problems as for graph matching is the different structures of the matrix $\mathbf{M}$. In the case of graph matching, this matrix contains a single strong cluster formed mainly by the correct assignments, while in the case of MAP problems, the matrix could contain several such clusters corresponding to completely different labelings. The idea of accidental alignment is not applicable to most MAP problems, and thus the learning algorithm could converge to several parameter vectors that would binarize the continuous solution, in which case supervised learning is required. Moreover, even in the case of supervised learning, training is sensitive to initialization in the case of MAP problems, a fact also observed by other researchers.

2.10 Experiments on Hypergraph Matching

We perform experiments on synthetic point sets, as well as real images. We compare our matching method with current state-of-the-art methods: the Probabilistic Method [24] (PM), the Higher Order Power Method [22] (HOPM), and the recently proposed algorithm [23] (FAM2). We also experimentally show that our learning method improves the performance of all hypergraph matching algorithms tested. Similar to earlier work on learning for graph matching [90], we found that the degree of supervision does not influence the quality of learning (see Sect. 2.10.4 for more details). We also experimentally show that our novel hypergraph matching algorithm (Algorithm 2.2) outperforms the state of the art, if used by itself or in combination.

In all our experiments, for triangulation, we used all triangles formed by connecting every point to its seven nearest neighbors. On top of that, we also added a Delaunay triangulation. This setup was used for both the model and the test images. For testing without learning, we use a uniform vector of parameters $\mathbf{w}$, with $w_i = 0.1$, for all experiments and all methods. The gradient ascent method for learning starts at a uniform vector of parameters $\mathbf{w}$ for all experiments.

2.10.1 Synthetic Data

Our synthetic experiments followed a setup similar to Chertok and Keller [23] and Zass and Shashua [24]. The model set I_m contains 20 points with 2D positions uniformly generated in the interval $[0, 1]^2$. The test set I_t is generated by randomly perturbing the positions of the model points with Gaussian noise of σ_p in the interval $[0.01, 0.1]$. Outliers were added only to the test set, with (x, y) positions from the same uniform distribution in the interval $[0, 1]^2$. The outliers are extra points in the test set that have no correct correspondence in the model set. The outliers rate o_r, which is the ratio of outliers to inliers, ranges from 0 to 1 (when the number of outliers is equal to the number of inliers). The test set was then scaled by a factor s and rotated by an angle α drawn from a Gaussian distribution with $\sigma_a \in [0, 45°]$.

As opposed to Chertok and Keller [23], we allowed every point in the model image to match any other point in the test image. In Chertok and Keller [23], the authors selected for each point in the model image only a few candidates from the test image and made sure that the correct match was always selected. As opposed to their setup, ours is invariant to changes in scale and rotation. We look at two types of third-order scores: scale and rotation-invariant and scale-invariant, but sensitive to rotation.

For each triangle (i, j, k), we define the following geometric descriptors (see Fig. 2.2 for details): $\mathbf{b}_{ijk} = [\beta_i, \beta_j, \beta_k]$, $\mathbf{t}_{ijk} = [\theta_{ij}, \theta_{jk}, \theta_{ki}]$ and $\mathbf{d}_{ijk} = [d_{ij}, d_{jk}, d_{ki}]/p_{ijk}$, where p_{ijk} is the perimeter of the triangle. Note that $\mathbf{b}_{ijk}$ and $\mathbf{d}_{ijk}$ are invariant to similarity transformations, whereas $\mathbf{t}_{ijk}$ is sensitive to rotations. For rotation-invariant matching, the deformation vector used by the third-order score (Eq. 2.5) is $\mathbf{g}_{ia;jb;kc} = [\sum |\mathbf{b}_{ijk} - \mathbf{b}_{abc}|, \max |\mathbf{b}_{ijk} - \mathbf{b}_{abc}|, \sum |\mathbf{d}_{ijk} - \mathbf{d}_{abc}|, \max |\mathbf{d}_{ijk} - \mathbf{d}_{abc}|]$. For rotation sensitive scores, we extend the vector $\mathbf{g}_{ia;jb;kc}$ with two more entries: $[\sum |\mathbf{t}_{ijk} - \mathbf{t}_{abc}|, \max |\mathbf{t}_{ijk} - \mathbf{t}_{abc}|]$, for a total of six dimensions.

For training, we used two different learning setups, by introducing (or not) the global rotation angle α in our third-order scores. We used five training point set pairs (I_m, I_t), for $\sigma_p = 0.03$, $o_r = 0$ and $\sigma_r = 5°$. In Fig. 2.15, notice that for rotation sensitive matching (plot on the right) learning was faster and the matching rates were generally higher than in the rotation-invariant case. Matching accuracy increased during learning, for all algorithms. Notice the superior convergence of our method (initialized with a noninformative uniform solution $\mathbf{x}_0$) versus the convergence of the other methods tested.

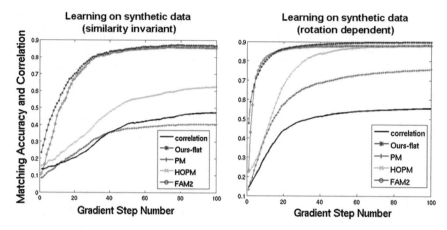

Fig. 2.15 Learning on synthetic data. Note how the correlation between the eigenvector and the ground truth is maximized during training. At the same time, the matching accuracy of all algorithms on training data increases during learning. The plots show average values over 60 different learning experiments. No outliers were generated during training; positional noise $\sigma_p = 0.03$; rotation angle noise $\sigma_a = 5°$

Table 2.5 Comparisons in objective scores. The results are averages over 412 different types of synthetic matching experiments, by uniformly varying $\sigma_p \in [0.01, 0.1]$, $\sigma_r \in [0, 45°]$ and $o_r \in [0, 1]$. For each type, 100 experiments were randomly generated, for a total of 41,200 experiments. First row: the average ratio of objective score S of other algorithms to the score S obtained by our method starting from a flat solution (Ours-flat). Second row: average ratio of S obtained by our algorithm initialized with other methods to the objective score of Ours-flat. On the last row: frequency with which the score of Ours-flat is greater than the score S of other algorithms. Note that Ours gives significantly better scores on all types of experiments either by itself or in combination with other algorithms

Algorithm	FAM2	HOPM	PM
S(Alg)/S(Ours-flat)	0.83	0.49	0.37
S(Alg + Ours)/S(Ours-flat)	0.99	0.96	0.97
Freq. S(Ours-flat) > S(Alg)	100%	100%	100%

For testing, we generated pairs of point sets (I_m, I_t), by uniformly varying $\sigma_p \in [0.01, 0.1]$, $\sigma_r \in [0, 45°]$ and $o_r \in [0, 1]$. For results, see Tables 2.5 and 2.6 and Fig. 2.17. Figure 2.16 illustrates how rotation sensitive matching can be useful when the global rotation angle α is small. During learning, the rotation parameters adapted to small rotations, which results in superior performance for the rotation-dependent matching when α is small. In contrast, the rotation-independent matching maintains high accuracy even when α is large.

Table 2.6 Average performance on 124 different types of synthetic matching experiments, with and without learning, with and without rotation sensitivity. The model and test sets (I_m, I_t) were generated by uniformly drawing $\sigma_p \in [0.01, 0.1]$, $\sigma_r \in [0, 45°]$ and $o_r \in [0, 1]$. For each type, 100 experiments were randomly generated, for a total of 12,400 experiments

Algorithm	Ours-flat (%)	FAM2 (%)	HOPM (%)	PM (%)
No learning	**66.7**	51.1	34.8	35.4
+ Learning	69.9	**70.8**	61.5	53.2
+ Ours	69.9	**71.6**	66.3	67.6

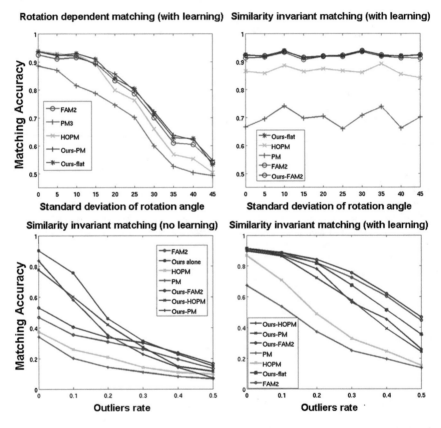

Fig. 2.16 The effect of learning. Upper left: matching using rotation-dependent scores. Upper right: matching using rotation-independent scores. Lower left: similarity-invariant matching without learning (all parameters set to $w_i = 0.1$). Lower right: similarity-invariant matching after learning. For the bottom row, $\sigma_p = 0.03$. All plots display averages over 100 experiments

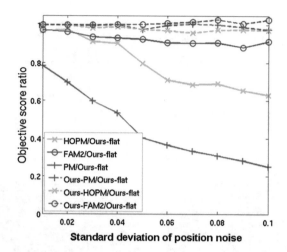

Fig. 2.17 Performance comparison of different algorithms as the positional noise is increased, w.r.t to the performance of our approach. Notice how our method, used as a final discretization step, significantly improves the accuracy of all methods

2.10.2   Experiments on Real Images

For the first set of real image experiments, we used the 50 image pairs of cars and motorbikes (20 motorbikes and 30 cars) from Leordeanu et al. [6]. In order to illustrate the capability of hypergraph matching to handle significant changes in scale, for each pair of model-test images, we down-scaled the test image by a factor of 2. The third-order scores are scale-invariant, so changes in scale do not alter matching accuracy. The actual scores used are of the same form as the rotation-dependent ones in our synthetic experiments, with a different set of parameters learned on these specific images (Fig. 2.19). Features consisted of 2D points and their normals, sampled on contours as in Leordeanu et al. [6].

We performed 30 different learning and testing experiments. For each experiment, we randomly chose five pairs of images of the same class for training and used the remaining ones belonging to the same class for testing. The results in Table 2.7 are averages over both classes over all 30 experiments. This matching task is difficult (see Fig. 2.18 for some examples) as reflected by the relatively low matching rates, but the results clearly confirm the usefulness of our learning and matching methods and their ability to improve the performance of current state-of-the-art methods.

Table 2.7 Average performance on the cars and motorbikes image pairs. The results are averages over test image pairs. Last row shows the average ratio of objective scores

Algorithm	Ours (%)	FAM2 (%)	HOPM (%)	PM (%)
Baseline	**27.5**	21.4	6.3	7.6
+ Learning	**54.9**	41.1	51.8	29.0
+ Ours	54.9	54.7	**56.3**	50.3
Alg/Ours	1	0.49	0.85	0.26

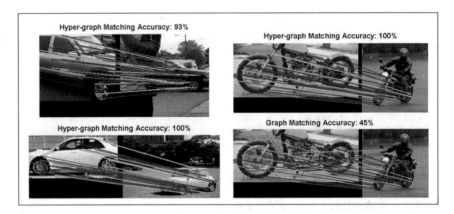

Fig. 2.18 Examples of scale-invariant matching on cars and motorbikes image pairs. Notice the robustness to changes in scale and viewpoint and the degree of background clutter

2.10.3 Matching People

People are a good example of deformable objects that require high-quality matching in a large variety of applications, including recognition and tracking of fast motions with large inter-frame displacement. We tested our algorithms on pairs of images containing people from videos captured with the recently released PrimeSense RGB-depth cameras. For every frame, we first register the depth with the color image, by using a linear transformation that was computed once, from manual correspondences and the ICP algorithm on a single color-depth frame. For all points of depth larger than 1000, we set their depth equal to 1000. After preprocessing the depth image in this way, we applied a Canny edge detector on the depth image and extracted occlusion boundaries that proved to be relatively stable and accurate.

For training (see Fig. 2.19) and Evaluation, we select 30 pairs of frames containing the same person in different poses about half a second apart. We manually selected ground truth correspondences between the two frames, on the depth boundaries obtained. For the test frames, we added outliers uniformly sampled from the depth boundaries. As in the previous experiments, we down-scaled the test image by a factor of 2.

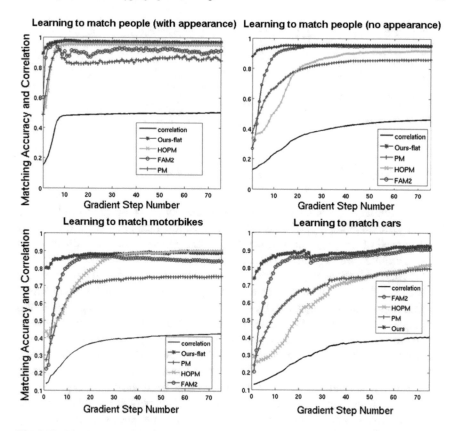

Fig. 2.19 Learning to match cars, motorbikes, and people. Notice that learning using appearance and geometry is faster and leads to better matching accuracy for all algorithms. Results are averages over 30 different learning experiments

We used two types of third-order scores: (i) the rotation-dependent ones from the previous experiments, with **w** learned from RGB-d images pairs; (ii) geometry-only scores augmented with appearance information (chi-squared distances between 40-dimensional histograms in the HSV color space) computed on the interior of the triangles defined by triplets of features. We performed 30 different experiments, using 5 image pairs for training and the remaining ones for testing (See Table 2.8). See Figs. 2.20 and 2.1 for testing examples.

We further tested our geometry and appearance hypergraph model on hundreds of pairs of RGB-d frames for which we do not have ground truth correspondences (Fig. 2.21). We implemented our method in C++, and the optimized code is able to match 2 frames in a scale-invariant manner in less than a second—total time that includes extracting features, constructing the tensor **H**, and matching.

Table 2.8 Matching accuracy over 30 image pairs containing people

Algorithm	Ours (%)	FAM2 (%)	HOPM (%)	PM (%)
No learning	73.4	**73.9**	23.2	24.6
+ Learning	**84.2**	83.5	81.5	65.4
+ Appearance	**91.0**	86.3	89.7	84.7
+ Ours	**91.0**	90.1	89.9	89.0

2.10.4 Supervised Versus Unsupervised Learning

In our experiments, the amount of supervision does not affect the quality of learning. Basically, the learning algorithm estimates almost identical parameters $\mathbf{w}$ at a very similar speed of convergence even when no correct, ground truth matches are given during learning (see Fig. 2.22). These findings are similar to the ones obtained on graph matching [90]. In Fig. 2.22, we show how the matching rate and the correlation between $\mathbf{v}$ and the ground truth increase during learning for different rates of supervision, ranging from unsupervised (no correct matches are known during training) to completely supervised. The values plotted are averages over 10 learning experiments. In each experiment, seven pairs of images of people are chosen randomly and used for training. In these plots, we used only three values for the third-order scores on triangles: deformations on gray-level appearance, gradient appearance, and geometry, normalized by some constants (kept fixed during all experiments) such that their average values were close to 1 (for numerical stability). The initial parameters $\mathbf{w}$ were all equal to $[0.1, 0.1, 0.1]$, and the ones learned were very similar for all degrees of supervision, e.g., for the unsupervised case the average learned $\mathbf{w} = [3.0, 2.2, 1.6]$, for 40% supervision $\mathbf{w} = [3.0, 2.2, 1.8]$ and for completely supervised $\mathbf{w} = [2.9, 2.1, 1.7]$.

2.11 Conclusions and Future Work

In this chapter, we have presented our efficient learning and optimization methods for graph/hypergraph matching and MAP inference problems, based on Quadratic Integer Program formulations. We demonstrated through numerous extensive experiments that our algorithms have the potential to improve matching and inference accuracy significantly. We brought valuable insights and methods to the area of graph and hypergraph matching, to hopefully open new roads for further exploration, especially in the directions of efficient matching and unsupervised learning. We handled two main aspects of matching: (1) finding optimal solutions to the relaxed problem in the continuous domain, and (2) obtaining a high-quality discrete solution with important theoretical properties, unlike the traditional approaches to obtaining the final hard assignment, based on greedy algorithms or the classical Hungarian method.

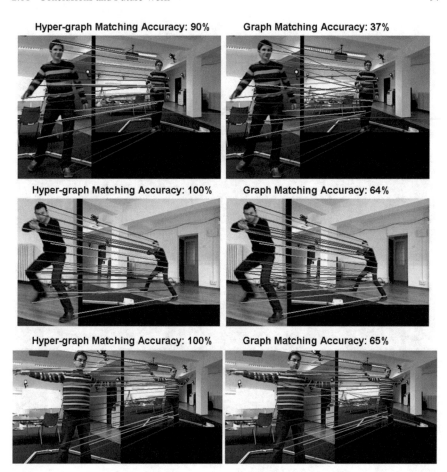

Fig. 2.20 Examples of matching people at different scales and poses. Notice the significant change in scale and the relatively large displacement of body parts

There are many avenues for future work in matching with second- or higher order constraints. They bring a significant boost in matching performance, as they are more robust to changes in scale, rotations, perspective or affine transformations, as well as diverse appearance changes and non-rigid deformations. Please see, for example, in Fig. 2.23, how hypergraph matching can easily outperform, in terms of matching accuracy, the well-known, popular approach to matching using SIFT features in the case of global affine transformations.

Some of the problems and directions that we want to explore further are enumerated below.

Fig. 2.21 Matching people at different scales and poses. Notice the significant change in scale and the relatively large displacements of body parts

2.11.1   Efficient Optimization

The introduction of our spectral matching algorithms contributed to the growth of computer vision applications based on graph matching, from medical imaging, 2D and 3D registration and reconstruction, object tracking, modeling, discovery, and learning. The applications of graph matching was extended to even higher levels of interpretation, in the realm of action recognition and video understanding. We foresee that the use of second- or higher order geometric and categorical relationships will be of essential interest in future applications. Along this line of thought, one of the main technical problems to solve will be that of efficiency and computational cost. Even though spectral matching is fast, as an optimization method, the construction

Correlation between v and ground truth

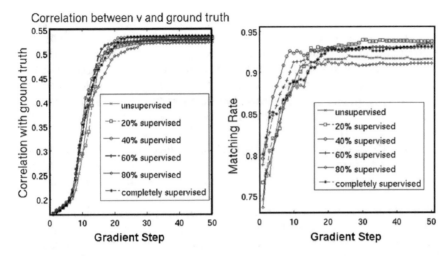

Fig. 2.22 Varying the degree of supervision during training. Note that there is almost no correlation on average between the degree of supervision and the speed of convergence. For each degree of supervision, the training pairs of images were randomly chosen

Hyper-graph Matching Accuracy: 100% (using only appearance)

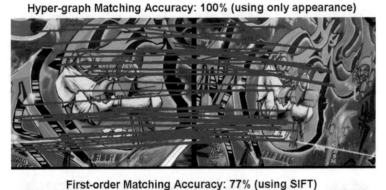

First-order Matching Accuracy: 77% (using SIFT)

Fig. 2.23 Hypergraph matching versus SIFT feature matching under affine transformations

of the assignment matrix or the tensor (for the higher order case) is slow. We believe that further substantial progress, in this direction, is needed. Our future plans include looking in more depth at this computational aspect, to study how to construct/set up the problem as fast as possible, without losing accuracy. The success of Deep Learning for Neural Networks influenced many other areas in which many *deep* methods are currently being proposed. We are not surprised to see a growing trend of deep learning and convolutional neural networks being applied to graph matching as well [21, 91–97].

In fact, the recent approaches for optical flow estimation, DeepFlow [98] as well as our recurrent graph neural network model and the more general object recognition system presented in Chap. 8 are also deeply rooted in models for matching using graphs.

As the capabilities of the current deep convolutional networks models are starting to become more clear and well understood, it is not a surprise that the classic graph models are returning with an increasing popularity in machine learning and computer vision. The standard deep learning approach has a single, feed-forward pass, without the possibility to return and correct mistakes to ensure that an equilibrium is reached. Finding an equilibrium solution between different processing stages usually involves the use of an iterative message passing procedure, which should naturally converge to a limit solution. Such iterative methods are at the core of learning and inference in the more classical probabilistic graphical models and graph matching methods. Thus, there should be expected that in the immediate future we will start seeing more and more approaches that combine the advantages of feed-forward deep nets— able to learn powerful features, with classical iterative graph methods, which favor consensual solutions.

2.11.2   Higher Order Relationships

Higher order terms are more powerful than second-order ones, as our experiments on third-order matching have shown. However, this benefit comes at a great cost: fast approximate algorithms for optimization with good matching accuracy are very hard to find, and setting up the problem, building the tensor is even more computational and memory demanding. That is one of the main reasons why higher order matching did not enjoy the level of success of graph matching, despite its theoretical attractiveness. In Leordeanu et al. [81], we present a powerful method for motion and occlusion estimation based on an elegant closed-form solution to the locally affine model that reduces the higher order geometric terms to a quadratic energy that is much easier to compute and optimize. We believe that in the next years we will witness a wonderful growth in the research dedicated to higher order matching, detection, and reasoning, with focus on both efficient approximations and higher order computational models.

The intelligent learning or manual design of such energies should consider appearance, geometry, segmentation, and occlusion cues, as well as higher level categorical information, in order to achieve strong performance in real-world images.

There are many other directions that could be explored in graph matching during the next period, in terms of modeling, their theoretical and optimality properties, as well as practical efficiency and applications. The ideas are many and the possible ways to apply them even more numerous. As the field of computer vision is enjoying a tremendous growth and a fast transition from research to industry, computer vision, and robotics, researchers will get real chances to develop many interesting and useful algorithms for this relatively young and exciting field.

References

1. Gold S, Rangarajan A (1996) A graduated assignment algorithm for graph matching. PAMI 18(4):377–388
2. Berg A, Berg T, Malik J (2005) Shape matching and object recognition using low distortion correspondences. In: CVPR
3. Cour T, Srinivasan P, Shi J (2006) Balanced graph matching. In: NIPS
4. Schellewald C, Schnorr C (2005) Probabilistic subgraph matching based on convex relaxation. In: International conference on energy minimization methods in computer vision and pattern recognition
5. Torr P (2003) Solving Markov random fields using semidefinite programming. In: International conference on artificial intelligence and statistics
6. Leordeanu M, Hebert M, Sukthankar R (2009) An integer projected fixed point method for graph matching and map inference. In: NIPS
7. Cho M, Lee J, Lee KM (2010) Reweighted random walks for graph matching. In: ECCV
8. Egozi A, Keller Y, Guterman H (2013) A probabilistic approach to spectral graph matching. Pattern Anal Mach Intell 35(1)
9. Zhou F, la Torre FD (2012) Factorized graph matching. In: Computer vision and pattern recognition
10. Cho M, Lee K (2012) Progressive graph matching: making a move of graphs via probabilistic voting. In: Computer vision and pattern recognition
11. Tolias G, Avrithis Y (2011) Speeded-up, relaxed spatial matching. In: ICCV
12. Escolano F, Hancock E, Lozano M (2011) Graph matching through entropic manifold alignment. In: Computer vision and pattern recognition
13. Torresani L, Kolmogorov V, Rother C (2013) A dual decomposition approach to feature correspondence. Pattern Anal Mach Intell 35(2)
14. Liu Z, Qiao H, Yang X, Hoi S (2014) Graph matching by simplified convex-concave relaxation procedure. Int J Comput Vis 1–18
15. Zhou F, la Torre FD (2013) Deformable graph matching. In: Computer vision and pattern recognition, pp 2922–2929
16. Leordeanu M, Hebert M (2005) A spectral technique for correspondence problems using pairwise constraints. In: ICCV
17. Caetano T, Cheng L, Le Q, Smola AJ (2007) Learning graph matching. In: ICCV
18. Leordeanu M, Hebert M (2008) Smoothing-based optimization. In: CVPR
19. Leordeanu M, Sukthankar R, Hebert M (2009) Unsupervised learning for graph matching. IJCV 96(1)
20. Cho M, Alahari K, Ponce J (2013) Learning graphs to match. In: International conference on computer vision, pp 25–32

21. Zanfir A, Sminchisescu C (2018) Deep learning of graph matching. In: Proceedings of the IEEE conference on computer vision and pattern recognition, pp 2684–2693
22. Duchenne O, Bach F, Kweon I, Ponce J (2009) A tensor-based algorithm for high-order graph matching. In: CVPR
23. Chertok M, Keller Y (2010) Efficient high order matching. IEEE Trans Pattern Anal Mach Intell 32:2205–2215
24. Zass R, Shashua A (2008) Probabilistic graph and hypergraph matching. In: CVPR
25. Çeliktutan O, Wolf C, Sankur B, Lombardi E (2014) Fast exact hyper-graph matching with dynamic programming for spatio-temporal data. J Math Imaging Vis 1–21
26. Wang J, Chen S, Fuh C (2014) Attributed hypergraph matching on a Riemannian manifold. Mach Vis Appl 25(4)
27. Yan J, Zhang C, Zha H, Liu W, Yang X, Chu SM (2015) Discrete hyper-graph matching. In: Proceedings of the IEEE conference on computer vision and pattern recognition, pp 1520–1528
28. Nie WZ, Liu AA, Gao Z, Su YT (2015) Clique-graph matching by preserving global and local structure. In: Proceedings of the IEEE conference on computer vision and pattern recognition, pp 4503–4510
29. Liu AA, Nie WZ, Gao Y, Su YT (2016) Multi-modal clique-graph matching for view-based 3d model retrieval. IEEE Trans Image Process 25(5):2103–2116
30. Adamczewski K, Suh Y, Mu Lee K (2015) Discrete Tabu search for graph matching. In: Proceedings of the IEEE international conference on computer vision, pp 109–117
31. Jiang B, Tang J, Ding C, Luo B (2017) Binary constraint preserving graph matching. In: Proceedings of the IEEE conference on computer vision and pattern recognition, pp 4402–4409
32. Shi X, Ling H, Hu W, Xing J, Zhang Y (2016) Tensor power iteration for multi-graph matching. In: Proceedings of the IEEE conference on computer vision and pattern recognition, pp 5062–5070
33. Yan J, Cho M, Zha H, Yang X, Chu SM (2015) Multi-graph matching via affinity optimization with graduated consistency regularization. IEEE Trans Pattern Anal Mach Intell 38(6):1228–1242
34. Leordeanu MD (2010) Spectral graph matching, learning, and inference for computer vision. PhD Thesis, Carnegie Mellon University
35. Leordeanu M, Zanfir A, Sminchisescu C (2011) Semi-supervised learning and optimization for hypergraph matching. In: IEEE international conference on computer vision
36. Leordeanu M, Zanfir A, Sminchisescu C (2011) Semi-supervised learning and optimization for hypergraph matching. In: ICCV
37. Leordeanu M, Hebert M (2006) Efficient map approximation for dense energy functions. In: International conference on machine learning
38. Cour T, Shi J (2007) Solving Markov random fields with spectral relaxation. In: International conference on artificial intelligence and statistics
39. Ravikumar P, Lafferty J (2006) Quadratic programming relaxations for metric labeling and Markov random field map estimation. In: International conference on machine learning
40. Yan P, Khan SM, Shah M (2008) Learning 4D action feature models for arbitrary view action recognition. In: CVPR
41. Gaur U, Zhu Y, Song B, Roy-Chowdhury A (2011) A string of feature graphs model for recognition of complex activities in natural videos. In: International conference on computer vision, pp 2595–2602
42. Liu J, Luo J, Shah M (2009) Recognizing realistic actions from videos "in the wild". In: CVPR
43. Nayak N, Sethi R, Song B, Roy-Chowdhury A (2011) Motion pattern analysis for modeling and recognition of complex human activities. In: Guide to video analysis of humans: looking at people
44. Zhu Y, Nayak N, Gaur U, Song B, Roy-Chowdhury A (2013) Modeling multi-object interactions using string of feature graphs. Comput Vis Image Underst 117(10)
45. Kim G, Faloutsos C, Hebert M (2008) Unsupervised modeling of object categories using link analysis techniques. In: CVPR

46. Kim G, Faloutsos C, Hebert M (2008) Unsupervised modeling and recognition of object categories with combination of visual contents and geometric similarity links. In: ACM international conference on multimedia information retrieval

47. Leordeanu M, Collins R, Hebert M (2005) Unsupervised learning of object features from video sequences. In: CVPR

48. Parikh D, Chen T (2007) Unsupervised learning of hierarchical semantics of objects (hSOs). In: CVPR

49. Parikh D, Chen T (2007) Unsupervised identification of multiple objects of interest from multiple images: discover. In: Asian conference on computer vision

50. Leordeanu M, Hebert M, Sukthankar R (2007) Beyond local appearance: category recognition from pairwise interactions of simple features. In: CVPR

51. de Aguiar E, Stoll C, Theobalt C, Ahmed N, Seidel HP, Thrun S (2008) Performance capture from sparse multi-view video. In: SIGGRAPH

52. Huhle B, Jenke P, Straer W (2008) On-the-fly scene acquisition with a handy multi-sensor system. Int J Intell Syst Technol Appl 5(3–4):255–263

53. Ren X (2007) Learning and matching line aspects for articulated objects. In: CVPR

54. Chertok M, Keller Y (2010) Spectral symmetry analysis. PAMI 32(7):1227–1238

55. Hays J, Leordeanu M, Efros A, Liu Y (2006) Discovering texture regularity as a higher-order correspondence problem. In: ECCV

56. Yi L, Kim VG, Ceylan D, Shen I, Yan M, Su H, Lu C, Huang Q, Sheffer A, Guibas L et al (2016) A scalable active framework for region annotation in 3d shape collections. ACM Trans Graph (TOG) 35(6):210

57. Im K, Raschle NM, Smith SA, Ellen Grant P, Gaab N (2015) Atypical sulcal pattern in children with developmental dyslexia and at-risk kindergarteners. Cereb Cortex 26(3):1138–1148

58. Im K, Pienaar R, Lee JM, Seong JK, Choi YY, Lee KH, Grant PE (2011) Quantitative comparison and analysis of sulcal patterns using sulcal graph matching: a twin study. Neuroimage 57(3):1077–1086

59. Im K, Guimaraes A, Kim Y, Cottrill E, Gagoski B, Rollins C, Ortinau C, Yang E, Grant PE (2017) Quantitative folding pattern analysis of early primary sulci in human fetuses with brain abnormalities. Am J Neuroradiol 38(7):1449–1455

60. Sotiras A, Davatzikos C, Paragios N (2013) Deformable medical image registration: a survey. IEEE Trans Med Imaging 32(7):1153

61. Singh R, Xu J, Berger B (2008) Global alignment of multiple protein interaction networks with application to functional orthology detection. Proc Natl Acad Sci 105(35):12,763–12,768

62. Jiang ZP, Xi W, Li X, Tang S, Zhao JZ, Han JS, Zhao K, Wang Z, Xiao B (2014) Communicating is crowdsourcing: Wi-Fi indoor localization with CSI-based speed estimation. J Comput Sci Technol 29(4):589–604

63. Hu Y, Liao S, Lei Z, Yi D, Li S (2013) Exploring structural information and fusing multiple features for person re-identification. In: Proceedings of the IEEE conference on computer vision and pattern recognition workshops, pp 794–799

64. Zhang S, Zhu Y, Roy-Chowdhury AK (2016) Context-aware surveillance video summarization. IEEE Trans Image Process 25(11):5469–5478

65. Mirzaalian H, Lee TK, Hamarneh G (2016) Skin lesion tracking using structured graphical models. Med Image Anal 27:84–92

66. Kang U, Hebert M, Park S (2013) Fast and scalable approximate spectral graph matching for correspondence problems. Inf Sci 220:306–318

67. Park S, Park SK, Hebert M (2013) Fast and scalable approximate spectral matching for higher order graph matching. IEEE Trans Pattern Anal Mach Intell 36(3):479–492

68. Yang X, Liu CK, Liu ZY, Qiao H, Wang BF, Wang ZD (2014) A probabilistic spectral graph matching algorithm for robust correspondence between lunar surface images. In: Proceeding of the 11th world congress on intelligent control and automation. IEEE, pp 385–390

69. Brendel W, Todorovic S (2010) Segmentation as maximum-weight independent set. In: NIPS

70. Jain A, Gupta A, Rodriguez M, Davis L (2013) Representing videos using mid-level discriminative patches. In: Computer vision and pattern recognition, pp 2571–2578

71. Semenovich D (2010) Tensor power method for efficient map inference in higher-order MRFs. In: ICPR
72. Monroy A, Bell P, Ommer B (2014) Morphological analysis for investigating artistic images. Image Vis Comput 32(6)
73. He L, Yang X, Liu ZY (2018) Generalizing integer projected graph matching algorithm for outlier problem. In: 2018 IEEE 3rd international conference on image vision and computing (ICIVC). IEEE, pp 50–54
74. Lu Y, Huang K, Liu CL (2016) A fast projected fixed-point algorithm for large graph matching. Pattern Recognit 60:971–982
75. Zaslavskiy M, Bach F, Vert JP (2008) A path following algorithm for graph matching. In: ICSP
76. Zaslavskiy M (2010) Graph matching and its application in computer vision and bioinformatics. PhD thesis, lEcole nationale superieure des mines de Paris
77. Besag J (1986) On the statistical analysis of dirty pictures. J R Stat Soc 48(5):259–302
78. Frank M, Wolfe P (1956) An algorithm for quadratic programming. Nav Res Logist Q 3(1–2):95–110
79. LeBlanc LJ, Helgason RV, Boyce DE (1985) Improved efficiency of the frank-wolfe algorithm for convex network programs. Transp Sci 19(4):445–462
80. Mirzaalian H, Lee T, Hamarneh G (2012) Uncertainty-based feature learning for skin lesion matching using a high order MRF optimization framework. In: Medical image computing and computer-assisted intervention, pp 98–105
81. Leordeanu M, Zanfir A, Sminchisescu C (2013) Locally affine sparse-to-dense matching for motion and occlusion estimation. In: ICCV
82. Semenovich D, Sowmya A (2010) Tensor power method for efficient map inference in higher-order MRFs. In: ICPR
83. Neudecker Magnus (1999) Matrix differential calculus with applications in statistics and econometrics. Wiley
84. Cour T, Shi J, Gogin N (2005) Learning spectral graph segmentation. In: International conference on artificial intelligence and statistics
85. Bach F, Jordan M (2003) Learning spectral clustering. In: NIPS
86. Baratchart L, Berthod M, Pottier L (1998) Optimization of positive generalized polynomials under lp constraints. J Convex Anal 5(2):133
87. Kofidis E, Regallia PA (2002) On the best rank-1 approximation of higher-order supersymmetric tensors. SIAM J Matrix Anal Appl 23(3)
88. Belongie S, Malik J, Puzicha J (2002) Shape matching and object recognition using shape context. PAMI 24(4):509–522
89. Kumar S (2005) Models for learning spatial interactions in natural images for context-based classification. PhD thesis, The Robotics Institute, Carnegie Mellon University
90. Leordeanu M, Hebert M (2009) Unsupervised learning for graph matching. In: CVPR
91. Guo M, Chou E, Huang DA, Song S, Yeung S, Fei-Fei L (2018) Neural graph matching networks for Fewshot 3d action recognition. In: Proceedings of the European conference on computer vision (ECCV), pp 653–669
92. Li Y, Gu C, Dullien T, Vinyals O, Kohli P (2019) Graph matching networks for learning the similarity of graph structured objects. arXiv:190412787
93. Martineau M, Raveaux R, Conte D, Venturini G (2018) Learning error-correcting graph matching with a multiclass neural network. Pattern Recognit Lett
94. Wang R, Yan J, Yang X (2019) Neural graph matching network: learning Lawler's quadratic assignment problem with extension to hypergraph and multiple-graph matching. arXiv:191111308
95. Xu K, Wang L, Yu M, Feng Y, Song Y, Wang Z, Yu D (2019) Cross-lingual knowledge graph alignment via graph matching neural network. arXiv:190511605
96. Nowak A, Villar S, Bandeira AS, Bruna J (2018) Revised note on learning quadratic assignment with graph neural networks. In: 2018 IEEE data science workshop (DSW). IEEE, pp 1–5

97. Wang W, Song H, Zhao S, Shen J, Zhao S, Hoi SC, Ling H (2019) Learning unsupervised video object segmentation through visual attention. In: Proceedings of the IEEE conference on computer vision and pattern recognition, pp 3064–3074
98. Weinzaepfel P, Revaud J, Harchaoui Z, Schmid C (2013) DeepFlow: large displacement optical flow with deep matching. In: ICCV

Chapter 3
Unsupervised Learning of Graph and Hypergraph Clustering

3.1 Introduction

Clustering is a fundamental problem in data mining, machine learning, and computer vision [1]. There is no definitive formulation of the clustering problem. However, it is generally believed that objects belonging to the same cluster should exhibit agreement relationships among each other, whereas objects that do not belong to a cluster should not exhibit such relationships. Many popular clustering methods are partition based [2–6]. They make the assumption that every data point belongs to a cluster. While they have a good performance on problems where such assumptions are valid, in most real applications there is a large number of outliers that do not belong to any cluster. Deciding on whether a point is an outlier or not is a challenging task, which is why most partition-based methods have poor performance when outliers abound, as also noted by other authors Liu et al. [7], Bulo and Pellilo [8]. Another limitation of most classical data clustering methods is their handling of only pairwise relationships between data points, given by the weights of an affinity graph. Since, in some applications, pairwise relations are not sufficient, it is important to develop algorithms that can handle higher order relationships among data points. Since the problem becomes increasingly harder as the order increases, efficiency is a crucial factor.

Current research on hypergraph clustering takes on several directions. One is to transform the hypergraph into a graph by mapping the higher order affinities to pairwise relationships [9–11]. Another direction is to generalize the methods from pairwise clustering, such as normalized cuts [3] and non-negative matrix factorization to hypergraphs and their corresponding tensors [12, 13]. Our formulation is related to more recent hypergraph clustering methods [7, 8]. In [8], the authors propose a formulation based on game theory and optimize the objective function using the results of Baum-Eagon growth transformation [14]. Their algorithm, like ours, iteratively updates the cluster membership for all nodes in parallel and converges relatively fast. We show experimentally that our method, by taking larger steps towards a maximum,

---

The material presented in this chapter is based in large part on the following paper:
Leordeanu, Marius and Cristian Sminchisescu. "Efficient hypergraph clustering." In Artificial Intelligence and Statistics (AISTATS), 2012.

has significantly better speed of convergence, with slightly better accuracy. In turn, the algorithm of Liu et al. [7] performs node-wise updates and converges slowly, and in our experiments is at least an order of magnitude slower than the method we propose.

There are many problems in computer vision and machine learning using higher order clustering, even though the number of papers devoted to hypergraph processing and interpretation is limited. Among those, we enumerate here a few representative ones, such as work on modeling motion using higher order models and spectral clustering [15], video object segmentation using hypergraph cuts [16], object matching using hypergraphs [17–20] (presented in Chap. 2), automatic music and news recommendation based on hypergraph learning, by combining different sources of information, such as social media information, user behavior, news and musical content [21, 22], image retrieval with hypergraph ranking [23], image categorization by hypergraph partitioning [24], and social image search by hypergraph learning of combined visual and text models [25]. Hypergraph learning and modeling is powerful due to the possibility of integrating several sources of information and features into unified higher order potentials. Its applications are diverse; therefore, it is important to design efficient algorithms that function in real-world scenarios. Most hypergraph-based problems are special cases of clustering, which motivates our work in this area. As we will see later in this chapter, hypergraph matching is strongly related to hypergraph clustering and could be seen as a particular case of the more general clustering problem.

In this chapter, we present our efficient approach to clustering [26] that is not partition based and can handle second- or higher order relationships among sets of data points. Our algorithm has important theoretical properties such as convergence and satisfaction of first-order necessary optimality conditions. In our experiments, it significantly outperforms the current state-of-the-art methods in terms of computational speed, while being at least as accurate.

3.2 Problem Formulation

In Chap. 1, we presented a mathematical formulation for the more classical graph clustering problem, where the affinities have second order. Here, we give a general formulation, where we define clustering with k-th order relationships on a hypergraph, also known as a k-graph, following a formulation similar to Bulo and Pellilo [8], Liu et al. [7]. A k-graph $G = (V, E, w)$ is formed by a set of n vertices $V = \{1, \ldots, n\}$, a set of hyperedges $E \subseteq V^k$, and a k-th order real-valued affinity function $w : W \rightarrow R$. The affinity function captures the strength of relationships on hyperedges, namely, the stronger the affinity associated with a hyperedge $\{v_1, \ldots, v_k\}$, the larger the function value $w(\{v_1, \ldots, v_k\})$. We define the associated super-symmetric tensor W of the k-graph as follows: $W(v_1, \ldots, v_k) = w(\{v_1, \ldots, v_k\})$ if the hyperedge $\{v_1, \ldots, v_k\}$ is in E and 0 otherwise. The tensor W is super-symmetric since vertices within a given hyperedge can be considered in any order without changing the corresponding affinity value, and each hyperedge $\{v_1, \ldots, v_k\}$ has $k!$ duplicate entries in the tensor W.

Strong clusters $C \subseteq V$ are sets of vertices with high corresponding hyperedge affinities. Similar to Liu et al. [7], we describe the cluster score as the average over the hyperedge affinities within that cluster. Therefore, if a set C has m vertices, the cluster score can be written as

$$S_C = \frac{1}{m^k} \sum_{v_1, \ldots, v_k \in C} W(v_1, \ldots, v_k). \tag{3.1}$$

We define the vector $\mathbf{x}$, which acts as an indicator function, such that $x_i = 1/m$ if vertex i is in the cluster C and $x_i = 0$ otherwise. The cluster score can be rewritten as

$$S(\mathbf{x}) = \sum_{v_1, \ldots, v_k \in V} W(v_1, \ldots, v_k) \prod_{i=1}^{k} x_{v_i}. \tag{3.2}$$

Finding a good cluster means finding a subset of features with a high cluster score S_C. Even if we knew the number of elements in the cluster, maximizing S_C optimally would be an expensive combinatorial problem. For practical applications, we want to find a solution efficiently, so we rely on approximations. We relax the problem, by allowing x to take values in the continuous domain $[0, \varepsilon]$; ε acts as an upper bound of the cluster membership probability:

$$\mathbf{x}^* = \arg \max \ S(\mathbf{x}) \ \text{s.t.} \ \sum x_i = 1, \ \mathbf{x} \in [0, \varepsilon]^n. \tag{3.3}$$

When $\varepsilon = 1$, this hypergraph clustering formulation is equivalent to that of Bulo and Pellilo [8]. In the case of $\varepsilon = 1$ and pairwise affinities in $\{0, 1\}$ (corresponding to unweighted graphs), the problem is identical to the classical computation of maximal cliques [27, 28].

The L1 norm constraint in Problem 3.3 favors in practice sparse solutions and biases towards categorical values for the membership assignments. This makes it easier to discretize and find actual clusters. Multiple clusters can be found in different ways, such as the ones proposed in Bulo and Pellilo [8], Liu et al. [7]. One idea [8] is to remove the points belonging to a cluster that was found and restart the problem on the remaining points. The other approach [7] is to start from different initial solutions that are close to different clusters and locally maximize the score (Eq. 3.3).

3.3 Algorithm: IPFP for Hypergraph Clustering

Our proposed method (Algorithm 3.2) follows directly from the integer projected fixed point method (IPFP) presented in Chaps. 1 and 2. IPFP for hypergraph clustering is an iterative procedure that, at each iteration t, approximates the higher order score $S(\mathbf{x_t})$ by its first-order Taylor expansion around the current solution $\mathbf{x_t}$. This transforms the higher order optimization problem into a sequence of linear programs, each defined in the neighborhood of the solutions $\mathbf{x_t}$, at each time step t. Note that the first-order approximations can be globally optimized efficiently on the continuous domain $\sum x_i = 1$, $\mathbf{x} \in [0, \varepsilon]^n$. We successfully took a similar optimization approach to the problems of MAP inference, and graph and hypergraph matching [29, 30].

Before presenting our method, we first introduce some notation. Given a possible solution $\mathbf{x}$ in the continuous domain, let vector $\mathbf{d}(\mathbf{x})$ be a function of $\mathbf{x}$, obtained by marginalizing the tensor $\mathbf{W}$ as follows:

$$d_i(\mathbf{x}) = \sum_{v_1, \ldots, v_{k-1}} W(v_1, \ldots, v_{k-1}, i) \prod_{j=1}^{k-1} x_j. \tag{3.4}$$

We can now write the first-order Taylor approximation of the clustering score around solution $\mathbf{x_t}$ (Eq. 3.2), in the following form:

$$S(\mathbf{x}) \approx (1 - k)S(\mathbf{x_t}) + k\mathbf{d}(\mathbf{x_t})^\top \mathbf{x}. \tag{3.5}$$

Maximizing the first-order approximation in the continuous domain of (Eq. 3.3) results in the following linear program defined using the current solution $\mathbf{x_t}$:

$$\mathbf{y}^* = \arg\max \mathbf{d}(\mathbf{x_t})^\top \mathbf{y} \text{ s.t. } \sum y_i = 1, \ \mathbf{y} \in [0, \varepsilon]^n. \tag{3.6}$$

This linear program can be optimally solved using the following algorithm:

Algorithm 3.1 Optimize Problem 3.6

$\mathbf{d} \leftarrow \mathbf{d}(\mathbf{x_t})$
$c \leftarrow \lfloor 1/\varepsilon \rfloor$
Sort $\mathbf{d}$ in decreasing order $d_{i_1} \geq \cdots \geq d_{i_c} \geq \cdots \geq d_{i_n}$
$y_{i_l} \leftarrow \varepsilon$, for all $l \leq c$
$y_{i_{c+1}} \leftarrow 1 - c\varepsilon$
$y_{i_l} \leftarrow 0$, for all $l > c + 1$
return y

It is relatively easy to show that the above method returns a global optimum of Eq. (3.6). Once we obtain $\mathbf{y}^*$, we continue by searching for the global optimum of the original clustering score on the line segment between the current solution $\mathbf{x_t}$ and $\mathbf{y}^*$, an approach related to the classical Frank–Wolfe method [31]. If the hypergraph order, k, is less than or equal to 3, then the global optimum on the line segment can be computed in closed form since the score becomes a quadratic (or a cubic) one-dimensional function. For higher order scores, an efficient line search algorithm can be applied. Note that, since both $\mathbf{x_t}$ and $\mathbf{y}^*$ obey the constraints $\sum x_i = 1$, $\mathbf{x} \in [0, \varepsilon]^n$, every sample on the line between the two also obeys the constraints, so the maximizer will also be in the domain. Our algorithm is summarized in Algorithm 3.2 in pseudocode.

Algorithm 3.2 IPFP for efficient hypergraph clustering

Initialize $\mathbf{x_0}, t \leftarrow 0$
repeat
   Step 1: $\mathbf{y}^* \leftarrow \arg\max \mathbf{d(x_t)}^\top \mathbf{y}$ s.t. $\sum y_i = 1$, $\mathbf{y} \in [0, \varepsilon]^n$. If $\mathbf{d(x_t)}^\top (\mathbf{y} - \mathbf{x_t}) = 0$ stop.
   Step 2: $\alpha^* \leftarrow \arg\max S((1 - \alpha)\mathbf{x_t} + \alpha\mathbf{y}^*)$, $\alpha \in [0, 1]$.
   Step 3: $\mathbf{x_{t+1}} \leftarrow (1 - \alpha^*)\mathbf{x_t} + \alpha^*\mathbf{y}, t \leftarrow t + 1$
until convergence
return $\mathbf{x_t}$

3.4 Theoretical Analysis

Proposition 3.1 *The score $S(\mathbf{x}_t)$ increases at every step t of Algorithm 3.2 and the sequence $\mathbf{x}_t$ converges.*

Proof The algorithm does not stop at Step 2 if there is a $\mathbf{y}$ different from $\mathbf{x_t}$ such that $\mathbf{d(x_t)}^\top (\mathbf{y} - \mathbf{x_t}) > 0$. Since $\mathbf{d(x_t)}$ is proportional to the gradient of the original clustering score S at the current $\mathbf{x_t}$, it follows that there exists a point on the line segment between $\mathbf{x_t}$ and $\mathbf{y}$ with a score greater than $S(\mathbf{x_t})$. Such a point is found in Step 3 during line search (maximizer). Therefore, the score increases at every step. Since S is also bounded above, the algorithm will converge to a limit score. This must happen in the limit at step 2, when $\mathbf{d(x_t)}^\top (\mathbf{y} - \mathbf{x_t}) = 0$ and the solution $\mathbf{x_t}$ is returned. □

Proposition 3.2 *For a point of convergence $\mathbf{x}$, if $x_i < x_j$ then $d_i(\mathbf{x}) \leq d_j(\mathbf{x})$.*

Proof We will use a proof by contradiction. Let us assume that there exist i, j such that $x_i < x_j$ and $d_i(\mathbf{x}) > d_j(\mathbf{x})$. Let $r = (x_j - x_i)/2 > 0$ and $\mathbf{y}$ be a vector of the same size as $\mathbf{x}$, having all elements equal to those of $\mathbf{x}$ except for the i-th and j-th elements, which are $y_i = x_i + r$ and $y_j = x_j - r$. It can be easily verified that $\mathbf{y}$ lies in the valid continuous domain, since $\mathbf{x}$ is also in the domain. It can also be verified that $\mathbf{d(x)}^\top \mathbf{y} - \mathbf{d(x)}^\top \mathbf{x} = \mathbf{d(x)}^\top (\mathbf{y} - \mathbf{x}) = r(d_i(\mathbf{x}) - d_j(\mathbf{x})) > 0$, which contradicts the assumption that $\mathbf{x}$ is a point of convergence, since there exists at least one $\mathbf{y}$ which gives a better result in Step 2 of the algorithm. Hence, $\mathbf{x}$ cannot be a maximizer at Step 2. □

Proposition 3.3 *The points of convergence of Algorithm 3.2 satisfy the Karush–Kuhn–Tucker (KKT) necessary optimality conditions for Problem 3.3.*

Proof The Lagrangian function of (3.3) is

$$L(\mathbf{x}, \lambda, \mu, \beta) = S(\mathbf{x}) - \lambda \left(\sum x_i - 1 \right) + \sum \mu_i x_i + \sum \beta_i (\varepsilon - x_i), \qquad (3.7)$$

where $\beta_i \geq 0$, $\mu_i \geq 0$ and λ are the Lagrange multipliers. Since $kd_i(\mathbf{x})$ is the partial derivative of $S(\mathbf{x})$ w.r.t x_i, the KKT conditions at a point $\mathbf{x}$ are

$$kd_i(\mathbf{x}) - \lambda + \mu_i - \beta_i = 0.$$
$$\sum_{i=1}^{n} \mu_i x_i = 0.$$
$$\sum_{i=1}^{n} \beta_i (\varepsilon - x_i) = 0.$$

As the elements of $\mathbf{x}$ and the Lagrange multipliers are non-negative, it follows that if $x_i > 0 \Rightarrow \mu_i = 0$ and $x_i < \varepsilon \Rightarrow \beta_i = 0$. Then there exists a constant $\delta = \lambda/k$ such that the KKT conditions can be rewritten as

$$d_i(\mathbf{x}) \begin{cases} \leq \delta, \ x_i = 0, \\ = \delta, \ x_i \in (0, \varepsilon), \\ \geq \delta, \ x_i = \varepsilon. \end{cases}$$

If these KKT conditions are not met, then the conclusion of Proposition 3.2 does not hold, implying that $\mathbf{x}$ is not a point of convergence, which gives a contradiction. Therefore, Proposition 3.3 must be true. $\qquad\qquad\square$

3.4.1 Computational Complexity

In the general case, when the number of hyperedges is of order $O(N^k)$ (N—number of data points, k—the order of hyperedges), the overall complexity of each iteration of our algorithm is also $O(N^k)$ (linear in the number of hyperedges). It is important to note that [7, 8] have the same $O(N^k)$ complexity per step.

In Figs. 3.1 and 3.2, we present an evaluation of the computational costs of the three methods on synthetic problems with hyperedges of second order ($k = 2$). The results are averages over 100 different synthetic problems with varying number of data points. In Fig. 3.1, left plot, we show the average computational time per iteration (in Matlab) of each method versus the squared number of data points. Note that the computation time per step for each method varies almost linearly with the squared number of data points, confirming the theoretical $O(N^k)$ complexity (for $k = 2$). As expected, the least expensive method per step is Liu et al. [7] which performs sequential updates, while ours and Bulo and Pellilo [8] are based on parallel updates. The drawback of Liu et al. [7] (right plot) is the large number of iterations to convergence. It is interesting to note that Liu et al. [7] needs approximately the same number of iterations to converge as the dimensionality of the problem (number of data points). On the other hand, both ours and Bulo and Pellilo [8] are relatively

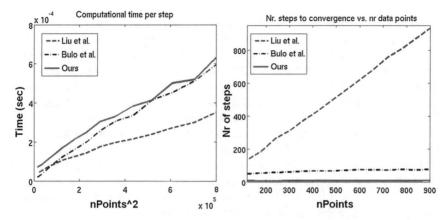

Fig. 3.1 Average computational cost over 100 different experiments as a function of the problem size. Left: the linear dependency of the runtime versus $nPoints^2$ shows experimentally that all algorithms have $O(N^2)$ complexity for hyperedges of order 2. Right: the method of Liu et al. [7] converges in a number of iterations that is directly proportional to the problem size, while the convergence of ours and Bulo and Pellilo [8] is relatively stable w.r.t. the number of points. In particular, ours reaches a high objective score in about 10 iterations

Fig. 3.2 Average total time to convergence over 100 experiments. We let ours run for 50 iterations, but it usually converges after about 10 iterations

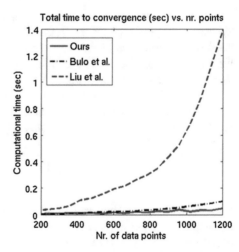

stable with respect to the number of points. Ours converges the fastest, needing on average less than 10 iterations. In terms of overall computational time, the method of Liu et al. [7] takes 1–2 orders of magnitude longer than ours to converge for 200–1200 points (Fig. 3.2).

3.5 Learning Graph and Hypergraph Clustering

The main difference, mathematically, between clustering and matching (presented in Chaps. 1 and 2) are the mapping constraints imposed on the indicator solution.

In the supervised learning schemes proposed for graph and hypergraph matching (Chap. 2), these constraints are not used during the gradient ascent optimization. The objective maximized during learning is the expected dot product between the given ground truth solution and the principal eigenvector of the matrix or tensor, containing the second- or higher order terms. Let us remember that the elements of the principal eigenvector are important for finding the cluster of correct assignments; they are indeed correlated with the elements of a main strong cluster of such assignments. Principal eigenvectors have been commonly used for clustering before, and thus such a learning scheme should also be suitable for clustering, even more so than for matching. Here, we propose a similar method for learning graph and hypergraph clustering, for the supervised, semi-supervised, and unsupervised cases, using gradient ascent and the efficient computation of the derivative of the eigenvector w.r.t to the parameters of the second- or higher order scores.

As discussed previously, in the case of graph matching (Chaps. 1 and 2), the objective function maximized during learning is the sum of the dot products between each $\mathbf{v}$ and the corresponding $\mathbf{b}$ over the training set of image pairs:

$$\mathbf{w}^* = \arg\max_{\mathbf{w}} \sum_i \mathbf{b}_i^T \mathbf{v}_i(\mathbf{w}).$$

Note that in the clustering case, as in the graph and hypergraph matching cases (Chap. 2), $\mathbf{b}$ could be either the ground truth discrete solution if it is known (the supervised case) or the projection of the eigenvector $\mathbf{v}(\mathbf{w})$ on the discrete domain defined by the clustering constraints. The learning approach from matching is thus transferred to the case of clustering with minimal modification.

As before in the case of graph and hypergraph matching, $\mathbf{v}(\mathbf{w})$ is a function of the parameters $\mathbf{w}$, since it depends on the elements of the tensor $\mathbf{H}$, which are all functions of $\mathbf{w}$. We optimize the objective by a gradient ascent update:

$$w_j \leftarrow w_j + \eta \sum_{i=1}^{N} \mathbf{b}_i(\mathbf{w})^T \frac{\partial \mathbf{v}_i(\mathbf{w})}{\partial w_j}. \tag{3.8}$$

Learning for graph and hypergraph clustering is strongly related to the one on learning for graph and hypergraph matching, since that is only a special case of clustering. In that case, the cluster is formed by nodes that represent candidate assignments and are linked and grouped through strong agreement edges or hyperedges. Below we reiterate the two main algorithms needed for computing the gradient ascent update from above (Eq. 3.8).

Algorithm 3.3 Symmetric higher order power method (S-HOPM)

$\mathbf{v} \leftarrow \mathbf{1}$
repeat
$\quad v_{ia} \leftarrow \sum_{ia,jb,kc} H_{ia;jb;kc} v_{jb} v_{kc}$
$\quad \mathbf{v} \leftarrow \mathbf{v}/\|\mathbf{v}\|$
until convergence
return $\mathbf{v}$

In Algorithm 3.3, we present the procedure for symmetric higher order power method (S-HOPM), to compute $\mathbf{v}(\mathbf{w})$, the equivalent of the eigenvector for third-order tensors. This method is the basis for estimating the gradient of the eigenvector $\mathbf{v}(\mathbf{w})$ computed using S-HOPM. Computation of this gradient is presented in Algorithm 3.4. Note that this algorithm is almost identical (with minimal differences due to the order of the matrix/tensor) for all learning problems addressed so far (Chaps. 1 and 2): graph and hypergraph matching, MAP inference in Markov random fields, as well as graph and hypergraph matching (presented in this chapter).

Since each iteration of S-HOPM computes the update $\mathbf{v} \leftarrow \mathbf{v}/\|\mathbf{v}\|$, it follows that the gradient of the estimate of $\mathbf{v}$ at iteration $k + 1$ can be expressed as a function of the gradient at the previous step. Hence, the gradient can be computed recursively, jointly with the computation of $\mathbf{v}$. Algorithm 3.4 applies these computations recursively.

Algorithm 3.4 Compute the gradient **dv** of **v**

$\mathbf{v} \leftarrow 1$
$\mathbf{dv} \leftarrow 0$
$\mathbf{dH} \leftarrow \frac{\partial \mathbf{H}}{\partial w_j}$
repeat
  $h_{ia} \leftarrow \sum_{jb;kc} dH_{ia;jb;kc}v_{jb}v_{kc} + 2H_{ia;jb;kc}v_{jb}dv_{kc}$
  $v_{ia} \leftarrow \sum_{jb;kc} H_{ia;jb;kc}v_{jb}v_{kc}$
  $\mathbf{h} \leftarrow \frac{\mathbf{h}}{\|\mathbf{v}\|}$
  $\mathbf{v} \leftarrow \frac{\mathbf{v}}{\|\mathbf{v}\|}$
  $\mathbf{dv} \leftarrow \mathbf{h} - (\mathbf{v}^T\mathbf{h})\mathbf{v}$
until convergence
return dv

3.6 Experiments on Third-Order Hypergraph Clustering

We present two types of experiments that are relevant for data clustering and compare the performance of our algorithm with that of current state-of-the-art methods. The first experiment is on 2D line fitting, similar to experiments from Liu et al. [7] and Bulo and Pellilo [8]. The second type of experiments we show is on higher order matching, where we perform comparisons with both clustering methods and two state-of-the-art hypergraph matching algorithms [17, 18]. This experiment reveals an interesting connection between hypergraph clustering and hypergraph matching.

3.6.1 Line Clustering

This experiment consists of finding lines as clusters of 2D points. Since any two points lie on a line, third-order affinity measures are needed. Similar to Liu et al. [7] and Bulo and Pellilo [8], for any triplet of points $\{i, j, k\}$, we use as dissimilarity measure the mean distance to the best fitting line $d(i, j, k)$. We define the similarity function $w(\{i, j, k\})$ using a Gaussian kernel $w(\{i, j, k\}) = \exp(-d(i, j, k)^2/\sigma_d^2)$, where σ_d is a parameter that controls the sensitivity to fitting errors.

We randomly generate points belonging to three lines in the range $[-0.5, 0.5]^2$, with 30 points per line perturbed with Gaussian noise $N(0, \sigma)$ and add a number of outliers in the same region of the 2D space. In Fig. 3.3, we present two examples of generated point sets, with the three lines shown in red, blue, and green and outliers shown with magenta. The figure on the right presents the case of 60 outliers with a relatively large amount of noise $\sigma = 0.04$. Note the difficulty of recovering the correct clusters.

We are interested in two important aspects: accuracy and speed of convergence, under varying levels of noise and number of outliers. For each level of noise and number of outliers, we test all methods in 50 random trials and average the results (Fig. 3.4). For each clustering problem, all algorithms are initialized with the same uniform initial solution, and the performance is evaluated using the F-measure. Our method and [7] use the same $\varepsilon = 1/30$. The performance of all methods in our experiments was relatively stable w.r.t σ_d, so we fixed $\sigma_d = 0.02$.

In Fig. 3.4, plots b and d, we show the average F-measure of all three algorithms at convergence, ours performing slightly better than the rest. An important aspect in practice is how fast a method can reach a good solution. This issue is particularly important in graph and hypergraph clustering problems that have combinatorial complexity. Our algorithm obtains accurate solutions much faster than both Liu et al. [7] and Bulo and Pellilo [8]. If our method takes on average between 5 and 10 steps

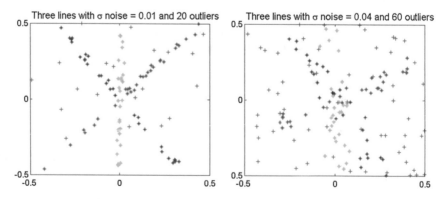

Fig. 3.3 Generated 3D points from three lines with outliers. Notice the difficulty of finding the correct clusters in the example on the right, when both the noise variance and the number of outliers are relatively large. Best viewed in color

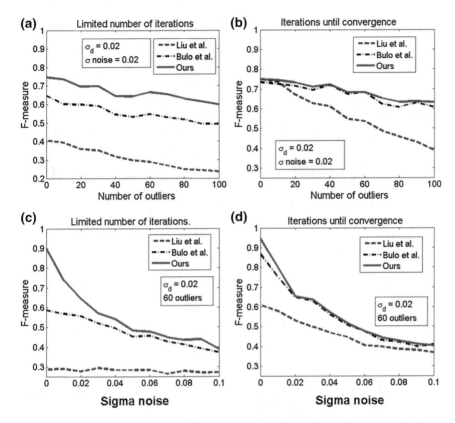

Fig. 3.4 Performance comparison of different methods on 2D line fitting. Left: our method has a far superior performance when all algorithms are allowed to run for the same, limited amount of computational time: ours for 5 iterations (less time), the competitors for 10 iterations. Comparison with right plot: notice that our method has almost converged in only 5 iterations. Best viewed in color

to reach a good solution, the algorithm of Liu et al. [7] needs more than 150 steps ($\approx$ the number of data points). Since the computation time per step can be shown to be comparable for all algorithms (ours is about 1.4 times slower per step than the others on these problems), we present in Fig. 3.4, plots *a* and *c*, the performance of all algorithms when ours is stopped after 5 iterations, whereas the other two methods are stopped after 10 steps. Empirically, we find that our proposed method converges about one order of magnitude faster.

In Fig. 3.5, we further study the convergence properties of all three algorithms. Plots *a*, *b*, and *c* show the solutions on one experiment for all methods after 5 and 50 iterations and at convergence. The case presented is a good example of how the methods behave in general. Our method, after only 5 iterations, is already close to its optimum, reaching a sparse solution. Note that the first 30 elements correspond to the points of the first line, so our method is already close to a perfect solution. The method

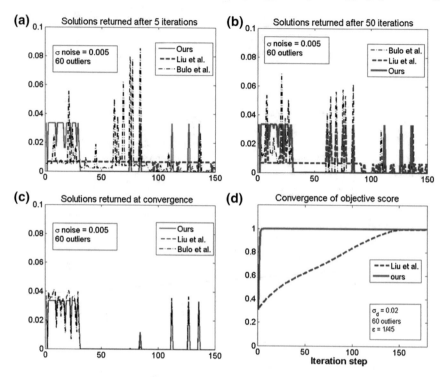

Fig. 3.5 Solutions achieved after different numbers of iterations on the same problem for all algorithms. For plots **a–c** the x-axis indexes the elements of the solution vectors. Plot **d**: average objective scores of our method versus [7] per iteration number. Notice the fast convergence of our method. Best viewed in color

of Liu et al. [7] is still very close to its initial solution, due to its site-wise updates. The method of Bulo and Pellilo [8] based on Baum-Eagon [14] performs parallel updates per step, but it takes smaller steps than ours. Our method uses the same problem formulation as Liu et al. [7] so we directly compare the objective scores obtained per iteration averaged over 300 experiments (Fig. 3.5, plot d: average scores normalized by the maximum average). Our algorithm converges very fast (<10 iterations), while Liu et al. [7] take almost 200 steps until it converges to an objective score slightly less on average than ours.

3.6.2 Affine-Invariant Matching

Affine-invariant point matching is an important problem in computer vision, with applications to object matching and recognition. Distant or planar objects seen from different viewpoints undergo transformations that can be well approximated by an affine or a piecewise affine transformation. We perform experiments on affine-invariant point matching and compare our method with hypergraph clustering as well as hypergraph matching methods [18] (TM), [17] (PM). Since the connection between hypergraph clustering and hypergraph matching is not well established yet, we believe that such experiments are relevant.

Hypergraph matching has a similar tensor formulation as hypergraph clustering (Eq. 3.3), the main difference being the constraints on the solution. Usually, assignment problems impose one-to-one matching constraints on a solution $\mathbf{x}$, where each element x_i corresponds to a candidate match $i = (u, v)$. Here, u is the index of a feature from one image and v is the index of the feature from the other image. The acceptable solutions are indicator vectors such that x_i is 1 if u is matched to v and 0 otherwise. A tensor, similar to W, is constructed such that its elements of type $(i_1, \ldots, i_k)$ represent matching similarities between the k-tuple of features $(u_1, \ldots, u_k)$ from one image and the corresponding k-tuple $(v_1, \ldots, v_k)$ from the other image, where $i_q = (u_q, v_q)$.

Although the constraints are different for hypergraph matching, it is important to note that the two state-of-the-art hypergraph matching methods [17, 18] ignore the one-to-one constraints during optimization and impose them only during a final, post-processing step. This makes them very efficient in practice. The method of Duchenne et al. [18] is based on computing the stationary point of the higher order power method applied to the tensor W, that is, the higher order extension of the principal eigenvector of matrices, well known for its usefulness in clustering. We therefore expect other clustering methods to be useful for matching.

Since affine transformations cannot be recovered from pairs of points, we use third-order clustering. Given three points (u_1, u_2, u_3) in one image and their corresponding matches (v_1, v_2, v_3) in the other image, transformed by the affine transformation $\mathbf{T}$, the areas of the corresponding triangles A_{u_1,u_2,u_3} and A_{v_1,v_2,v_3} are related by the formula $|\det \mathbf{T}| = A_{u_1,u_2,u_3}/A_{v_1,v_2,v_3}$. We use third-order matching scores of the form $W(i_1, \ldots, i_k) = \exp((1 - \sqrt{|\det \mathbf{T}| A_{v_1,v_2,v_3}/A_{u_1,u_2,u_3}})^2/\sigma^2)$.

We generated 2D point sets from 420 real images containing 7 different persons collected with an RGB-depth Kinect camera. For each image, we first obtained the occlusion boundaries of the person and extracted around 40–50 points per image equally spaced on the boundary. Using this set of points, we obtained the second set by transforming the points based on a randomly generated affine transformation $\mathbf{T} = [1 + N(0, 0.2), N(0, 0.2); N(0, 0.2), 1 + N(0, 0.2)]$, and then perturbed their position with Gaussian noise $N(0, \sigma)$. We also added outliers to the second set of points (Fig. 3.6). Since $|\det \mathbf{T}|$ is unknown at testing time, we estimated its distribution (from the distribution of $\mathbf{T}$), and for each matching example, we sampled a few values and picked the one giving the best objective score.

In clustering, a sparse solution is always preferred. In the case of matching, in order to obtain a sparse solution, it is better to use the one-to-one matching constraints. If clustering already provides a sparse solution, it is not clear how to use the one-to-one constraints in the final post-processing step. The three clustering methods from the previous experiment tend to give sparse solutions. In turn, hypergraph

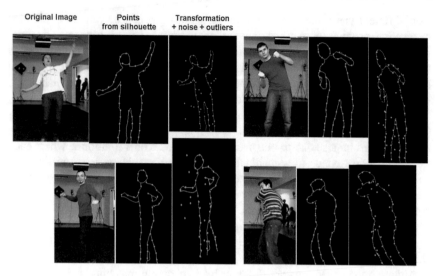

Fig. 3.6 Examples from our dataset: original image (left), points sampled on the boundaries of the person (middle), transformed points with added noise and outliers (right). Best viewed in color

matching algorithms [17, 18] almost always give continuous solutions. They could be later discretized by the Hungarian algorithm that imposes the one-to-one constraints. We therefore expect the matching algorithms to benefit more from the Hungarian method than the clustering algorithms. For this reason, we evaluate algorithms in two ways. First, we measure the matching rate by applying the Hungarian method to the raw output $\mathbf{x}$ of each algorithm, namely, solve the problem $\mathbf{y}^* = \arg\max \mathbf{x}^\top \mathbf{y}$, given the one-to-one matching constraints. Second, we measure the closeness of the raw output to the sparse ground truth, by computing the normalized correlation between the two. Note that the second measure should favor methods that give sparse clustering solutions, which are close to the matching ground truth without the knowledge of the one-to-one constraints. We present the results in Fig. 3.7. Note that our method outperforms on average all the others in terms of both matching and normalized correlation scores against ground truth. The fact that our method gives sparse solutions close to ground truth, without using the one-to-one matching constraints, could be useful in matching applications where the one-to-one constraints do not hold. For example, in cases of large changes in scale or when matching subparts of objects to full parts, the exact many-to-one or many-to-many matching constraints are unknown. In those cases, a robust clustering method could recover the correct matches without using any matching constraints.

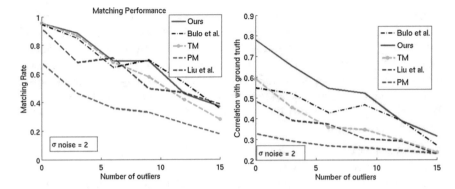

Fig. 3.7 Results on affine-invariant matching. Note that our method gives solutions that are already close to the sparse ground truth without using the one-to-one matching constraints. Best viewed in color

3.7   Conclusions and Future Work

We presented an original hypergraph clustering method with important theoretical properties and state-of-the-art performance, applicable to hyperedges of any order, including the usual case of second-order edges used in classical graph clustering. Our algorithm is based on an efficient iterative procedure directly derived from the initial integer projected fixed point algorithm for graph matching, with significantly better convergence speed than existing competitive methods, without any loss of accuracy. We have also tested our method on matching problems and showed excellent performance. In Chaps. 4 and 5, we will apply this method on different tasks, such as unsupervised feature selection, classifier, and descriptor learning. The applicability of clustering so many problems related to unsupervised learning proves once again that clustering is fundamental to learning in the unsupervised case. After all, clustering is at the core of finding mutual agreements between many entities, classifiers, or HPP features, which is also at the core of our principles for unsupervised learning proposed in Chap. 1.

Hyperedges consider higher level constraints and relationships, by accumulating information over larger groups of features than in the pairwise case of graph clustering. Such higher order scores can compute geometric-invariant quantities that are more powerful in cases of affine or perspective transformations, or appearance-based functions that could be robust to illumination changes, deformations or intra-category variations. The bottom line is as follows: the more the data points are processed jointly, the easier it is to decide whether they are part of the same cluster or not. The main bottleneck of hypergraph clustering is computation and storage cost. The higher the degree of the hyperedges, the more costly it is and the increase in cost is exponential as a function of the degree. We think that one possible way to overcome this limitation is to move towards sampling-based probabilistic formulations, where the groups of features, that is, the hyperedges, could be randomly sampled, at each iteration, depending on the available storage and computation capabilities and speed requirements. This would require dropping the deterministic convergence and climbing theoretical properties and seek stochastic properties that are likely to hold, with high probability, when the data is sufficiently large. The idea of stochastic clustering is not novel in vision [32], but our proposal to adapt and extend current hypergraph clustering approaches to the stochastic case is novel. Below, we present the sketch of a possible stochastic approach to higher order clustering, to be further developed and studied:

1. Step 1: At each iteration t, sample a subset of hyperedges S_t, given the memory and computation power available.
2. Step 2: Compute their potential functions and obtain the overall objective clustering score C_t as a function of the potentials over the sampled hyperedges and the current solution x_t.
3. Step 3: Find the next soft clustering solution x_t that produces a better clustering score $C_{t+1} > C_t$, for the hyperedges sampled so far.
4. Iterate Steps 1–3 until convergence of x_t, for a sufficiently large t.

Note that the algorithm above does not have in general a deterministic theoretical guarantee over all possible hyperedges, but we could hope to design a method with probabilistic properties and excellent performance in practice. In the case of large-scale data we should not avoid probabilistic solutions, they might be the only practical way to go in order to achieve efficiency and robustness to real-world noise in vision. After all, numerous studies in neuroscience reveal the noisy, stochastic underlying, neurological functioning of the brain [33].

The algorithmic design, theoretical study, and implementation of efficient hypergraph methods for learning and clustering, along with the connection between clustering, matching, discovery, and recognition, could further fortify the bridge between graph theory and the developing fields of computer vision and machine learning. In particular, it would be worth exploring in more depth the connections between the iterative projected fixed point method presented in the first chapter on the task of feature matching, and the related clustering method introduced in this chapter. These related problems and methods should, in future work, be integrated further into a unified approach for efficient graph-based modeling, learning, and optimization for computer vision.

References

1. Jain A, Murty M, Flynn P (1999) Data clustering: a review. ACM Comput Surv
2. Kanungo T, Mount D, Netanyahu N, Piatko C, Silverman R, Wu A (2002) An inequality with applications to statistical estimation for probabilistic functions of Markov processes and to a model for ecology. PAMI
3. Shi J, Malik J (2000) Normalized cuts and image segmentation. PAMI 22(8). http://www.citeseer.ist.psu.edu/shi97normalized.html

4. Dhillon I, Guan Y, Kulis B (2004) Kernel k-means: spectral clustering and normalized cuts. In: ACM international conference on knowledge discovery and data mining
5. Ng A, Jordan M, Weiss Y (2002) On spectral clustering: analysis and an algorithm. In: NIPS
6. Frey B, Dueck D (2007) Clustering by passing messages between data points. Science
7. Liu H, Latecki L, Yan S (2010) Robust clustering as ensembles of affinity relations. In: NIPS
8. Bulo S, Pellilo M (2009) A game-theoretic approach to hypergraph clustering. In: NIPS
9. Zien J, Schlag M, Chan P (1999) Multilevel spectral hypergraph partitioning with arbitrary vertex sizes. IEEE Trans Comput-Aided Des Integr Circuits Syst
10. Agarwal S, Lim J, Zelnik-Manor L, Perona P, Kriegman D, Belongie S (2005) Beyond pairwise clustering. In: CVPR
11. Rodriguez J (2003) On the Laplacian spectrum and walk-regular hypergraphs. Linear Multilinear Algebr
12. Zhou D, Huang J, Scholkopf B (2007) Learning with hypergraphs: clustering, classification and embedding. In: NIPS
13. Shashua A, Zass R, Hazan T (2003) Multi-way clustering using super-symmetric non-negative tensor factorization. In: ECCV
14. Baum L, Eagon J (1967) An inequality with applications to statistical estimation for probabilistic functions of Markov processes and to a model for ecology. Bull Am Math Soc
15. Ochs P, Brox T (2012) Higher order motion models and spectral clustering. Comput Vis Pattern Recognit
16. Huang Y, Liu Q, Metaxas D (2009) Video object segmentation by hypergraph cut. Comput Vis Pattern Recognit
17. Zass R, Shashua A (2008) Probabilistic graph and hypergraph matching. In: CVPR
18. Duchenne O, Bach F, Kweon I, Ponce J (2009) A tensor-based algorithm for high-order graph matching. In: CVPR
19. Leordeanu M, Zanfir A, Sminchisescu C (2011) Semi-supervised learning and optimization for hypergraph matching. In: ICCV
20. Chertok M, Keller Y (2010) Efficient high order matching. IEEE Trans Pattern Anal Mach Intell 32:2205–2215
21. Bu J, Tan S, Chen C, Wang C, Wu H, Zhang L, He X (2010) Music recommendation by unified hypergraph: combining social media information and music content. In: Proceedings of the international conference on multimedia
22. Li L, Li T (2013) News recommendation via hypergraph learning: encapsulation of user behavior and news content. In: Proceedings of the sixth ACM international conference on web search and data mining
23. Huang Y, Liu Q, Zhang S, Metaxas D (2010) Image retrieval via probabilistic hypergraph ranking. Comput Vis Pattern Recognit
24. Huang Y, Liu Q, Lv F, Gong Y, Metaxas D (2011) Unsupervised image categorization by hypergraph partition. Pattern Anal Mach Intell 33(6)
25. Gao Y, Wang M, Luan H, Shen J, Yan S, Tao D (2011) Tag-based social image search with visual-text joint hypergraph learning. In: Proceedings of the 19th ACM international conference on multimedia
26. Leordeanu M, Sminchisescu C (2012) Efficient hypergraph clustering. In: International conference on artificial intelligence and statistics
27. Ding C, Li T, Jordan M (2008) Nonnegative matrix factorization of combinatorial optimization: spectral clustering, graph matching, and clique finding. In: IEEE international conference on data mining
28. Motzkin T, Straus E (1965) Maxima for graphs and a new proof of a theorem of Turan. Canad J Math
29. Leordeanu M, Hebert M, Sukthankar R (2009) An integer projected fixed point method for graph matching and map inference. In: NIPS
30. Leordeanu M, Zanfir A, Sminchisescu C (2011) Semi-supervised learning and optimization for hypergraph matching. In: IEEE international conference on computer vision

31. Frank M, Wolfe P (1956) An algorithm for quadratic programming. Nav Res Logist Q 3(1–2):95–110
32. Gdalyahu Y, Weinshall D, Werman M (2001) Self-organization in vision: stochastic clustering for image segmentation, perceptual grouping, and image database organization. Pattern Anal Mach Intell 23(10)
33. Rolls E, Deco G (2010) The noisy brain: stochastic dynamics as a principle of brain function, vol 34. Oxford University Press, Oxford

Chapter 4
Feature Selection Meets Unsupervised Learning

4.1 Introduction

The problem of feature selection is naturally linked to learning individual classifiers or classifier ensembles, as it is able to answer a fundamental problem in machine learning research: which cues in the vast ocean of signals are relevant for learning new concepts at the next level of abstraction and semantics. For that matter, feature selection has been heavily researched over the last three decades, with many papers marking the machine learning and computer vision roadmap [1, 1–7]. Applications of feature selection techniques abound in all areas of computer vision and machine learning, ranging from detection of specific categories [8] to more general recognition tasks. Each year, new visual features and classifiers are discovered or automatically learned, stored, and often made publicly available. As the vast pool of powerful features continues to grow, efficient feature selection mechanisms must be devised.

Classes are often *triggered* by only a few key input features, as shown in Fig. 4.1, which presents a case that has been revealed in our experiments detailed in Sect. 4.4. Since feature selection is NP-hard [9, 10], most previous work focused on greedy methods, such as sequential search [11] and boosting [5], relaxed formulations with l^1- or l^2-norm regularization, such as ridge regression [12] and the Lasso [13, 14], or heuristic genetic algorithms [15].

We formulate feature selection as a problem of discriminant linear classification [16] with novel affine constraints on the solution and the features. We put an upper bound on the solution weights and require it to be an affine combination of soft-categorical features, which should have on average stronger outputs on the positive class versus the negative one. We term these *signed features*. We present both a supervised and an almost unsupervised approach. We call it an almost unsupervised case because we still require the knowledge or good guessing of a specific feature sign. The supervised method is a convex constrained minimization problem, which

The material presented in this chapter is based in large part on the following paper:
Leordeanu, Marius, Alexandra Radu, Shumeet Baluja, and Rahul Sukthankar. "Labeling the features not the samples: Efficient video classification with minimal supervision." In Thirtieth AAAI conference on artificial intelligence (AAAI). 2016.

© Springer Nature Switzerland AG 2020
M. Leordeanu, *Unsupervised Learning in Space and Time*,
Advances in Computer Vision and Pattern Recognition,
https://doi.org/10.1007/978-3-030-42128-1_4

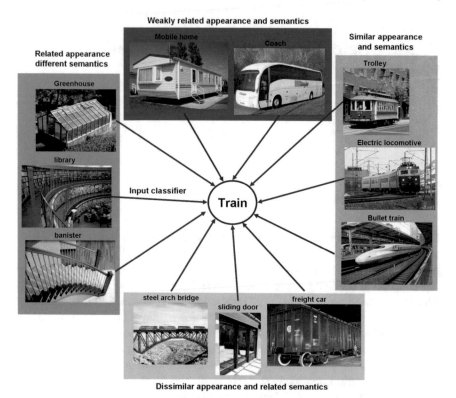

Fig. 4.1 What classes can *trigger* the idea of a "train"? Many classes have similar appearance but are semantically less related (blue box); others are semantically closer but visually less similar (green box). There is a complex discrete-continuum that relates appearance, context, and semantics. Can we find a group of classifiers that combined signal the presence of a specific, higher level concept, which are together robust to outliers, over-fitting, and missing features? Here, we show classifiers (e.g., library, steel arch bridge, freight car, mobile home, trolley, etc.) that are consistently selected by our method from limited training data as giving valuable input to the class "train"

we extend it to the case of almost unsupervised learning, with a concave minimization formulation, in which, as mentioned, the only bits of supervised information required are the feature signs. However, the training samples are completely unlabeled in the unsupervised case. We also present and test a combined semi-supervised formulation, in which a few labeled examples are used together with large amounts of unlabeled data. An interesting observation, which will be discussed in more detail in later sections, is that the unsupervised case is formulated as a concave minimization problem (which is NP-hard), whereas the supervised one has a convex minimization form (which can be globally optimized efficiently).

Both formulations have the desired sparsity and optimality properties as well as strong generalization capabilities in practice. The proposed schemes also serve as feature selection mechanisms, such that the majority of features with zero weights can be safely ignored while the remaining ones form a powerful classifier ensemble. Consider Fig. 4.1: here we use image-level CNN classifiers [17], pretrained on ImageNet, to recognize trains in video frames from the YouTube-Objects dataset [18]. Our method rapidly discovers the relevant feature group from a large pool of 6000 classifiers. Since each classifier corresponds to one ImageNet concept, we show a few of the classifiers that have been consistently selected by our method over 30 experimental trials.

The material presented in this chapter is based on our original paper [19]. Here, we present a more in-depth experimental analysis, to better understand the properties of our approach, its ability to learn from limited training data, and how it compares in such cases to well-established powerful algorithms such as support vector machines.

The key contributions presented in this chapter are the following:

1. We describe an efficient method for joint linear classifier learning and feature selection that directly relates to the integer projected fixed point method (IPFP), which was previously presented (Chaps. 1, 2 and 3) in the context of graph matching and clustering. Note that more recent research [20] has also revealed the interesting connection between feature selection and clustering. Here, we show that, both in theory and practice, our solutions to the classifier learning problem are sparse, and thus the task becomes equivalent to feature selection. The number of features selected can be set to k and the non-zero weights are equal to $1/k$. The simple solution enables good generalization and learning in an almost unsupervised setting, with minimal supervision, as explained next. This is very different from classical regularized approaches such as the Lasso.

2. Our formulation requires minimal supervision in the following sense. It is required to know or be able to estimate the *signs* of features with respect to the target class (defined in Sect. 4.2). A feature has a positive sign if on average it has higher values for the positive class than for the negative class. Thus, features can always be *flipped* to become positive, if their initial sign is known. These signs can be estimated from a small set of labeled samples, and once determined, our methods can handle large quantities of unlabeled data with excellent accuracy and generalization in practice. The algorithms presented are also robust to large errors in feature sign estimation.

3. Our feature selection by clustering approach demonstrates superior performance in terms of learning time and accuracy when compared to established algorithms such as AdaBoost, Lasso, Elastic Net, and SVM, especially in the case of limited supervision. Besides being able to efficiently tackle the very actual problem of learning in the case of video data with minimal supervision, the comparative experiments provide a broader view of our approach in the context of classic machine learning approaches, which have been extensively researched in the last decades.

4.1.1 Relation to Principles of Unsupervised Learning

The idea of learning a new concept in an unsupervised way, by using the co-firing of classifiers, specialized for other concepts, as supervisory signal, is mostly related to the following principle for unsupervised learning (introduced in Chap. 1).

> **Principle 6**
> When several weak independent classifiers consistently fire together, they signal the presence of a higher level class for which a stronger classifier can be learned.

This general principle is supported, in the context of this chapter, by the following fact: the covariance matrix of the outputs of many independent classifiers will often contain at least one strong cluster (corresponding to a sub-block of the covariance matrix), which is formed by an unknown subset of the classifiers (remaining to be discovered), which are strongly dependent given the current unsupervised (unlabeled) input data. However, such classifiers are usually independent and become dependent only when *this specific data* is given. This unlikely correlation becomes strong HPP supervisory signal indicating the presence of a hidden, unknown class. We also see the relation to other principles of unsupervised learning (Principles 2 and 3 in Chap. 1), on which the general unsupervised learning approach taken in this book is based.

Thus, a specific subset of independent classifiers become strongly correlated, whenever the unknown class is present in the data. The correlation, captured within the overall covariance matrix, from all classifiers or features, could be detected. Then, the subset of strongly correlated features could be discovered by a clustering procedure, as shown next in this chapter. While the intuition behind the co-firing of different classifiers for the discovery of a new class (on which they are conditioned) is pretty clear, we still need to discuss the original idea of the *feature sign*.

Before we motivate the introduction of this concept, we reiterate here what the sign of a feature means: a feature (or classifier output) has a positive sign if it is, in loose terms, positively correlated with the presence of the (hidden) class of interest and has a negative sign otherwise. More specifically, a positive feature is one that has, on average, larger values when the class of interest is present (the true positive cases) than when the class of interest is not present (true negative cases). If we know in advance the true signs of features, then we could easily make them all positive, by a simple *flip* operation. Then, the cluster (subset) of strongly correlated classifier outputs (conditioned on the presence of that specific class) would become even stronger, as more relevant features will become strongly positively correlated. Consequently, they would be easier to discover and the unsupervised learning process would be more efficient and robust.

In Sect. 4.2, we provide a more in-depth mathematical justification for this fact. For now, it is important to first understand the intuitive ideas behind it. Next, we go a bit deeper and motivate the idea of using the feature signs instead of the true labels of the data samples, in order to achieve (almost) unsupervised learning.

4.1.2   Why Labeling the Features and Not the Samples

Features are functions of data samples. We could then imagine using the output of these functions to indirectly infer about the class of the samples. If flipped correctly, when a feature has a strong positive value (w.r.t to its overall mean value in time), then we could expect the data sample, which triggered it, to also be positive. In the opposite case, if the positively flipped feature has a low value, then we could also expect the specific sample that triggered it to be a negative sample.

Thus, features could play the role of weak classifiers: when they are correctly flipped, their high output values indicate (on average) positive samples and their low output values indicate (on average) negative samples. On one hand, features represent transformations of the data, so they are data. On the other hand, they could represent weak classifiers when they are correctly flipped: in some sense, signed features are both data and classifiers. Then, as explained previously, if I have many of such weak classifiers (features) and I also know their correct signs, I could use them for unsupervised learning—that is training of a strong classifier without knowing the true labels of data samples. In the next section (Sect. 4.2), we also provide the mathematical justification that naturally leads to these conclusions.

One could argue that by doing so, we simply push the learning problem to the flipping of features instead of labeling samples. We argue that correctly flipping the features is much easier and could be done by using the unsupervised learning principles proposed in the book. For example, HPP supervisory signal could be used to correctly infer the correct flipping of features and their transformations into classifiers. Moreover, many features are correctly and naturally flipped from the start: for example, the presence of a light source is in general positively correlated with the presence of an object (of any type) and not with the absence of one. So light is a positive feature, almost always, for learning about most classes.

Let us now synthesize the main ideas, which justify why labeling the features might be a better idea than labeling the samples.

1. We have a limited number of features (e.g., neurons in the brain), whereas we have virtually unlimited number of data samples. Once I know the correct sign of each of my finite number of features, I could be done.
2. Labeling individual features might be simpler and could require less data than learning a very complex class from lots of data samples. Let us think of the light example: figuring out that a light source is a positive feature for many unknown classes could be much easier than learning the correct classifier for those classes. Light is a simple feature but those unknown classes could be rather complex and very different concepts.
3. Things get even better when we realize that, in practice, we might not need to flip features at all. We could use them as they are. As we show in our mathematical

derivations, nothing requires that we have all features flipped correctly. We only discover a subset of features that are already flipped correctly anyway. We really do not need to know a priori which or how many features are flipped correctly. Let us imagine we have features that are outputs of some weak classifiers, which are already trained to respond to the presence of some known classes. We could simply consider them already flipped correctly w.r.t some still unknown class, which is yet to be discovered. Then we could use these existing weak classifiers (assuming we have many already), without flipping them, in order to learn about classes that are naturally positively correlated with them, as they are. Thus, we will be able to learn about unknown classes for which we already have sufficiently many correctly flipped features. Interestingly enough, it is also true that we would be blind to classes for which we do not have sufficient support from existing positively correlated features.

The ideas discussed above immediately suggest examples in real life, in which we could learn in an unsupervised way, by using the existing correlations between our pretrained features and new, unknown classes: for example, by using such pretrained features we could *learn about new types of cars from features of old types of cars*; or, as another example, we could *learn to detect new types of activities (squash) from features (patterns) of old activities (tennis)*. In principle, we could learn about any new variation of a class, evolution of a class or new classes that are strongly and positively related to old ones. Other examples include *learning about sky scrappers from already acquired knowledge (positively correlated features/classifiers) about houses; learning about airplanes from knowledge (positively correlated features/classifiers) about birds*, and so on. At the same time, we might not be able to learn, without strong supervision, about classes that are a combination of positively and negatively correlated features, for which our a priori positive correlation guess is wrong. But it is somewhat expected to get confused and unable to learn when there is no consensus of prior knowledge with respect to a novel and unusual concept.

These ideas and examples also relate to the more general idea of a universal unsupervised learning machine, which is discussed in the last chapter (Chap. 8). Next, we proceed by looking more closely at the mathematics that lays behind the ideas of labeling the features and the approach to unsupervised learning as feature selection by clustering. We will return to intuition throughout the chapter, when needed in order to better understand the mathematics and the experimental results.

4.2 Mathematical Formulation

We address the case of binary classification and apply the one versus all strategy to the multi-class scenario. We have a set of N samples, with each ith sample expressed as a column vector $\mathbf{f}_i$ of n features with values in $[0, 1]$; such features could themselves be outputs of classifiers. We want to find vector $\mathbf{w}$, with elements in $[0, 1/k]$ and unit l^1-norm, such that $\mathbf{w}^T \mathbf{f}_i \approx \mu_P$ when the ith sample is from the positive class and

$\mathbf{w}^T \mathbf{f}_i \approx \mu_N$ otherwise, with $0 \le \mu_N < \mu_P \le 1$. For a labeled training sample i, we fix the ground truth target $t_i = \mu_P = 1$ if positive and $t_i = \mu_N = 0$ otherwise. Our novel constraints on $\mathbf{w}$ limit the impact of each individual feature f_j, encouraging the selection of features that are powerful in combination, with no single one strongly dominating. This produces solutions with good generalization power. In Sect. 4.3, we show that k is equal to the number of selected features, all with weights $= 1/k$. The solution we look for is a weighted feature average with an ensemble response that is stronger on positives than on negatives. For that, we want any feature f_j to have expected value $E_P(f_j)$ over positive samples greater than its expected value $E_N(f_j)$ over negatives. We estimate its sign $sign(f_j) = E_P(f_j) - E_N(f_j)$ from labeled samples and if it is negative we simply *flip* the feature: $f_j \leftarrow 1 - f_j$.

4.2.1 Supervised Learning

We begin with the supervised learning task, which we formulate as a least squares constrained minimization problem. Given the $N \times n$ feature matrix $\mathbf{F}$ with $\mathbf{f}_i^T$ on its ith row and the ground truth vector $\mathbf{t}$, we look for $\mathbf{w}^*$ that minimizes $\|\mathbf{Fw} - \mathbf{t}\|^2 = \mathbf{w}^T(\mathbf{F}^T\mathbf{F})\mathbf{w} - 2(\mathbf{F}^T\mathbf{t})^T\mathbf{w} + \mathbf{t}^T\mathbf{t}$, and obeys the required constraints. We drop the last constant term $\mathbf{t}^T\mathbf{t}$ and obtain the following convex minimization problem:

$$\mathbf{w}^* = \arg\min_w J(\mathbf{w}) \qquad (4.1)$$

$$= \arg\min_w \mathbf{w}^T(\mathbf{F}^T\mathbf{F})\mathbf{w} - 2(\mathbf{F}^T\mathbf{t})^T\mathbf{w}$$

$$\text{s.t.} \sum_j w_j = 1, \ w_j \in [0, 1/k].$$

The least squares formulation is related to Lasso, Elastic Net, and other regularized approaches, with the distinction that in our case individual elements of $\mathbf{w}$ are restricted to $[0, 1/k]$. This leads to important properties regarding sparsity and directly impacts generalization power (Sect. 4.3).

4.2.2 Unsupervised Case: Labeling the Features Not the Samples

Consider a pool of signed features correctly flipped according to their signs, which could be known a priori, or estimated from a small set of labeled data. We make the simplifying assumption that the signed features' expected values for positive and negative samples, respectively, are close to the ground truth target values (μ_P, μ_N). The assumption could be realized in practice, in an approximation sense, by appropriate normalization, estimated from the small supervised set.

Then, for a given sample i, and any $\mathbf{w}$ obeying the constraints, the expected value of the weighted average $\mathbf{w}^\top \mathbf{f}_i$ is also close to the ground truth target t_i: $E(\mathbf{w}^\top \mathbf{f}_i) = \sum_j w_j E(\mathbf{f}_i(j)) \approx (\sum_j w_j) t_i = t_i$. Then, for all samples we have the expectation $E(\mathbf{F}\mathbf{w}) \approx \mathbf{t}$, such that any feasible solution will produce, on average, approximately correct answers. Thus, we can regard the supervised learning scheme as attempting to reduce the variance of the feature ensemble output, as their expected value is close to the ground truth target. If we approximate $E(\mathbf{F}\mathbf{w}) \approx \mathbf{t}$ into the objective $J(\mathbf{w})$, we get a new ground-truth-free objective $J_u(\mathbf{w})$ with the following learning scheme, which is unsupervised once the feature signs are known. Here $\mathbf{M} = \mathbf{F}^\top \mathbf{F}$:

$$
\begin{aligned}
\mathbf{w}^* &= \arg\min_w \; J_u(\mathbf{w}) \\
&= \arg\min_w \; \mathbf{w}^\top (\mathbf{F}^\top \mathbf{F})\mathbf{w} - 2(\mathbf{F}^\top (\mathbf{F}\mathbf{w}))^\top \mathbf{w} \\
&= \arg\min_w \; (-\mathbf{w}^\top (\mathbf{F}^\top \mathbf{F}\mathbf{w})) = \arg\max_w \; \mathbf{w}^\top \mathbf{M}\mathbf{w} \\
&\text{s.t.} \sum_j w_j = 1 \; , \; w_j \in [0, 1/k].
\end{aligned}
\tag{4.2}
$$

Interestingly, while the supervised case is a convex minimization problem, the unsupervised learning scheme is a concave minimization problem, which is NP-hard. This is due to the change in sign of the matrix $\mathbf{M}$. As $\mathbf{M}$ in the (almost) unsupervised case could be created from larger quantities of unlabeled data, $J_u(\mathbf{w})$ could in fact be less noisy than $J(\mathbf{w})$ and produce significantly better local optimal solutions—a fact confirmed by experiments.

We should note the difference between our formulation and other, much more costly semi-supervised or transductive learning approaches based on label propagation with quadratic criterion [21] (where the quadratic term is very large, being computed from pairs of data samples, not features) or on transductive support vector machines [22]. There are also methods for unsupervised feature selection, such as the regularization scheme of [23], but they do not simultaneously learn a discriminative classifier, as it is the case here.

4.2.3   Intuition

Let us consider two terms involved in our objectives, the quadratic term: $\mathbf{w}^\top \mathbf{M}\mathbf{w} = \mathbf{w}^\top (\mathbf{F}^\top \mathbf{F})\mathbf{w}$ and the linear term: $(\mathbf{F}^\top \mathbf{t})^\top \mathbf{w}$. Assuming that feature outputs have similar expected values and then minimizing the linear term in the supervised case will give more weight to features that are strongly correlated with the ground truth and are good for classification, even independently. Things become more interesting when looking at the role played by the quadratic term in the two cases of learning. The positive definite matrix $\mathbf{F}^\top\mathbf{F}$ contains the dot products between pairs of feature responses over the samples. In the supervised case, minimizing $\mathbf{w}^\top(\mathbf{F}^\top\mathbf{F})\mathbf{w}$ should find *groups of features* that are as uncorrelated as possible. Thus, they should be individually relevant due to the linear term, but not redundant with respect to each other due to the quadratic term. They should be *conditionally independent* given the class, an observation that is consistent with earlier research (e.g., Dieterich [3], Rolls and Deco [24]). This idea is also related to the recent work on discovering discriminative groups of HOG filters [25].

In the (almost) unsupervised case, the task seems reversed: maximize the same quadratic term $\mathbf{w}^\top\mathbf{M}\mathbf{w}$, with no linear term involved. We could interpret this apparently opposite approach as transforming the learning problem into a special case of clustering with pairwise constraints, related to methods such as spectral clustering with l^2-norm constraints [26] and robust hypergraph clustering with l^1-norm constraints [27, 28]. The transformation of the discriminative learning task into a clustering problem is exactly what relates this approach to the clustering formulation presented in Chap. 3. And, as we will see later in the chapter, the same Integer Projected Fixed Point (IPFP) algorithm will be effectively applied for optimization.

Therefore, the unsupervised learning case is addressed by finding the group of features with strongest intra-cluster score that is the largest amount of covariance. In the absence of ground truth labels, if we assume that features in the pool are, in general, correctly signed and not redundant, then the maximum covariance is attained by those whose collective average varies the most as the hidden class labels also vary.

Thus, we hope that the (almost) unsupervised variant seeks features that respond in a united, combined manner, which is sensitive to the natural variation in the distributions of the two classes. In other words, the variation in the total second-order signal (captured by the covariance matrix) of the good feature responses, caused by the natural differences between the two classes should be the strongest when compared to noisy variations that are more or less independent and often cancel each other out.

4.3 Feature Selection and Learning by Clustering with IPFP

We first need to determine the *sign* for each feature, as defined before. Once it is estimated, we can set up the optimization problems to find $\mathbf{w}$. In Algorithm 4.1, we present the unsupervised method, with the supervised variant being constructed by modifying the objective appropriately. The supervised case is a convex problem, with efficient global optimization possible in polynomial time. The unsupervised scheme is a concave minimization problem, which is NP-hard. There are many possible fast methods for approximate optimization. Here, we adapted the integer projected fixed point (IPFP) approach [29, 30], described in detail in Chap. 2. IPFP has proved efficient in practice on all tasks we have tested it on, from graph matching and clustering to feature selection, descriptor learning, and object segmentation in video (Fig. 4.2c). Note that IPFP is immediately applicable to the supervised as well as the unsupervised case. The method converges to a stationary point—the global optimum in the supervised case. We remind the reader that at each iteration IPFP approximates the original objective with a linear, first-order Taylor approximation that can be optimized immediately in the feasible domain. That step is followed by a *line search* with rapid closed-form solution, and the process is repeated until convergence.

In practice, 10–20 iterations bring us close to the stationary point; nonetheless, for thoroughness, we use 100 iterations in all tests. See, for example, comparisons to Matlab's *quadprog* runtime for the convex supervised learning case in Fig. 4.2 and to other learning methods in Fig. 4.5. The quadratic formulations for feature selection that we propose here have the advantage that once the linear and quadratic terms are set up, the learning problems are independent of the number of samples and only dependent on the number n of features considered, since $\mathbf{M}$ is $n \times n$ and $\mathbf{F}^\top \mathbf{t}$ is $n \times 1$.

Algorithm 4.1 Feature selection and classifier learning by clustering with IPFP

Learn feature signs from a small set of labeled samples.
Create $\mathbf{F}$ with flipped features from unlabeled data.
Set $\mathbf{M} \leftarrow \mathbf{F}^\top \mathbf{F}$,
Find $\mathbf{w}^* = \arg\max_w \mathbf{w}^\top \mathbf{M} \mathbf{w}$
     s.t. $\sum_j w_j = 1$, $w_j \in [0, 1/k]$.
return $\mathbf{w}^*$

4.3.1 Theoretical Analysis

First, we show that the solutions are sparse with equal non-zero weights (Proposition 4.5), also observed in practice (Fig. 4.2b). This property makes the classifier learning by clustering also an excellent feature selection mechanism. Next, we show that simple equal weight solutions are likely to minimize the output variance over samples of a given class (Proposition 4.6) and minimize the error rate. This explains the good generalization power of our method. Then, we show how the error rate is expected to go towards zero when the number of considered non-redundant features increases (Proposition 4.7), which explains why a large diverse pool of features is beneficial. Let $J(\mathbf{w})$ be our objective for either the supervised or semi-supervised case:

Proposition 4.5 *Let* $\mathbf{d}(\mathbf{w})$ *be the gradient of* $J(\mathbf{w})$. *The partial derivatives* $d(\mathbf{w})_i$ *corresponding to those elements* w_i^* *of the stationary points with non-sparse, real values in* $(0, 1/k)$ *must be equal to each other.*

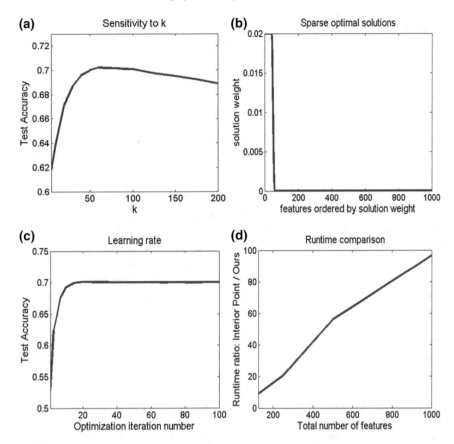

Fig. 4.2 Optimization and sensitivity analysis: **a** sensitivity to k. Performance improves as features are added, is stable around the peak $k = 60$, and falls for $k > 100$ as useful features are exhausted. **b** Features ordered by weight for $k = 50$ confirming that our method selects equal weights up to the chosen k. **c** The method almost converges in 10–20 iterations. **d** Runtime of interior point method divided by ours, both in Matlab and with 100 max iterations. All results are averages over 100 random runs

Proof The stationary points for the Lagrangian satisfy the Karush–Kuhn–Tucker (KKT) necessary optimality conditions. The Lagrangian is $L(\mathbf{w}, \lambda, \mu, \beta) = J(\mathbf{w}) - \lambda(\sum w_i - 1) + \sum \mu_i w_i + \sum \beta_i (1/k - w_i)$. From the KKT conditions at a point $\mathbf{w}^*$, we have

$$\mathbf{d}(\mathbf{w}^*) - \lambda + \mu_i - \beta_i = 0,$$

$$\sum_{i=1}^{n} \mu_i w_i^* = 0,$$

$$\sum_{i=1}^{n} \beta_i (1/k - w_i^*) = 0.$$

Here, $\mathbf{w}^*$ and the Lagrange multipliers have non-negative elements, so if $w_i > 0 \Rightarrow \mu_i = 0$ and $w_i < 1/k \Rightarrow \beta_i = 0$. Then, there must exist a constant λ such that

$$d(\mathbf{w}^*)_i = \begin{cases} \leq \lambda, \ w_i^* = 0, \\ = \lambda, \ w_i^* \in (0, 1/k), \\ \geq \lambda, \ w_i^* = 1/k. \end{cases}$$

This implies that all w_i^* that are different from 0 or $1/k$ correspond to partial derivatives $d(\mathbf{w})_i$ that are equal to some constant λ; therefore, those $d(\mathbf{w})_i$ must be equal to each other, which concludes our proof.   □

From Proposition 4.5, it follows that in the general case, when the partial derivatives of the objective error function at the Lagrangian stationary point are unique, the elements of the solution $\mathbf{w}^*$ are either 0 or $1/k$. As $\sum_j w_j^* = 1$, it follows that the number of non-zero weights is exactly k, in the general case. Therefore, our solution is not just a simple linear separator (hyperplane), but also a sparse representation and a feature selection procedure that effectively averages the selected k (or close to k) features. The algorithm is robust to the choice of k (Fig. 4.2a) and seems to be less sensitive to the number of features selected than Lasso (see Fig. 4.3). In terms of memory cost, compared to the Lasso solution that in general has different real weights for all features, whose storage requires $32n$ bits in floating-point representation, our averaging of k selected features needs only $k \log_2 n$ bits—select k features out of n possible and automatically set their weights to $1/k$. The good generalization from very limited training data, which the algorithms presented here display in practice, seem to follow the Occam's Razor principle [31]. The form of the classifier learned has the simplest form, with all non-zero values being equal to a predefined value. Thus, it becomes simultaneously a linear separator between positive and negative cases and a feature selector.

Next, for a better statistical interpretation, we assume the somewhat idealized case when all features have equal means (μ_P, μ_N) and equal standard deviations (σ_P, σ_N) over positive (P) and negative (N) training sets, respectively. This effectively induces a bimodal distribution for the average response of n features.

Proposition 4.6 *If we assume that the input soft classifiers are independent and better than random chance, the error rate converges towards 0 as their number n goes to infinity.*

Proof Given a classification threshold θ for $\mathbf{w}^T \mathbf{f}_i$, such that $\mu_N < \theta < \mu_P$, then, as n goes to infinity, the probability that a negative sample will have an average response greater than θ (a false positive) goes to 0. This follows from Chebyshev's inequality. By a similar argument, the chance of a false negative also goes to 0 as n goes to infinity.   □

Proposition 4.7 *The weighted average $\mathbf{w}^T \mathbf{f}_i$ with smallest variance over positives (and negatives) has equal weights.*

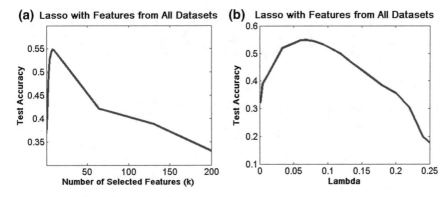

Fig. 4.3 Sensitivity analysis for Lasso: Left: sensitivity to number of features with non-zero weights in the solution. Note the higher sensitivity when compared to ours. Lasso's best performance is achieved for fewer features, but the accuracy is worse than in our case. Right: sensitivity to lambda λ, which controls the L1-regularization penalty

Proof We consider the case when $\mathbf{f}_i$'s are from positive samples, the same being true for the negatives. Then $\mathrm{Var}(\sum_j w_j \mathbf{f}_i(j)/\sum_j w_j) = \sum w_j^2/(\sum w_j)^2 \sigma_P^2$. We minimize $\sum w_j^2/(\sum w_j)^2$ by setting its partial derivatives to zero and get $w_q(\sum w_j) = \sum w_j^2, \forall q$. Then $w_q = w_j, \forall q, j$. $\qquad\square$

4.4 Experimental Analysis

We evaluate our method's ability to generalize and learn quickly from limited data, in both the supervised and the almost unsupervised case when only the feature signs are known. We also explore the possibility of transferring and combining knowledge from different datasets, containing video or low- and medium-resolution images of many potentially unrelated classes. We focus on video classification and compare to established methods for selection and classification and report classification accuracy per frame. We test on the large-scale YouTube-Objects V1 video dataset [18], with difficult sequences from ten categories (aeroplane, bird, boat, car, cat, cow, dog, horse, motorbike, train) taken *in the wild*. The training set contains about 4200 video shots, for a total of 436,970 frames, and the test set has 1284 video shots for a total of over 134,119 frames. The videos have significant clutter, with objects coming in and out of foreground focus, undergoing occlusions, extensive changes in scale and viewpoint. This set is difficult because the *intra*-class variation is large and sudden between video shots. Given the very large number of frames and variety of shots, their complex appearance, and variation in length, presence of background clutter with many distracting objects, changes in scale, viewpoint and drastic intra-class variation, the task of learning the main category from only a few frames presents a significant challenge. We used the same training/testing split as described in [18].

Diversifying features by learning classifiers on different sub-parts (clusters) of the input space

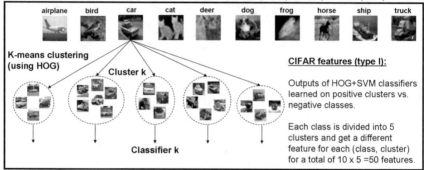

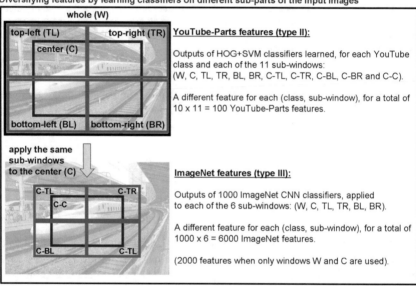

Fig. 4.4 We encourage feature diversity by taking classifiers trained on three datasets and by looking at different parts of the input space (Type I) or different locations within the image (Types II and III)

In all our tests, we present results averaged over 30–100 randomized trials, for each method. We generate a large pool of over 6000 different features (see Fig. 4.4), computed and learned from three different datasets: CIFAR10 [32], ImageNet [33], and a hold-out part of the YouTube-Objects training set:

CIFAR10 features (Type I): The CIFAR10 dataset contains 60,000 32 × 32 color images in 10 classes (airplane, automobile, bird, cat, deer, dog, frog, horse, ship, truck), with 6000 images per class. There are 50,000 training and 10,000 test images. We randomly chose 2500 images per class to create features. They are HOG + SVM classifiers trained on data obtained by clustering images from each class into five groups using k-means applied to their HOG descriptors. Each classifier was trained to separate its own cluster from the others. We hoped to obtain, for each class, diverse and relatively independent classifiers that respond to different, naturally clustered, parts of the input space. Each of the $5 \times 10 = 50$ such classifiers becomes a different feature, which we compute on all training and test images from YouTube-Objects. It is important to note that the CIFAR10 images are statistically different from the YouTube ones. More specifically, the CIFAR10 classes coincide only partially (7 out of 10) with the YouTube-Objects classes and generally cover a larger variety of scene contexts per class. Also, CIFAR10 images are statistically very different from the YouTube images, especially due to their very small resolution of 32×32. In many cases, it is hard even for human observers to tell the correct class of a given image.

YouTube-parts features (Type II): We formed a separate dataset with 25,000 images from video, randomly selected from a subset of YouTube-Objects training videos, not used in subsequent recognition experiments. Features are outputs of linear SVM classifiers using HOG applied to the different parts of each image. Each classifier is trained and applied to its own dedicated sub-window as shown in Fig. 4.4. To speed up training and remove noise, we also applied PCA to the resulted HOG and obtained descriptors of 46 dimensions, before passing them to SVM. Experiments with a variety of SVM kernels and settings showed that linear SVM with default parameters for libsvm worked best, and we kept that fixed in all experiments. For each of the 10 classes, we have 11 classifiers, one for each sub-window, and get a total of 110 type II features.

ImageNet features (Type III): We considered the soft feature outputs (before soft-max) of the pretrained ImageNet CNN features using Caffe [17], each of them over six different sub-windows: *whole, center, top left, top right, bottom left*, and *bottom right*, as presented in Fig. 4.4. There are 1000 such outputs, one for each ImageNet category, for each sub-window, for a total of 6000 features. In some of our experiments, when specified, we used only 2000 ImageNet features, restricted to the *whole* and *center* windows.

Short discussion on the choice of features: The types of features chosen for our experiments vary and include diverse features that come from different types of images (low-resolution, video frames, ImageNet images). The classes that these classifier features are trained on are strongly related in the case of YouTube features, but less related in the case of CIFAR10 and ImageNet features. It is important to note here that all these types of features have some limitations with respect to the fully supervised case when classifiers have access to full images belonging to the classes that we are interested in. In the case of CIFAR10 features, the classifiers (acting as features in our experiments) are trained to predict anonymous clusters, which divide the larger class groups. They also see only low-resolution images and have relatively simpler HOG descriptors. In the case of YouTube-Objects features, they are trained on a separate set of images not included in the actual training set used to train the final classifiers. Moreover, most of them have access only to subparts of images. The third group of features is pretrained on ImageNet, a dataset with completely different visual characteristics and wide range of very different image classes. While one might argue that our set of pretrained feature classifiers is very strong, the experiments presented next show that in fact they are still weak, with all linear classifiers tested having a hard time distinguishing between image classes. As we will see in the next section, these results suggest the following: (1) The problem of determining the class of a video from which an individual frame is extracted is notoriously hard—the object belonging to the target class might be missing and the context is not always strong enough. In some sense, we push the limits of classification by indirectly exploiting context as much as possible, under the given models and tests. (2) There is a hidden but strong difference in the statistical visual characteristics of different datasets. There is a strong case of dataset bias, since, for example, ImageNet higher level features are not capable of always predicting the correct class. (3) As we discuss in the final parts of our experimental section, following next, it will be interesting to see what features are actually selected to form the final classifiers. A closer look at these features will reveal some interesting findings that relate appearance, context, and semantics.

4.4.1 Comparative Experiments

We evaluated seven methods: ours, SVM on all input features, Lasso, AdaBoost on all input features, ours with SVM (applying SVM only to features selected by our method, idea related to [34–36]), forward–backward selection (FoBa), [37] and simple averaging of all signed features, with values in [0, 1] and flipped as discussed before. While most methods work directly with the signed features provided, AdaBoost further transforms each feature into a weak binary classifier by choosing the threshold that minimizes the expected exponential loss at each iteration (this explains why AdaBoost is much slower). For SVM, we used the latest LIBSVM [38] implementation, with kernel and parameter C validated separately for each type of experiment. For the Lasso, we used the latest Matlab library and validated the L1-regularization parameter λ for each experiment. Additionally, we compared to Elastic Net (L1 + L2 regularization) [39]. Our supervised variant outperformed Elastic Net by over 10% when learning from large feature pools (Fig. 4.5, last two rows). The results (Fig. 4.5) show that our method has a constant optimization time (after creating $\mathbf{F}$, and then computing $\mathbf{F}^\top \mathbf{F}$). It is significantly less expensive to use than SVM, AdaBoost (time too large to show in the plot), FoBa, and even the latest Matlab's Lasso. We outperform most other methods, especially in the case of limited labeled training data, when our selected feature averages are even stronger than in combination with SVM or SVM by itself, due to a superior generalization power. In

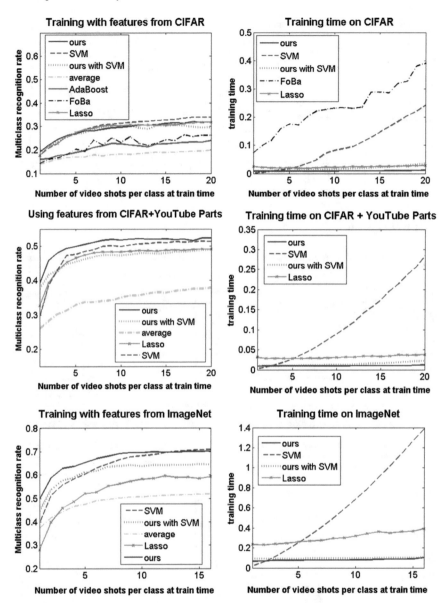

Fig. 4.5 Accuracy and training time on YouTube-Objects, with varying training video shots (10 frames per shot and results averaged over 30 runs). Input feature pool, row 1: 50 type I features on CIFAR10; row 2: 110 type II features on YouTube-Parts + 50 CIFAR10; row 3: 2000 type III features in ImageNet; row 4: 2160 all features. Ours outperforms SVM, Lasso, AdaBoost, and FoBa

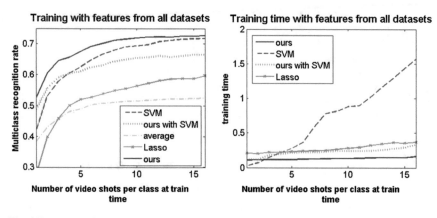

Fig. 4.5 (continued)

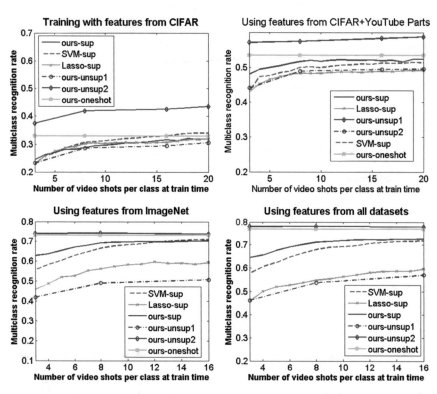

Fig. 4.6 Comparison of our almost unsupervised approach to the supervised case for different methods. In our case, unsup1 uses training data in $J_u(\mathbf{w})$ only from the shots used by all supervised methods; unsup2 also includes frames from testing shots, with unknown test labels; one shot is unsup2 with a single-labeled image per class used only for feature sign estimation. This truly demonstrates the ability of our approach to efficiently learn with minimal supervision

Table 4.1 Improvement in recognition over the supervised case, by using only unlabeled training data features signs learned from the same (1, 3, 8, 16) shots as the supervised case. The first column presents the one-shot learning case, when the almost unsupervised method uses a single-labeled image per shot and also per class to estimate the feature signs. Results are averages over 30 random runs

Training # shots	3	8	16
Feature I (%)	+16.85	+13.92	+13.98
Feature I + II (%)	+10.15	+6.2	+6.12
Feature III (%)	+11.21	+4.9	+3.33
Feature I + II + III (%)	+13.43	+6.73	+5.36

Table 4.2 Mean accuracy per class, over 30 random runs of unsupervised learning with 16 labeled training shots

Class name	Aeroplane	Bird	Boat	Car	Cat	Cow	Dog	Horse	Motorbike	Train
Accuracy (%)	91.53	91.61	99.11	86.67	70.02	78.13	61.63	53.65	72.66	83.37

the case of unlabeled learning with signed features, we outperformed all other methods by a very large margin, up to over 20% (Fig. 4.6 and Table 4.1). Of particular note is when only a single-labeled image per class was used to estimate the feature signs, with all the other data being unlabeled (Fig. 4.6).

Next, we present some representative correct and incorrect frame classifications for each category in YouTube-Objects. For each class, we show the mean per frame class recognition accuracy, over 30 random runs of unsupervised learning with 16 labeled training shots (Table 4.2). We also present lists (Figs. 4.7, 4.8, and 4.9) of ten classified frames per category, for which the ratio of correct to incorrect samples matches the class accuracy. Note that we considered all frames in a video shot as belonging to a single category—even though sometimes a significant amount of frames did not contain any of the above categories. Therefore, often our results look qualitatively better than the quantitative evaluation.

4.4.2 Additional Comparisons with SVM

In our experiments, our supervised method generalized significantly better, on average, than SVM or in combination with SVM for very limited training data. We believe that this is due to the power of feature averages, as also indicated by our theoretical results. Our formulation is expected to discover features that are independent and strong as a group, not necessarily individually. That is why we prefer to give all features selected equal weight than to put too much faith into a single strong feature, especially in the case of limited training data. As shown in Fig. 4.10, our approach generalizes better than SVM or in combination with SVM, as reflected by the perfor-

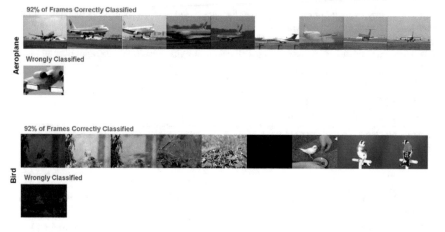

Fig. 4.7 Lists of ten classified frames per category, for which the ratio of correct to incorrect samples matches the mean class recognition accuracy. Note that we considered all frames in a video shot as belonging to a single category—even though sometimes a significant amount of frames did not contain any of the above categories. Therefore, often our results look qualitatively better than the quantitative evaluation. Please also see the next two figures

mance differences between the testing and training case. Note that we have used the latest SVM library with kernel and parameter C validated separately for each type of experiment—in our experiments the linear kernel performed the best. We can also see that our method often generalizes from just one frame per video shot, for a total of eight positive training frames per class in the experiments in Fig. 4.11.

4.5   The Effect of Limited Training Data

The accuracy of our almost unsupervised learning approach with signed features mainly depends on several key factors: (1) the ability to estimate the signs of features, (2) the access to large quantities of unsupervised data, and (3) the available pretrained features. Here, we present several experiments to evaluate the accuracy of the estimated signs with respect to the available labeled data, as well as the robustness of our method to errors in sign estimation. Our experiments show that our method is robust to noise in sign estimation, whose accuracy can be as low as 60–65%. Next, we also present experiments that measure how the accuracy depends on the quantity of unsupervised data available at training time. Our experiments indicate that the performance increases with the quantity of the unlabeled data available, as expected, but it almost reaches a plateau when a limited fraction of the unlabeled frames used for training are available. We also look at the generalization capability of our approach, when learning with limited data and show that it is less prone to over-fitting than SVM in such cases. We also give more intuitions behind our approach and conclude with a few qualitative results from actual videos, to demonstrate the difficulty of the task.

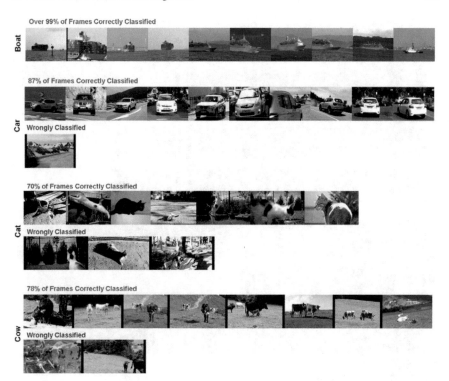

Fig. 4.8 Ten classified frames per class, for which the ratio of correct to incorrect samples matches the mean class recognition accuracy

4.5.1 Estimating Feature Signs from Limited Data

For each feature type used in our experiments, we varied the number of labeled training shots that were used to estimate the feature signs. We have compared the estimated feature signs with the ones estimated from the entire unlabeled test set of the database and present estimation *accuracies*, where the signs estimated from the test set were considered the empirical ground truth. Given the large size of the test set compared to the limited labeled training data, it is important to evaluate the ability to guess the right sign from very limited data. The results are shown in Fig. 4.12 and Table 4.3. The first two rows in the table and values in the plots correspond to the cases where a single-labeled video shot was used for sign estimation: the *1 labeled training frame per class* case (the same as the *one-shot* case) and the *10 frames per shot per class* case, as before, in which 10 evenly space frames were taken from each labeled training shot.

An interesting direction for future work is to explore the possibility of borrowing feature signs from classes that are related in meaning, shape, or context. We have performed some experiments and compared the estimated feature signs between classes (see Fig. 4.13). Does the plane share more feature signs with the bird, or with

Fig. 4.9 Ten classified frames per class, for which the ratio of correct to incorrect samples matches the mean class recognition accuracy

another man-made class, such as the train? The possibility of sharing or borrowing feature signs from other classes could pave the way for a more unsupervised type of learning, where we would not need to estimate the signs from labeled data of the specific class. The results in Fig. 4.13 indicate that, indeed, classes that are closer in meaning share more signs than classes that mean very different things. For example, the class *aeroplane* shares most signs with *boat, motorbike, bird, train, bird* with *cat, dog, motorbike, aeroplane, cow, boat* with *train, car aeroplane*, and *car* with *train, boat, motorbike*. We also have *cat: dog, bird, cow, horse, cow: horse, dog, cat, bird, dog: cow, horse, cat, bird, horse: cow, dog, cat, motorbike: aeroplane, bird, car*, and *train: car, boat, aeroplane*, for the remaining classes. We notice that indeed classes that are similar in meaning, appearance, or context, such as animals, or man-made categories, share more signs among themselves than classes that are very different. This experiment confirms the intuition that these signs could be a good indicator of a class, and they are also similar between similar classes.

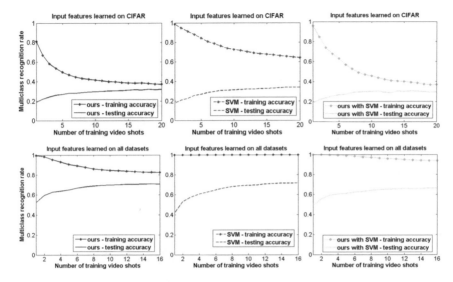

Fig. 4.10 Supervised learning: Our method generalizes better (training and test errors are closer) compared to SVM or in combination with SVM on limited training data

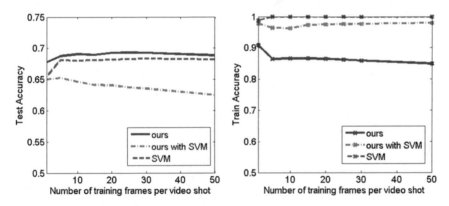

Fig. 4.11 Supervised learning: avg. test recognition accuracy over 30 independent experiments of our method as we vary the number of training frames uniformly sampled from 8 random training video shots. Note how well our method generalizes from just 1 frame per video shot, for a total of 8 positive training frames per class

4.5.2 Varying the Amount of Unsupervised Data

We also evaluated the influence on the performance of varying amounts of unsupervised (unlabeled) data used for unsupervised learning. We present the following experimental setup: first, we randomly split the unlabeled test data into two equal sets of frames, S_A and S_B. We have used one set of unlabeled frames S_A for unsupervised learning and the other set S_B for testing. While keeping the test frames set S_B con-

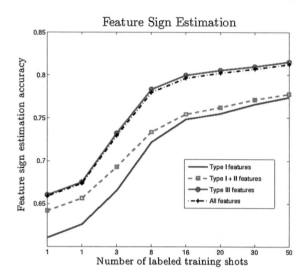

Fig. 4.12   For a feature sign estimation accuracy (agreement with sign estimation from the test set) of roughly 70% our method manages to significantly outperform the supervised case, demonstrating its robustness and solid practical value. The first value corresponds to one-shot-one-frame case (please also see Table 4.3)

Table 4.3   Accuracy of feature sign estimation for different numbers of labeled shots: note that the sign estimation is not accurate in the case of limited data, proving that our method is remarkably robust to noises in these sign estimations. This brings our method closer to a fully unsupervised approach. In the one-shot learning case, signs are correctly estimated only 65% of the time, with chance being %50

Number of shots	Features I (%)	Features I + II (%)	Features III (%)	Features I + II + III (%)
1 (1 frame)	61.06	64.19	66.03	65.89
1 (10 frames)	62.61	65.64	67.53	67.39
3 (10 frames)	66.53	69.31	73.21	72.92
8 (10 frames)	72.17	73.36	78.33	77.97
16 (10 frames)	74.83	75.44	79.97	79.63
20 (10 frames)	75.51	76.23	80.54	80.22
30 (10 frames)	76.60	77.15	80.98	80.70
50 (10 frames)	77.41	77.79	81.52	81.24

stant, and 8 labeled video shots per class for feature sign estimation from the training set (with 10 evenly spaced frames per shot), we varied the amount of unlabeled data used for unsupervised learning, by varying the percentage of unlabeled frames used from S_A. We present results over 30 random runs in Fig. 4.14 and Tables 4.4 and 4.5.

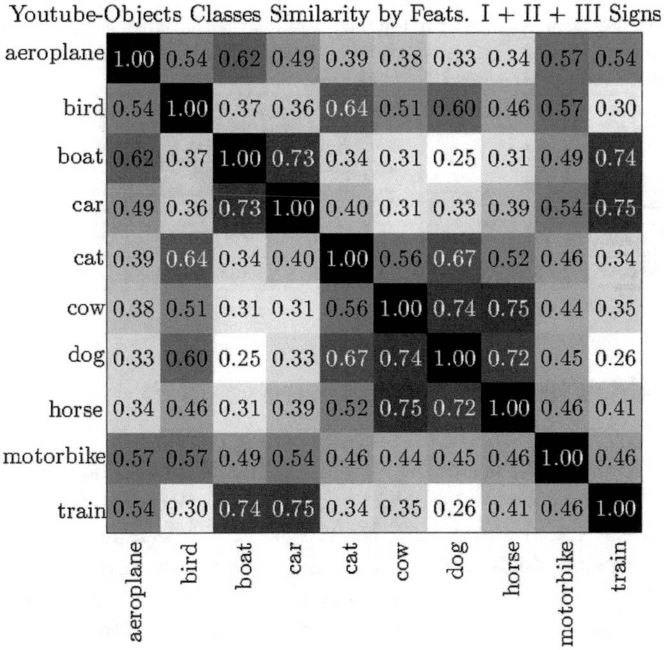

Youtube-Objects Classes Similarity by Feats. I + II + III Signs
aeroplane 1.00 0.54 0.62 0.49 0.39 0.38 0.33 0.34 0.57 0.54
bird 0.54 1.00 0.37 0.36 0.64 0.51 0.60 0.46 0.57 0.30
boat 0.62 0.37 1.00 0.73 0.34 0.31 0.25 0.31 0.49 0.74
car 0.49 0.36 0.73 1.00 0.40 0.31 0.33 0.39 0.54 0.75
cat 0.39 0.64 0.34 0.40 1.00 0.56 0.67 0.52 0.46 0.34
cow 0.38 0.51 0.31 0.31 0.56 1.00 0.74 0.75 0.44 0.35
dog 0.33 0.60 0.25 0.33 0.67 0.74 1.00 0.72 0.45 0.26
horse 0.34 0.46 0.31 0.39 0.52 0.75 0.72 1.00 0.46 0.41
motorbike 0.57 0.57 0.49 0.54 0.46 0.44 0.45 0.46 1.00 0.46
train 0.54 0.30 0.74 0.75 0.34 0.35 0.26 0.41 0.46 1.00
aeroplane
bird
boat
car
cat
cow
dog
horse
motorbike
train

Fig. 4.13 YouTube-Objects classes similarity based on their estimated signs of features. For each pair of features, we present the percent of signs of features that coincide. Note that classes that are more similar in meaning, shape, or context have, on average, more signs that coincide. These signs were estimated from all training data

Performance vs. quantity of unsupervised data
0.8
0.7
0.6
0.5
0.4
0.3
0.2
0.1
Testing accuracy per frame
0%
25%
50%
75%
100%
Percentage of unsupervised data
Type I features
Type I+II features
Type III features
All features

Fig. 4.14 Testing accuracy vs. amount of unlabeled data used for unsupervised learning. Note that the performance almost reaches a plateau after a relatively small fraction of unlabeled frames are added

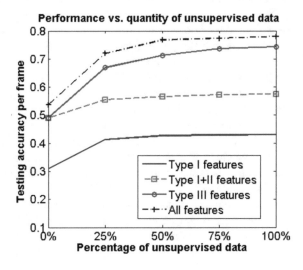

Table 4.4 Testing accuracy when varying the quantity of unlabeled data used for unsupervised learning. We used 8 video shots per class with 10 frames each, for estimating the signs of the features. Results are averages over 30 random runs

Unsupervised data	Features I (%)	Features I + II (%)	Features III	Features I + II + III (%)
Train + 0% test S_A	30.86	48.96	49.03	53.71
Train + 25% test S_A	41.26	55.50	66.90	72.01
Train + 50% test S_A	42.72	56.66	71.31	76.78
Train + 75% test S_A	42.88	57.24	73.65	77.39
Train + 100% test S_A	43.00	57.44	74.30	78.05

4.6   Intuition Regarding the Selected Features

Another interesting finding (see Fig. 4.15) is the consistent selection of diverse input features (classifiers) that are related to the target class in surprising ways: similar w.r.t. global visual appearance, but not semantic meaning—banister :: train, tiger shark :: plane, Polaroid camera :: car, scorpion :: motorbike, remote control :: cat's face, space heater :: cat's head; related in co-occurrence and context, but not in global appearance—helmet versus motorbike; connected through part-to-whole relationships—{grille, mirror and wheel} :: car; or combinations of the above—dock :: boat, steel bridge :: train, albatross :: plane. The relationships between the target class and the selected features could also hide combinations of many other factors. Meaningful relationships could ultimately join together correlations along many dimensions, from appearance to geometric, temporal, and interaction-like relations.

Since categories share shapes, parts, and designs, it is perhaps unsurprising that classifiers trained on semantically distant classes that are visually similar can help improve learning and generalization from limited data. Another interesting aspect is that the classes found are not necessarily central to the main category, but often peripheral, acting as guardians that separate the main class from the rest. This is where feature diversity plays an important role, ensuring both *separation* from nearby classes and robustness to missing values.

An additional possible benefit is the capacity to immediately learn novel concepts from old ones, by combining existing high-level concepts to recognize new classes. In cases where there is insufficient data for a particular new class, sparse averages of reliable classifiers can be an excellent way to combine previous knowledge. Consider the class *cow* in Fig. 4.15. Although "cow" is not present in the 1000 label set, our method is able to learn the concept by combining existing classifiers. Also, note that as categories borrow shapes, parts, and even designs from each other, it is not surprising that using input cues from distant classes (in meaning) but close in appearance could improve learning and generalization from limited data.

Fig. 4.15 For each training target class from YouTube-Objects videos (labels on the left), we present the most frequently selected ImageNet classifiers (input features), over 30 independent experiments, with 10 frames per shot and 10 random shots for training. We show the classes that were always selected by our method when $k = 50$. On the right, we show the probability of selection for the most important 50 features. Note how stable the selection process is. Also note the connection between the selected classes and the target class in terms of appearance, context, or geometric part-whole relationships. We find two aspects interesting: (1) the consistent selection of the same classes, even for small random training sets and (2) the fact that unrelated classes in terms of meaning can be useful for classification, based on their shape and appearance similarity

4.6.1 Location Distribution of Selected Features

Table 4.6 summarizes the location distribution of ImageNet features selected by our method for each category in YouTube-Objects. We make several observations. First, the majority of features for all classes consider the whole image (W), which suggests that the image background is relevant. Second, for several categories (e.g., car, motorbike, aeroplane), the center (C) is important. Third, some categories (e.g., boat) may be located off-center or benefit from classifiers that focus on non-central regions. Finally, we see that object categories that may superficially seem similar (cat vs. dog) exhibit rather different distributions: dogs seem to benefit from the whole image, while cats benefit from sub-windows; this may be because cats are smaller and appear in more diverse contexts and locations, particularly in YouTube videos.

Table 4.5 Recognition accuracy of our method on YouTube-Objects dataset, using 10 training shots (first row) and 20 training shots (second row), as we combine features from several datasets. The accuracy increases significantly (it doubles) as the pool of features grows and becomes more diverse. Features are type I: CIFAR; types I + II: CIFAR + YouTube-Parts; types I + II + III: CIFAR + YouTube-Parts + ImageNet (6000)

Accuracy	I (50)	I + II (160)	I + II + III (6160)
10 train shots (%)	29.69	51.57	69.99
20 train shots (%)	31.97	52.37	71.31

Table 4.6 The distribution of sub-windows for the input classifiers selected for each category. The most selected one is *whole*, indicating that information from the whole image is relevant, including the background. For some classes, the object itself is also important as indicated by the high percentage of *center* classifiers. The presence of classifiers at other locations indicates that the object of interest, or objects co-occurring with it might be present at different places. Note, for example, the difference in distribution between *cats* and *dogs*. The results suggest that looking more carefully at location and moving towards detection could benefit the overall classification

Locations	W	C	TL	TR	BL	BR
Aeroplane	65.6	30.2	0	0	2.1	2.1
Bird	78.1	21.9	0	0	0	0
Boat	45.8	21.6	0	0	12.3	20.2
Car	54.1	40.2	2.0	0	3.7	0
Cat	76.4	17.3	5.0	0	1.3	0
Cow	70.8	22.2	1.8	2.4	0	2.8
Dog	92.8	6.2	1.0	0	0	0
Horse	75.9	14.7	0	0	8.3	1.2
Motorbike	65.3	33.7	0	0	0	1.0
Train	56.5	20.0	0	2.4	12.8	8.4

Discussion: The additional tests further confirm the practical value of our approach, which is robust to limited labeled data and capable of good generalization from unsupervised samples. In our tests on the almost unsupervised case with labeled features, the limited supervision is only used to estimate feature signs, which are often wrongly classified as shown here. The effect of misclassified feature signs is mitigated, as our method is sufficiently robust to noise in sign estimation and still select good groups of features that are likely to have the sign correctly estimated. We also tested with varying the amount of unsupervised data used, which showed that performance tends to reach a plateau after adding 25–30% of the unlabeled frames. It is also important to note that our approach can handle large quantities of unlabeled data and improves as new unlabeled data is added, thus being an efficient and robust feature selection method with promise to be useful in many big data, practical, applications.

4.7 Concluding Remarks and Future Work

We present a fast feature selection and learning method with minimal supervision, and we apply it to video classification. It has strong theoretical properties and excellent generalization and accuracy in practice. The crux of our approach is its ability to learn from large quantities of unlabeled data once the feature signs are determined, and it is very robust to feature sign estimation errors. A key difference between our features signs and the weak features used by boosting approaches such as AdaBoost is that in our case, the sign estimation requires minimal labeling and that sign is the only bit of supervision needed. AdaBoost requires large amounts of training data to carefully select and weigh new features. This aspect reveals a key insight: being able to approximately label *the features* and *not the data* is sufficient for learning. For example, when learning to separate oranges from cucumbers, if we knew the features that are likely to be positively correlated with the "orange" class (roundness, redness, sweetness, temperature, latitude where image was taken) in the immense sea of potential cues, we could then employ huge amounts of unlabeled images of oranges and cucumbers, to find the best relatively small group of such features. With a formulation that permits very fast optimization and effective learning from large heterogeneous feature pools, our approach provides a useful tool for many other recognition tasks, and it is suited for real-time, dynamic environments. Thus it could open doors for new and exciting research in machine learning, with both practical and theoretical impacts.

One of the interesting directions that we wish to pursue in the future is to extend this approach to multiple layers of classifiers at different levels of abstraction as it happens in the case of deep neural networks. What we presented here is only a first layer of linear classifiers learned in an almost unsupervised way from pretrained features. Is it possible to learn deep hierarchical structures of classifiers with minimal training data, using this idea? One idea is to take a pretrained deep neural network with many filters (features) at each level of depth and apply our feature learning scheme to prune the network. We will achieve that by automatically selecting, at each network level, only the features that are most relevant and form the strongest block in the feature covariance matrix. We could go even further and use the joint co-occurrence signal of the selected features as supervisory signal for fine-tuning the initial features selected backpropagating gradient descent updates down the network. This idea of using the combined output of the group as supervision signal for a deep unsupervised learning strategy immediately relates to the material presented in the next chapters, especially the Visual Story Network concept presented in the last part of the book.

References

1. Hansen L, Salamon P (1990) Neural network ensembles. PAMI 12(10)
2. Vasconcelos N (2003) Feature selection by maximum marginal diversity: optimality and implications for visual recognition. In: CVPR
3. Dietterich T (2000) Ensemble methods in machine learning. Springer
4. Breiman L (1996) Bagging predictors. Mach Learn 24(2)
5. Freund Y, Schapire R (1995) A decision-theoretic generalization of on-line learning and an application to boosting. Comp Learn Theory
6. Kwok S, Carter C (1990) Multiple decision trees. In: Uncertainty in artificial intelligence
7. Criminisi A, Shotton J, Konukoglu E (2012) Decision forests: a unified framework for classification, regression, density estimation, manifold learning and semi-supervised learning. Found Trends Comput Graph Vis 7(2–3)
8. Viola P, Jones M (2004) Robust real-time face detection. IJCV 57(2)
9. Guyon I, Elisseeff A (2003) An introduction to variable and feature selection. J Mach Learn Res 3:1157–1182
10. Ng A (1998) On feature selection: learning with exponentially many irrelevant features as training examples. In: ICML
11. Pudil P, Novovičová J, Kittler J (1994) Floating search methods in feature selection. Pattern Recognit Lett 15(11)
12. Vogel CR (2002) Computational methods for inverse problems. Society for industrial and applied mathematics, vol 23
13. Tibshirani R (1996) Regression shrinkage and selection via the lasso. J R Stat Soc 267–288
14. Zhao P, Yu B (2006) On model selection consistency of lasso. J Mach Learn Res 7:2541–2563
15. Siedlecki W, Sklansky J (1989) A note on genetic algorithms for large-scale feature selection. Pattern Recognit Lett 10(5)
16. Duda R, Hart P (1973) Pattern classification and scene analysis. Wiley
17. Jia Y, Shelhamer E, Donahue J, Karayev S, Long J, Girshick R, Guadarrama S, Darrell T (2014) Caffe: convolutional architecture for fast feature embedding. In: ACM multimedia
18. Prest A, Leistner C, Civera J, Schmid C, Ferrari V (2012) Learning object class detectors from weakly annotated video. In: CVPR
19. Leordeanu M, Radu A, Baluja S, Sukthankar R (2015) Labeling the features not the samples: efficient video classification with minimal supervision. arXiv:151200517
20. Li Z, Cheong LF, Yang S, Toh KC (2018) Simultaneous clustering and model selection: algorithm, theory and applications. IEEE Trans Pattern Anal Mach Intell 40(8):1964–1978
21. Bengio Y, Delalleau O, Roux NL (2006) Label propagation and quadratic criterion. Semi-supervised learning
22. Joachims T (1999) Transductive inference for text classification using support vector machines. In: ICML

23. Yang Y, Shen HT, Ma Z, Huang Z, Zhou X (2011) L2, 1-norm regularized discriminative feature selection for unsupervised learning. In: IJCAI
24. Rolls E, Deco G (2010) The noisy brain: stochastic dynamics as a principle of brain function, vol 34. Oxford University Press, Oxford
25. Ahmed E, Shakhnarovich G, Maji S (2014) Knowing a good HOG filter when you see it: efficient selection of filters for detection. In: ECCV
26. Sarkar S, Boyer K (1998) Quantitative measures of change based on feature organization: eigenvalues and eigenvectors. CVIU 71(1):110–136
27. Bulo S, Pellilo M (2009) A game-theoretic approach to hypergraph clustering. In: NIPS
28. Liu H, Latecki L, Yan S (2010) Robust clustering as ensembles of affinity relations. In: NIPS
29. Leordeanu M, Sminchisescu C (2012) Efficient hypergraph clustering. In: International conference on artificial intelligence and statistics
30. Leordeanu M, Hebert M, Sukthankar R (2009) An integer projected fixed point method for graph matching and map inference. In: NIPS
31. Blumer A, Ehrenfeucht A, Haussler D, Warmuth M (1987) Occam's razor. Inf Process Lett 24(6)
32. Krizhevsky A, Hinton G (2009) Learning multiple layers of features from tiny images. Comp Sci Dep, Univ of Toronto, Tech Rep
33. Deng J, Dong W, Socher R, Li L, Li K, Fei-Fei L (2009) ImageNet: a large-scale hierarchical image database. In: CVPR
34. Nguyen M, De la Torre F (2010) Optimal feature selection for support vector machines. Pattern Recognit 43(3)
35. Weston J, Mukherjee S, Chapelle O, Pontil M, Poggio T, Vapnik V (2000) Feature selection for SVMs. In: NIPS, vol 12
36. Kira K, Rendell L (1992) The feature selection problem: traditional methods and a new algorithm. In: AAAI
37. Zhang T (2009) Adaptive forward-backward greedy algorithm for sparse learning with linear models. In: NIPS
38. Chang CC, Lin CJ (2011) LIBSVM: a library for support vector machines. ACM Trans Intell Syst Technol (TIST) 2(3):27
39. Zou H, Hastie T (2005) Regularization and variable selection via the elastic net. J R Stat Soc: Ser B (Stat Methodol) 67(2)

Chapter 5
Unsupervised Learning of Object Segmentation in Video with Highly Probable Positive Features

5.1 From Simple Features to Unsupervised Segmentation in Video

By now we should be convinced that unsupervised learning in video is a very challenging task. Fully solving this problem would shed new light on our understanding of intelligence from a scientific perspective. It would also have a strong impact in many real-world applications, as large datasets of unlabeled videos could be collected at a relatively low cost.

   We start from a simple idea, which reveals a powerful concept: often positive data samples could be detected with high precision in an unsupervised way by taking advantage of natural grouping properties of the real world. Then, such positive samples could be reliably used to learn robust classifiers, without human supervision. Next, ensembles of such classifiers could be applied to novel data and the outputs could become HPP features at the next iteration. From one iteration to the next, the learned classifiers become more and more powerful with better generalization properties. This line of thought follows precisely the principles of unsupervised learning, which we presented and briefly motivated in the introductory chapter. We reiterate them here, one below the other. At this point in the book, we start developing a full system for object discovery in video that puts together all these principles, following one after the other. These principles will be further exploited jointly in the following chapters, towards the creation of a general and fully unsupervised learning system.

The material presented in this chapter is based in large part on the following papers:
Stretcu, Otilia and Marius Leordeanu. "Multiple Frames Matching for Object Discovery in Video." British Machine Vision Conference (BMVC) 2015.
Haller, Emanuela and Marius Leordeanu. "Unsupervised object segmentation in video by efficient selection of highly probable positive features." IEEE International Conference on Computer Vision (ICCV) 2017.

© Springer Nature Switzerland AG 2020   157
M. Leordeanu, *Unsupervised Learning in Space and Time*,
Advances in Computer Vision and Pattern Recognition,
https://doi.org/10.1007/978-3-030-42128-1_5

The first seven principles of unsupervised learning:

Principle 1
Objects are local outliers in the global scene, of small size and with a different
appearance and movement than their larger background.

Principle 2
It is often possible to pick with high-precision (not necessarily high recall)
data samples that belong to a single object or category.

Principle 3
Objects display symmetric, coherent, and consistent properties in space and
time. Grouping cues based on appearance, motion, and behavior can be used
to collect positive object samples with high precision. Such cues, which are
very likely to belong to a single object or category are called Highly Probable
Positive (HPP) features.

Principle 4
Objects form strong clusters of motion trajectories and appearance patterns in
their space-time neighborhood.

Principle 5
Accidental alignments are rare. When they happen they usually indicate correct alignments between a model and an image. Alignments, which could be
geometric or appearance based, rare as they are, when they take place form a
strong cluster of agreements that re-enforce each other in multiple ways.

Principle 6
When several weak independent classifiers consistently fire together, they signal the presence of a higher level class for which a stronger classifier can be
learned.

Principle 7
To improve or learn new classes, we need to increase the quantity and level of
difficulty of the training data as well as the power and diversity of the classifiers.
In this way, we could learn in an unsupervised way over several generations, by
using the agreements between the existing classifiers as a teacher supervisory
signal for the new ones.

In this chapter, we focus on the following methods, one leading to the next:

1. We first present a simple and effective algorithm for segmenting foreground
   object in images using the contrasting properties of foreground versus background based on simple colors. We present an interesting theoretical result that
   will constitute the basis for learning from highly probable positive features.

2. Next, we show how the idea presented initially for foreground segmentation in images using simple colors could be extended to video by using Principal Component Analysis to differentiate between foreground and background in the spatiotemporal domain.
3. In the remaining of the chapter, we explore in more depth and develop on the idea of learning from highly probably positive features, to build a complete and highly efficient system for unsupervised foreground object segmentation in video.

There are several different published approaches for unsupervised learning and discovery of the salient object in video [1–4], but most have a high computational cost. In general, algorithms for unsupervised mining and clustering are expected to be computationally expensive due to the inherent combinatorial nature of the problem [5].

Here, we address the computational cost challenge and propose an algorithm that is both accurate and fast. We achieve our goal based on a key insight: we focus on selecting and learning from features that are highly correlated with the presence of the object of interest and can be rapidly selected and computed. **Note**: throughout the book, when referring to Highly Probable Positive (HPP) features, we use "feature" to indicate a feature vector sample (i.e., a data sample), not a feature type. While we do not require these features to cover all instances and parts of the object of interest (we could expect low recall), we show that it is possible to find, in the unsupervised case, positive features with high precision (a large number of those selected are indeed true positives). Then, we prove theoretically that we can reliably train an object classifier using sets of positive and negative samples, both selected in an unsupervised way, as long as the set of features considered to be positive has high precision, regardless of the recall, if certain conditions are met (and they are often met in practice). We present an algorithm that can effectively and rapidly achieve this task in practice, in an unsupervised way, with state-of-the-art results in difficult experiments, while being at least 10x faster than its competition. The proposed method outputs both the soft segmentation of the main object of interest and its bounding box. Two examples are shown in Fig. 5.1.

While we do not make any assumption about the type of object present in the video, we do expect the sequence to contain a single salient object, as our method performs foreground soft segmentation and does not expect videos with no salient

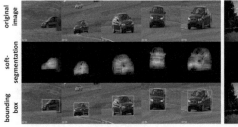

Fig. 5.1 Qualitative results of our method, which provides the soft segmentation of the main object of interest and its bounding box

object or with multiple objects of interest. The key insights that led to our formulation and algorithm are the following:

1. First, the foreground and background are complementary and in contrast to each other—they have different sizes, appearances, and movements. We observed that the more we can take advantage of these contrasting properties the better the results, in practice. While the background occupies most of the image, the foreground is usually small and has distinct color and movement patterns—it stands out against its background scene.
2. The second main idea of our approach is that we should use this foreground-background complementarity in order to automatically select, with high precision, foreground features, even if the expected recall is low. Then, we could reliably use those samples as positives, and the rest as negatives, to train a classifier for detecting the main object of interest. We present this formally in Sect. 5.4.1.

These insights lead to our two main contributions in this chapter: first, we show theoretically that by selecting features that are positive with high probability, a robust classifier for foreground regions can be learned. Second, we present an efficient method based on this insight, which in practice outperforms its competition on many different object classes, while being an order of magnitude faster than most other approaches.

The task of object discovery in video has been tackled for many years, with early approaches being based on local features matching [1, 2]. Current literature offers a wide range of solutions, with varying degrees of supervision, going from fully unsupervised methods [3, 4] to partially supervised ones [6–10]—which start from region, object, or segmentation proposals estimated by systems trained in a supervised manner [11–13]. Some methods also require user input for the first frame of the video [14]. Most object discovery approaches that produce a fine shape segmentation of the object also make use of off-the-shelf shape segmentation methods [15–19].

The material presented here is based on work that we previously published in Leordeanu [20], Stretcu and Leordeanu [10], and Haller and Leordeanu [21]. We start by presenting the initial method for object segmentation in images using simple color distributions. Then, we extend the idea to develop VideoPCA, a background subtraction method in video using Principal Component Analysis. Third, for the main and remaining part of the book, we develop, starting from the ideas and insights of the first two approaches, our full competitive system for unsupervised foreground object segmentation in video.

5.2 A Simple Approach to Unsupervised Image Segmentation

In this section, we introduce a simple, yet effective method for foreground object segmentation using color statistics that separate the foreground from the background. We show that color histograms alone are often powerful enough to segment the foreground from the background, without using any local intensity information or edges. Other work that uses color histograms to separate the foreground from the background, given minimal user input, includes GrabCut [22] and Lazy Snapping [23]. That work required more extensive user input, such as markings on the object and outside of it. In our case, we want to discover, in a soft way, the object mask relative to a given point on the foreground, without any other information given. We want to be able to use the power of color histograms without knowing the true mask (or bounding box) of the foreground. But this seems almost impossible. How can we compute the color histogram of the foreground if we have no clue about the foreground size and rough shape?

For now, let us assume that we have the bounding box of an object in an image. Color grouping should separate the foreground from the background in terms of certain global statistics using the bounding box given. In this case, we use color likelihoods derived from color histograms: the histogram for the foreground object is computed from the bounding box of the object, and the histogram of the background is computed from the rest of the image, similar to the idea that we previously used in object tracking [24]. This should be reasonable if we had a correct bounding box. Here, we show that even a completely wrong bounding box that meets certain assumptions will give a reasonably good result. Then, at any given pixel, if we choose a bounding box that meets those assumptions, we could potentially find which other points in the image are likely to be on the same object as that particular point. As we will show later a fairly good foreground mask can be obtained based on color distributions from a bounding box centered on the object, but of completely wrong shape, size, and location. This is explained theoretically if we make certain assumptions. In Fig. 5.2, the foreground is shown in blue and the bounding box in red. The bounding box's center is on the foreground, but its size and location are obviously wrong. We make the following assumptions that are easily met in practice, even without any prior knowledge about the shape and size of the foreground. We also give a visual explanation of the assumptions made in Fig. 5.3.

1. The area of the foreground is smaller than that of the background: this is true for most objects in images.

Fig. 5.2 Automatically discovering the foreground mask

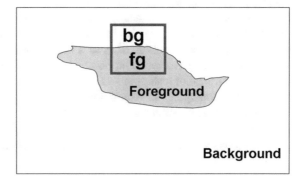

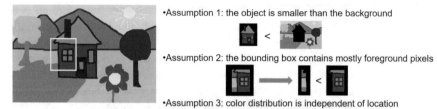

Fig. 5.3 A visual representation of the assumptions behind our simple segmentation algorithm. The first assumption is more general and states that the object is generally smaller than the background scene in the image. The second is particular to the bounding box: we require the bounding box to contain mostly foreground pixels, without necessarily being the true box. This case could be often achieved once we know a single point inside the object region and then taking a small bounding box around it. The third and strongest assumption is that if a color is more often found on the object than in the background, then this property is true everywhere in the image, independent of location. This idea intuitively makes sense and could be used effectively in practice, even though it is only an approximation. It is based on the observation that very often objects have particular, distinctive color patterns with respect to their background scene

2. The majority of pixels inside the bounding box belong to the true foreground: this is also easily met in practice since the center of the bounding box is considered to be on the foreground object by (our own) definition.
3. The color distributions of the true background and foreground are independent of position: this assumption is reasonable in practice, but harder to satisfy than the first two, since color is sometimes dependent on location (e.g., the head of a person has a different distribution than that person's clothes).

Even though the three assumptions above are not necessarily true all the time in practice, most of the time they do not need to be *perfectly true* for the following result to hold (they represent only loose sufficient conditions): let $p(c|obj)$ and $p(c|bg)$ be the true foreground and background color probabilities for a given color c, and $p(c|box)$ and $p(c|\neg box)$ the ones computed using the (wrong) bounding box satisfying the assumptions above. We want to prove that for any color c such that $p(c|obj) > p(c|bg)$ we must also have $p(c|box) > p(c|\neg box)$ and vice versa. This result basically shows that whenever a color c is more often found on the true object than in the background, it is also true that c will be more likely to be found inside the bounding box than outside of it, so a likelihood ratio test (> 1) would give the same result if using the bounding box instead of the true object mask. This result enables us to use color histograms as if we knew the true object mask, by using any bounding box satisfying the assumptions above. The proof is straight forward and it is based on those assumptions:

$$p(c|box) = p(c|obj, box)p(obj|box) + p(c|bg, box)p(bg|box), \qquad (5.1)$$

and

$$p(c|\neg box) = p(c|obj, \neg box)p(obj|\neg box) + p(c|bg, \neg box)p(bg|\neg box). \quad (5.2)$$

Assuming that the color distribution is independent of location (third assumption) for both the object and the background, we have $p(c|obj, box) = p(c|obj)$ and $p(c|bg, box) = p(c|bg)$. Then we have $p(c|box) = p(c|obj)p(obj|box) + p(c|bg)p(bg|box)$ and similarly $p(c|\neg box) = p(c|obj)p(obj|\neg box) + p(c|bg) p(bg|\neg box)$.

Since the object is smaller than the background (first assumption) but the main part of the bounding box is covered by the object (second assumption), we have $p(obj|box) > 0.5 > p(bg|box)$ and $p(obj|\neg box) < 0.5 < p(bg|\neg box)$. By also using $p(c|obj) > p(c|bg)$, $p(bg|box) = 1 - p(obj|box)$ and $p(bg|\neg box) = 1 - p(obj|\neg box)$, we finally get our result $p(c|box) = p(obj|box)(p(c|obj) - p(c|bg)) + p(c|bg) > p(obj|\neg box)(p(c|obj) - p(c|bg)) + p(c|bg) = p(c|\neg box)$ (since $p(obj|box) > p(obj|\neg box)$). The reciprocal result is obtained in the same fashion, by noticing that in order for the previous result to hold we must have $p(c|obj) - p(c|bg) > 0$, since $p(obj|box) > p(obj|\neg box)$.

5.2.1 A Fast Color Segmentation Algorithm

Next, we present a simple algorithm for computing soft masks of the foreground object by following the ideas presented previously. In Fig. 5.4, we present some results of soft masks obtained, as explained below in more detail, by computing the posterior probabilities of foreground using a small box centered at a specific location on the object (red dot). It is important that the masks obtained are relatively stable and not so sensitive to the location of the point chosen on the object. This suggests that by combining several such soft segmentations obtained from different potential locations on the object or by using different bounding boxes at different scales, we could obtain a more robust mask of higher quality.

Computing the soft object masks: The soft masks are computed as follows: each pixel of color c in the image is given the posterior value $p(c|box)/(p(c|box) + p(c|\neg box))$ in the mask. By the result obtained previously, we know that this posterior is greater than 0.5 whenever the true posterior is also greater than 0.5. Therefore, we expect that the soft mask we obtain to be similar to the one we would have obtained if we had used the true mask instead of the bounding box for computing the color distributions. We compute such masks at a given location over four different bounding boxes of increasing sizes (the sizes are fixed, the same for all images in all our experiments). At each scale, we zero out all pixels of value less than 0.5 and keep only the largest connected component that touches the inside of the bounding box. That is the soft mask for a given scale. Then, as a final result, we average the soft masks over all four scales. We term this procedure, the FG-BG color segmentation

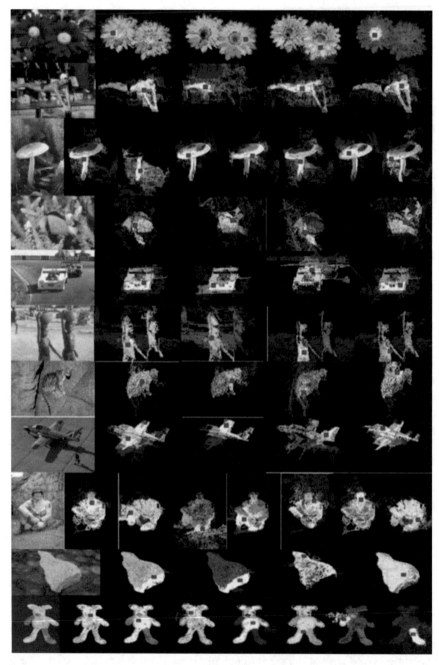

Fig. 5.4 Results obtained from the MSR database. We show a few representative segmentation results for each image, computed using a bounding box around the chosen point on the object (red dot). Note that the soft masks are relatively stable for different locations

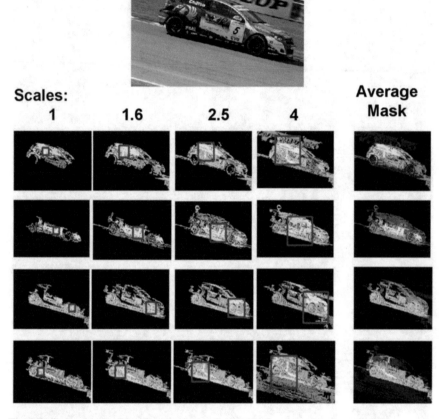

Fig. 5.5 Finding reasonable object masks is robust to the location of the bounding box, even for objects with a complex color distribution

approach. In Fig. 5.6, we present some representative visual results of the proposed soft segmentation algorithm. In Fig. 5.5, we show that the mask obtained is robust to the location of the bounding boxes, as long as the bounding boxes' centers are on the object of interest.

From class-specific object detection to segmentation: Class-specific or semantic segmentation is an important problem in computer vision that is enjoying a growing interest and has a wide range of real-world applications. Here, we show another simple and efficient method for this task, by the FG-BG segmentation approach with an off-the-shelf object class detector. The steps of this procedure are the following:

| Orig. Image | 1 | 1.6 | 2.5 | 4 | Avg. Mask |

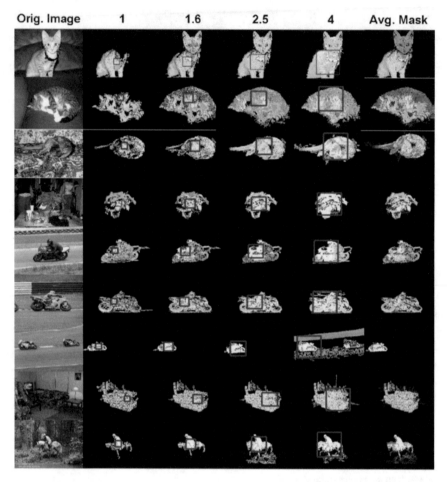

Fig. 5.6 Object masks obtained from arbitrary bounding boxes, centered on the object of interest. Averaging over several fixed scales improves the result. The masks shown are computed from color likelihood histograms based on the bounding boxes (completely wrong) shown, which are centered on the object of interest. The interior of the boxes is considered to be the foreground, while their exterior is the background. The posterior color likelihood image is obtained, thresholded at 0.5 and the largest connected component touching the interior of the bounding box is retained. We notice the even when the bounding boxes have wrong locations and sizes the masks obtained are close to the ground truth. Different bounding box sizes (which are fixed for every image, starting at 8 pixels and growing at a rate of 1.6) are tried and the average mask is shown in the rightmost column

1. Use any class-specific detector to get the approximate bounding box of objects from the category of interest.
2. Reduce in half the width and length of the bounding box returned by the detector, while keeping the same box center. This step will guarantee that most pixels in the reduced bounding box belong to the object of interest.

Fig. 5.7 Segmentation examples using the FG-BG color method on images of cars, motorcycles, and buses from VOC2012 dataset

3. Obtain the weighted grouping mask using the grouping method presented in this chapter. In this case, we do not need to compute masks from boxes at different scales, since we already have a good bounding box, centered on the object, of size that already has the same scale and spatial extent as the object.
4. Threshold the weighted mask to obtain the final hard segmentation of the object.

We used this simple idea for getting segmentations of objects from the VOC2012 dataset using the well-known, classical detector of Felzenszwalb et al. [25]. This dataset is very difficult for segmentation, objects of the same class varying greatly in appearance, size, and shape. We present some visual representative in Fig. 5.7. Note the high degree of variability in these images, with the object being of very different types and sizes, in images with complex backgrounds. Surprisingly, the simple color segmentation scheme, without having any semantic knowledge and without any pre-training, manages to produce results of relatively high quality, often comparable to the state-of-the-art methods in the literature that are much more computationally extensive and often require a high degree of pre-training.

5.3   VideoPCA: Unsupervised Background Subtraction in Video

We present our method based on Principal Component Analysis for rapidly estimating the frame pixels that are more likely to belong to the foreground object [10].[1] We make the observations that usually the object of interest has more complex and varied movements than its background scene, it often causes occlusions, it has a distinctive appearance, and it usually occupies less space. All these differences make the foreground more difficult to model with a simple PCA-based scheme than the background. Since the background contains the bulk of the information in the video and varies less than the foreground, we expect that it is better captured by the lower dimensional subspace of the frames from a given video shot. Several competitive methods for detecting potentially interesting, foreground objects as salient regions in images are also based on the general idea that objects are different from their backgrounds and that this foreground-background contrast can be best estimated by computing global image statistics over the input test image or by learning a background prior [26, 27]. For example, the successful spectral residual approach [28] is an efficient method that finds interesting regions in an image by looking at the difference between the average Fourier spectrum of the image, estimated using filtering, and the actual raw spectrum. The more recent Discriminative Regional Feature Integration (DRFI) approach [29] learns a background prior and finds objects that distinguish themselves from the global background using regression.

Different from the current literature, our method takes advantage of the spatiotemporal consistency that naturally exists in video shots and learns, in an unsupervised manner using PCA, a linear subspace of the background. It takes advantage of the redundancy and also of the rich information available in the whole video sequence. Relative to the main subspace of variation, the object is expected to be an outlier, an element of noise, harder to reconstruct. Note that every single change in appearance from one frame to the next, and every rapid movement, would be hard to capture by blindly using PCA on whole frames. We used this intuition to find pixels belonging to potential foreground objects and their occlusion regions, by a method related to background subtraction. In our case, the background is, in fact, the image reconstructed in the reduced subspace. Let the principal components be $\mathbf{u}_i$, $i \in [0 \dots n_u]$ (we used $n_u = 8$) and the reconstructed frame $\mathbf{f}$ be $\mathbf{f}_r \approx \mathbf{f}_0 + \sum_{i=1}^{n_u}((\mathbf{f} - \mathbf{f}_0)^\top \mathbf{u}_i)\mathbf{u}_i$. We obtained the error image $f_{\text{diff}} = |\mathbf{f} - \mathbf{f}_r|$. We notice that the difference image enhances the pixels belonging to the foreground object or to occlusions caused by the movement of this object (Fig. 5.8). By smoothing these regions with a large enough Gaussian and then thresholding, we obtain masks whose pixels tend to belong to objects rather than to background. Then, by applying another large and centered Gaussian to the obtained masks, we get a refined mask that is more likely to belong to the object of interest. Next, by accumulating such masks over the entire video shot, we can construct a relatively stable, robust foreground-background color model in

---

[1]Code available at: https://sites.google.com/site/multipleframesmatching/.

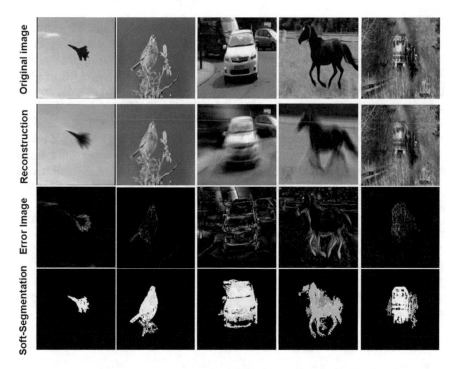

Fig. 5.8 First row: original images. Second row: reconstructed images with PCA and the first 8 principal components chosen. Third row: error image between the original and the reconstructed. Fourth row: final foreground segmentation computed with the SoftSeg method using color models obtained from foreground regions estimated with VideoPCA

order to estimate a soft segmentation, using the SoftSeg method presented next. While the mask is not optimal, it is computed at a high speed (50–100 fps), and, in our extensive experiments, it was always useful at obtaining high-quality candidate bounding boxes and initial foreground soft segmentations (see Fig. 5.8).

5.3.1 Soft Foreground Segmentation with VideoPCA

Foreground-background segmentation should separate well the object of interest from the background, based on statistical differences between the object and its surroundings. Here, we present a simple and effective way (termed SoftSeg) of producing soft object masks by capturing global object and background color properties, related to the method for soft foreground segmentation in static images presented in Leordeanu and Hebert [30]. For both the object, represented as a bounding box, and the background, considered as a border surrounding the bounding box (of thickness half the size of the bounding box), we estimate the empirical color distributions, such that for a given color c the foreground color likelihood is estimated as $p(c|F) = N_c^{(F)}/N^{(F)}$, where $N_c^{(F)}$ is the number of times the color c appeared inside the foreground region and $N^{(F)}$ is the total number of foreground pixels. Similarly, we compute the background color likelihood $p(c|B)$. Given the two distributions, we estimate the probability of foreground for each pixel of color c in the image, using Bayes' rule with equal priors: $p(F|c) = p(c|F)/(p(c|F) + p(c|B))$. In order to obtain the soft foreground segmentation mask, we simply estimate the foreground probability for each pixel with the above formula. In the case of multiple frames, when estimating the higher order terms $H_i(\mathbf{x}, \theta)$ these two distributions are computed from pixels accumulated from all frames considered. Segmentations obtained with probabilities estimated from multiple frames are of higher quality, less prone to accidental box misalignments and other noises.

The initial soft segmentation produced here is not optimal but it is computed fast (20 fps) and of sufficient quality to ensure the good performance of the subsequent stages. The first two steps of the method follow the algorithm VideoPCA first proposed in Stretcu and Leordeanu [10].

In Sect. 5.4.1, we present and prove our main theoretical result (Proposition 5.1), which explains in large part why our approach is able to produce accurate object segmentation in an unsupervised way.

5.4   Unsupervised Segmentation in Video Using HPP Features

Our method receives as input a video sequence, in which there is a main object of interest, and it outputs its soft segmentation masks and associated bounding boxes. The proposed approach has, as starting point, a processing stage based on Principal Component Analysis of the video frames, which provides an initial soft segmentation of the object—similar to the recent VideoPCA algorithm introduced as part of the object discovery approach of Stretcu and Leordeanu [10]. This soft segmentation usually has high precision but may have low recall. Starting from this initial stage that classifies pixels independently based only on their individual color, next we learn a higher level descriptor that considers groups of pixel colors and is able to capture higher order statistics about the object properties, such as different color patterns and textures. During the last stage, we combine the soft segmentation based on appearance with foreground cues computed from the contrasting motion of the main object versus its scene. The resulting method is accurate and fast (≈ 3 fps in Matlab, 2.60 GHz CPU—see Sect. 5.5.3). Our code is available online.[2]

Below, we summarize the steps of our approach (also see Fig. 5.9), in relation with Algorithm 5.1 (the pseudocode of our approach).

---

[2]https://goo.gl/2aYt4s.

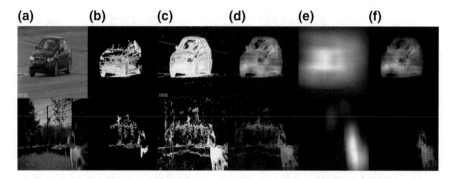

Fig. 5.9 Algorithm overview. **a** Original image, **b** first pixel-level appearance model, based on initial object cues (Steps 1 and 2), **c** refined pixel-level appearance model, built from the projection of soft segmentation (Steps 3 and 4), **d** patch-level appearance model (Step 5), **e** motion estimation mask (part of Step 6), **f** final soft segmentation mask (Step 6)

- **Step 1**: Select highly probable foreground pixels based on the differences between the original frames and the frames projected on their subspace with Principal Component Analysis (Sect. 5.3, Algorithm 5.1—lines [2, 5]).
- **Step 2**: Estimate empirical color distributions for foreground and background from the pixel masks computed in Step 1. Use these distributions to estimate the probability of foreground for each pixel independently based on its color (Sect. 5.3.1, Algorithm 5.1—line 6).
- **Step 3**: Improve the soft segmentation from Step 2, by projection on the subspace of soft segmentations (Sect. 5.4.1.1, Algorithm 5.1—lines [7, 9).
- **Step 4**: Re-estimate empirical color distributions for foreground and background from the pixel masks updated in Step 3. Use these distributions to estimate the probability of foreground for each pixel independently based on its color (Sect. 5.3.1, Algorithm 5.1—line 10).
- **Step 5**: Learn a discriminative classifier of foreground regions with regularized least squares regression on the soft segmentation real output $\in [0, 1]$. Use a feature vector that considers groups of colors that co-occur in larger patches. Run classifier at each pixel location in the video and produce improved per frame foreground soft segmentation (Sect. 5.4.2, Algorithm 5.1—lines [11, 15]).
- **Step 6**: Combine soft segmentation using appearance (Step 5) with foreground motion cues efficiently computed by modeling the background motion. Obtain the final soft segmentation (Sect. 5.4.3, Algorithm 5.1—lines [16, 23]).
- **Step 7**: Optional: Refine segmentation using GrabCut [15], by considering as potential foreground and background samples the pixels given by the soft segmentation from Step 6 (Sect. 5.4.3.1).

Selecting the initial highly probable object regions in video: We discover the initial foreground object regions in Steps 1 and 2 which represent the technique for video object segmentation using VideoPCA [10], presented in Sect. 5.3. Note that other approaches for soft foreground discovery could have been applied here, such as Zitnick and Dollár [31], Hou and Zhang [28], Jiang et al. [29], but we have found the direction using VideoPCA to be both fast and reliable and to fit perfectly with the later stages of our method. As discussed previously, the main idea behind VideoPCA is that the principal components will represent a linear subspace of the background, as the object is expected to be an outlier, not obeying the principal variation observed in the video, thus harder to reconstruct. At this step, we project the frames on the resulted subspace and compute reconstruction error images as differences between original frames and their PCA reconstructed counterparts. The first two steps of our approach represent the first stage of learning with highly probably positive features in a sequence of steps that apply the same strategy sequentially. Thus, at every stage of the algorithm, discussed in more detail next, we reiterate the principle of learning from HPP features and thus manage to start from a classifier that has high precision but low recall and continue improving the recall, stage by stage, while maintaining the precision high. The overall F1 score (which is an appropriate measure for evaluating object segmentation) of the overall system thus increases from one iteration to the next.

Key ideas: Our algorithm has at its core two main ideas, which come directly from the principles that we proposed in Chap. 1. The first key idea is that the object and the background have contrasting properties in terms of size, appearance, and movement. This insight leads to the ability of reliably selecting a few regions in the video that are highly likely to belong to the object. The following, second idea, which brings certain formal guarantees, is that if we are able to select, in an unsupervised manner, even a small portion of the foreground object, but with high precision, then, under some reasonable assumptions, we could train a robust foreground-background classifier that can be used for the automatic discovery of the object. In Table 5.1, we present the improvements in precision, recall, and F-measure between the different steps of our algorithm. Note that the arrows go from the precision and recall of the samples initially considered to be positive, to the precision and recall of the pixels finally classified as positive. The significant improvement in F-measure is explained by our theoretical result (stated in Proposition 5.1), which shows that under certain conditions, a reliable classifier will be learned even if the recall of the corrupted

Table 5.1 Evolution of precision, recall, and F-measure of the feature samples considered as positives (foreground) at different stages of our method (SegTrack dataset). We start with a corrupted set of positive samples with high precision and low recall, and improve both precision and recall through the stages of our method. Thus, the soft masks become more and more accurate from one stage to the next

	Steps 1 and 2	Steps 3 and 4	Step 5
Precision	$66 \rightarrow 70$	$62 \rightarrow 60$	$64 \rightarrow 74$
Recall	$17 \rightarrow 51$	$45 \rightarrow 60$	$58 \rightarrow 68$
F-measure	$27 \rightarrow 59$	$53 \rightarrow 60$	$61 \rightarrow 72$

Table 5.2 Performance analysis and execution time for all stages of our method

	Steps 1 and 2	Steps 3 and 4	Step 5	Step 6
F-meas. (SegTrack)	59.0	60.0	72.0	74.6
F-meas. (YTO)	53.6	54.5	58.8	63.4
Runtime (s/frame)	0.05	0.03	0.25	0.02

positive samples is low, as long as the precision is relatively high. In Table 5.2, we introduce quantitative results of the different stages of our method, along with the associated execution times.

Algorithm 5.1 Video object segmentation

1: get input frames $\mathbf{F}^i$
2: PCA($\mathbf{A}_1$) $=> \mathbf{V}_1$ eigenvectors; $\mathbf{A}_1(i, :) = \mathbf{F}^i(:)$
3: $\mathbf{R}_1 = \bar{\mathbf{A}}_1 + (\mathbf{A}_1 - \bar{\mathbf{A}}_1) * \mathbf{V}_1 * \mathbf{V}_1^T$ - reconstruction
4: $\mathbf{P}_1^i = \mathbf{d}(\mathbf{A}_1(i, :), \mathbf{R}_1(i, :))$
5: $\mathbf{P}_1^i = \mathbf{P}_1^i \otimes \mathbf{G}_{\sigma_1}$
6: $\mathbf{P}_1^i =>$ pixel-level appearance model $=> \mathbf{S}_1^i$
7: PCA($\mathbf{A}_2$) $=> \mathbf{V}_2$ eigenvectors; $\mathbf{A}_2(i, :) = \mathbf{S}_1^i(:)$
8: $\mathbf{R}_2 = \bar{\mathbf{A}}_2 + (\mathbf{A}_2 - \bar{\mathbf{A}}_2) * \mathbf{V}_2 * \mathbf{V}_2^T$ - reconstruction
9: $\mathbf{P}_2^i = \mathbf{R}_2^i \otimes \mathbf{G}_{\sigma_2}$
10: $\mathbf{P}_2^i =>$ pixel-level appearance model $=> \mathbf{S}_2^i$
11: $\mathbf{D}$ - data matrix containing patch-level descriptors
12: $\mathbf{s}$ patch labels extracted from $\mathbf{S}_2^i$
13: select k features from $\mathbf{D} => \mathbf{D_s}$
14: $\mathbf{w} = (\lambda \mathbf{I} + \mathbf{D}_s^T \mathbf{D}_s)^{-1} \mathbf{D}_s^T \mathbf{s}$
15: evaluate $=>$ patch-level appearance model $=> \mathbf{S}_3^i$
16: **for** each frame i **do**
17:    compute $\mathbf{I}_x, \mathbf{I}_y$ and $\mathbf{I}_t$
18:    build motion matrix $\mathbf{D}_m$
19:    $\mathbf{w}_m = (\mathbf{D_m}^T \mathbf{D_m})^{-1} \mathbf{D_m}^T \mathbf{I}_t$
20:    compute motion model $\mathbf{M}^i$
21:    $\mathbf{M}^i = \mathbf{M}^i \otimes \mathbf{G}_\sigma^i$
22:    combine $\mathbf{S}_3^i$ and $\mathbf{M}^i => \mathbf{S}_4^i$
23: **end for**

5.4.1 Learning with Highly Probable Positive Features

In Proposition 5.1, we show that a classifier trained on corrupted sets of positive and negative samples can learn the right thing as if true positives and negatives were used for training, if the following condition is met: the set of corrupted positives should contain positive samples in a proportion that is greater than the overall proportion of true positives in the whole training set. This proposition is the basis for both stages

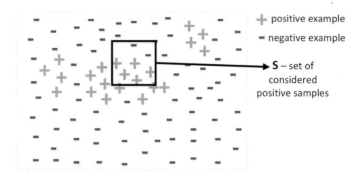

Fig. 5.10 Learning with HPP feature vectors. Essentially, Proposition 5.1 shows that we could learn a reliable discriminative classifier from a small set of corrupted positive samples, with the rest being considered negatives, if the corrupted positive set contains mostly good features such that the ratio of true positives in the corrupted positive set is greater than the overall ratio of true positives. This assumption can often be met in practice and efficiently used for unsupervised learning

of our method, the one that classifies pixels independently based on their colors and the second in which we consider higher order color statistics among groups of pixels.

Let us start with the example in Fig. 5.10, where we have selected a set of samples S (inside the box) as being positive. The set S has high precision (most samples are indeed positive), but low recall (most true positives are wrongly labeled). Next, we show that the sets S and $\neg S$ could be used reliably (as defined in Proposition 5.1) to train a binary classifier.

Let $p(E_+)$ and $p(E_-)$ be the true distributions of positive and negative elements, and $p(\mathbf{x}|S)$ and $p(\mathbf{x}|\neg S)$ be the probabilities of observing a sample inside and outside the considered positive set S and negative set $\neg S$, respectively.

Proposition 5.1 (learning from highly probable positive (HPP) features) *Considering the following hypotheses* $\mathbf{H_1} : p(E_+) < q < p(E_-)$, $\mathbf{H_2} : p(E_+|S) > q > p(E_-|S)$, *where* $q \in (0, 1)$, *and* $\mathbf{H_3} : p(\mathbf{x}|E_+)$ *and* $p(\mathbf{x}|E_-)$ *are independent of S, then, for any sample* $\mathbf{x}$ *we have* $p(\mathbf{x}|S) > p(\mathbf{x}|\neg S) <=> p(\mathbf{x}|E_+) > p(\mathbf{x}|E_-)$. *In other words, a classifier that classifies pixels based on their likelihoods w.r.t to S and $\neg S$ will take the same decision as if it was trained on the true positives and negatives, and we refer to it as a* reliable *classifier.*

Proof We express $p(E_-)$ as $\frac{(p(E_-) - p(E_-|S) \cdot p(S))}{(1-p(S))}$ (Eq. 1), using the hypothesis and the sum rule of probabilities. Considering (Eq. 1), hypothesis $\mathbf{H_1}$, $\mathbf{H_2}$, and the fact that $p(S) > 0$, we obtain that $p(E_-|\neg S) > q$ (Eq. 2). In a similar fashion, $p(E_+|\neg S) < q$ (Eq. 3). The previously inferred relations (Eqs. 2 and 3) generate $p(E_-|\neg S) > q > p(E_+|\neg S)$ (Eq. 4), which along with hypothesis $\mathbf{H_2}$ help us conclude that $p(E_+|S) > p(E_+|\neg S)$ (Eq. 5). Also, from $\mathbf{H_3}$, we infer that $p(\mathbf{x}|E_+, S) = p(\mathbf{x}|E_+)$ and $p(\mathbf{x}|E_-, S) = p(\mathbf{x}|E_-)$ (Eq. 6). Using the sum rule and hypothesis $\mathbf{H_3}$, we obtain that $p(\mathbf{x}|S) = p(E_+|S) \cdot (p(\mathbf{x}|E_+) - p(\mathbf{x}|E_-)) + p(\mathbf{x}|E_-)$ (Eq. 7). In a similar way, it results that $p(\mathbf{x}|\neg S) = p(E_+|\neg S) \cdot (p(\mathbf{x}|E_+) - p(\mathbf{x}|E_-)) + p(\mathbf{x}|E_-)$ (Eq. 8).

$p(\mathbf{x}|S) > p(\mathbf{x}|\neg S) \Rightarrow p(\mathbf{x}|E_+) > p(\mathbf{x}|E_-)$: using the hypothesis and previously inferred results (Eqs. 5, 7 and 8), it results that $p(\mathbf{x}|E_+) > p(\mathbf{x}|E_-)$.

$p(\mathbf{x}|E_+) > p(\mathbf{x}|E_-) \Rightarrow p(\mathbf{x}|S) > p(\mathbf{x}|\neg S)$: from the hypothesis we can infer that $p(\mathbf{x}|E_+) - p(\mathbf{x}|E_-) > 0$, and using (Eq. 5) we obtain $p(\mathbf{x}|S) > p(\mathbf{x}|\neg S)$. $\square$

5.4.1.1 Object Proposals Refinement

During this stage, the soft segmentations obtained so far are improved using a projection on their PCA subspace. In contrast to VideoPCA, now we select the probable object regions as the PCA projected versions of the soft segmentations computed in previous steps. For the projection, we consider the first eight principal components, with the purpose of reducing the amount of noise that might be leftover from the previous steps. Further, color likelihoods are re-estimated to obtain the soft segmentation masks.

5.4.2 Descriptor Learning with IPFP

The foreground masks obtained so far were computed by treating each pixel independently, which results in masks that are not always correct, as first-order statistics, such as colors of individual pixels, cannot capture more global characteristics about object texture and shape. At this step, we move to the next level of abstraction by considering groups of colors present in local patches, which are sufficiently large to capture object texture and local shape. We define a patch descriptor based on local color occurrences, as an indicator vector $\mathbf{d}_W$ over a given patch window W, such that $\mathbf{d}_W(c) = 1$ if color c is present in window W and 0 otherwise (Fig. 5.11). Colors are indexed according to their values in HSV space, where channels H, S, and V are discretized in ranges $[1, 15]$, $[1, 11]$, and $[1, 7]$, generating a total of 1155 possible colors. The descriptor does not take into consideration the exact spatial location of a given color in the patch, nor its frequency. It only accounts for the presence of c in the patch. This leads to invariance to most rigid or non-rigid transformations, while preserving the local appearance characteristics of the object. Then, we take a classification approach and learn a classifier (using regularized least squares regression, due to its considerable speed and efficiency) to separate between Highly Probable Positive (HPP) descriptors and the rest, collected from the whole video according to the soft masks computed at the previous step. The classifier is generally robust to changes in viewpoint, scale, illumination, and other noises, while remaining discriminative (Fig. 5.9).

Not all 1155 colors are relevant to our classification problem. Most object textures are composed of only a few important colors that distinguish them against the background scene. Effectively reducing the number of colors in the descriptor and selecting only the relevant ones can improve both speed and performance. We use the efficient selection algorithm presented in Leordeanu et al. [32]. The method

Fig. 5.11 Initial patch
descriptors encoding color
occurrences (*n* number of
considered colors)

proceeds as follows. Let *n* be the total number of colors and $k < n$ the number of relevant colors we want to select. The idea is to identify the group of *k* colors with the largest amount of covariance—they will be the ones most likely to select well the foreground versus the background (see Leordeanu et al. [32] for details). Now consider **C** the covariance matrix of the colors forming the rows in the data matrix **D**. The task is to solve the following optimization problem:

$$\mathbf{w}^* = \arg \max_{\mathbf{w}} \ \mathbf{w}^T \mathbf{C} \mathbf{w}$$

$$s.t. \sum_{i=1}^{n} w_i = 1, w_i \in \left[0, \frac{1}{k}\right]. \tag{5.3}$$

The non-zero elements of $\mathbf{w}^*$ correspond to the colors we need to select for creating our descriptor used by the classifier (based on regularized least squares regression model), so we define a binary mask $\mathbf{w}_s \in \mathbb{R}^{n \times 1}$ over the colors (that is the descriptor vector) as follows:

$$\mathbf{w_s}(i) = \begin{cases} 1 & \text{if } \mathbf{w}^*(i) > 0 \\ 0 & \text{otherwise} \end{cases} \tag{5.4}$$

The problem above is NP-hard, but a good approximation can be efficiently found by the IPFP method, originally introduced in Leordeanu et al. [32] and presented in this book (Chaps. 1–4) in the context of various optimization and learning tasks, with similar formulations as the one in Eq. 5.4. The optimal number of selected colors is a relatively small fraction of the total number, as expected. Besides the slight increase in performance, the real gain is in the significant decrease in computation time (see Fig. 5.12).

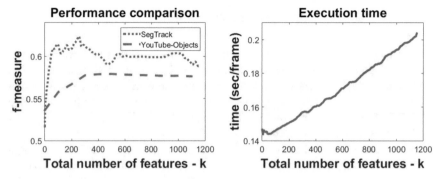

Fig. 5.12 Features selection—optimization and sensitivity analysis

Relation to graph clustering and feature selection: The reader will immediately notice the similarity between the descriptor learning formulation in Eq. 5.4 and the graph clustering formulations in Chaps. 1 and 2. Also, note that the exact formulation is also used for the classifier learning approach in Chap. 3. What also unites all these tasks, besides their similar mathematical representations is the algorithm we use for optimization, the Integer Projected Fixed Point (IPFP) method, which proves to be very effective in finding efficiently high-quality discrete solutions every time we apply it to such optimization problems. Also, note the interesting connection between finding the relevant colors that define a patch descriptor and feature selection, which is also related to classifier learning (Chap. 4). While all these unsupervised learning tasks seem different initially, they have a deep common core—that of finding co-occurrences of feature or classifier responses in space and time. They are unlikely by accident, which is why they can be reliably used for detecting the presence of meaningful causes.

Next, we define $\mathbf{D}_s \in \mathbb{R}^{m \times (1+k)}$ to be the data matrix, with a training sample per row, after applying the selection mask to the descriptor; m is the number of training samples and k is the number of colors selected to form the descriptor (we add a constant column of 1's for the bias term). Then, the weights $\mathbf{w} \in \mathbb{R}^{(1+k) \times 1}$ of the regularized regression model are learned very fast, in closed form:

$$\mathbf{w} = (\lambda \mathbf{I} + \mathbf{D}_s^T \mathbf{D}_s)^{-1} \mathbf{D}_s^T \mathbf{s}, \tag{5.5}$$

where $\mathbf{I}$ is the identity matrix, λ is the regularization term, and $\mathbf{s}$ is the vector of soft segmentation masks' values (estimated at the previous step) corresponding to the samples chosen for training of the descriptor. Then, the final appearance-based soft segmentation masks are generated by evaluating the regression model for each pixel.

5.4.3   Combining Appearance and Motion

The foreground and background have complementary properties at many levels, not just that of appearance. Here, we consider that the object of interest must distinguish itself from the rest of the scene in terms of its motion pattern. A foreground object that does not move in the image, relative to its background, cannot be discovered using information from the current video alone. We take advantage of this idea by the following efficient approach.

Let $\mathbf{I}_t$ be the temporal derivative of the image as a function of time, estimated as difference between subsequent frames $\mathbf{I}_{t+1} - \mathbf{I}_t$. Also let $\mathbf{I}_x$ and $\mathbf{I}_y$ be the partial derivatives in the image w.r.t x and y. Consider $\mathbf{D}_m$ to be the motion data matrix, with one row per pixel p in the current frame corresponding to $[\mathbf{I}_x, \mathbf{I}_y, x\mathbf{I}_x, x\mathbf{I}_y, y\mathbf{I}_x, y\mathbf{I}_y]$ at locations estimated as background by the foreground segmentation estimated so far. Given such a matrix at time t, we linearly regress $\mathbf{I}_t$ on $\mathbf{D}_m$. The solution would be a least square estimate of an affine motion model for the background using first-order Taylor expansion of the image w.r.t time: $\mathbf{w}_m = (\mathbf{D}_m{}^T\mathbf{D}_m)^{-1}\mathbf{D}_m{}^T\mathbf{I}_t$. Here, $\mathbf{w_m}$ contains the six parameters defining the affine motion (including translation) in 2D.

Then, we consider deviations from this model as potential good candidates for the presence of the foreground object, which is expected to move differently than the background scene. The idea is based on an approximation, of course, but it is very fast to compute and can be reliably combined with the appearance of soft masks. Thus, we evaluate the model in each location p and compute errors $|\mathbf{D}_m(p)\mathbf{w}_m - \mathbf{I}_t(p)|$. We normalize the error image and map it to [0, 1]. This produces a soft mask (using motion only) of locations that do not obey the motion model—they are usually correlated with object locations. This map is then smoothed with a Gaussian (with σ proportional to the distribution on x and y of the estimated object region).

At this point, we have a soft object segmentation computed from appearance alone, and one computed independently, based on motion cues. The two soft results are multiplied to obtain the final segmentation.

5.4.3.1   Optional Refinement of Video Object Segmentation

Optionally, we can further refine the soft mask by applying an off-the-shelf segmentation algorithm, such as GrabCut [15] and feeding it our soft foreground segmentation. **Note**: in our experiments, we used GrabCut only for evaluation on SegTrack, as we were interested in the fine details of the objects' shape. All other experiments are performed without this step.

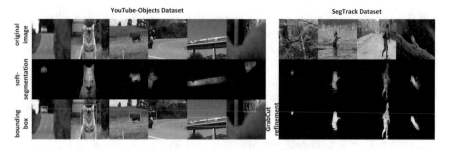

Fig. 5.13 Qualitative results on YouTube-Objects dataset and SegTrack dataset

5.5 Experimental Analysis

Our experiments were performed on two datasets: YouTube-Objects dataset and SegTrack v2 dataset. We first introduce some qualitative results of our method, on the considered datasets (Fig. 5.13). Note that for the final evaluation on the YouTube-Objects dataset, we also extract object bounding boxes that are computed using the distribution of the pixels with high probability of being part of the foreground. Both position and size of the boxes are computed using a mean shift approach. For the final evaluation on the SegTrack dataset, we have refined the soft segmentation masks, using the GrabCut algorithm [15]. In Table 5.2, we present evaluation results for different stages of our algorithm, along with the execution time, per stage. The F-measure is increased with each stage of our algorithm.

5.5.1 Tests on YouTube-Objects Dataset

Dataset: The YouTube-Objects dataset [33] contains a large number of videos filmed in the wild, collected from YouTube. It contains challenging, unconstrained sequences of ten object categories (aeroplane, bird, boat, car, cat, cow, dog, horse, motorbike, train). The sequences are considered to be challenging as they are completely unconstrained, displaying objects performing rapid movements, with difficult dynamic backgrounds, illumination changes, camera motion, scale and viewpoint changes, and even editing effects, like flying logos or joining of different shots. The ground truth is provided for a small number of frames and contains bounding boxes for the object instances. Usually, a frame contains only one primary object of the considered class, but there are some frames containing multiple instances of the same class of objects. Two versions of the dataset were released, the first (YouTube-Objects v1.0) containing 1407 annotated objects from a total of ≈570,000 frames, while the second (YouTube-Objects v2.2) contains 6975 annotated objects from ≈720,000 frames.

Metric: For the evaluation on the YouTube-Objects dataset, we have adopted the CorLoc metric, computing the percentage of correctly localized object bounding boxes. We evaluate the correctness of a box using the PASCAL criterion (intersection over union ≥ 0.5).

Results: We compare our method against [3, 6, 7, 10, 33]. We considered their results as originally reported in the corresponding papers. The comparison is presented in Table 5.3. From our knowledge, the other methods were evaluated on YouTube-Objects v1.0, on the training samples (the only exception would be [10], where they have considered the full v1.0 dataset). Considering this, and the differences between the two versions, regarding the number of annotations, we have reported our performances on both versions, in order to provide a fair comparison and also to report the results on the latest version, YouTube-Objects v2.2 (not considered for comparison). We report results of the evaluation on v1.0 by only considering the training samples, for a fair comparison with other methods. Our method, which is unsupervised, is compared against both supervised and unsupervised methods. In the table, we have marked state-of-the-art results for unsupervised methods (bold) and overall state-of-the-art results (underlined). We also mention the execution time for the considered methods, in order to prove that our method is one order of magnitude faster than others (see Sect. 5.5.3 for details).

Table 5.3 The CorLoc scores of our method and five other state-of-the-art methods, on the YouTube-Objects dataset (note that result for v2.2 of the dataset are not considered for comparison)

Method super-vised?	Jun Koh et al. [6] Y	Zhang et al. [7] Y	Prest et al. [33] N	Stretcu and Leordeanu [10] N	Papazoglou and Ferrari [3] N	Ours $v1.0$ N	Ours $v2.2$ N
Aeroplane	64.3	75.8	51.7	38.3	65.4	**76.3**	76.3
Bird	63.2	60.8	17.5	62.5	67.3	**71.4**	68.5
Boat	<u>73.3</u>	43.7	34.4	51.1	38.9	**65.0**	54.5
Car	68.9	<u>71.1</u>	34.7	54.9	**65.2**	58.9	50.4
Cat	44.4	46.5	22.3	64.3	46.3	**68.0**	59.8
Cow	<u>62.5</u>	54.6	17.9	52.9	40.2	**55.9**	42.4
Dog	<u>71.4</u>	55.5	13.5	44.3	65.3	**70.6**	53.5
Horse	52.3	<u>54.9</u>	**48.4**	43.8	**48.4**	33.3	30.0
Motorbike	<u>78.6</u>	42.4	39.0	41.9	39.0	**69.7**	53.5
Train	23.1	35.8	25.0	<u>45.8</u>	25.0	42.4	60.7
Avg.	60.2	54.1	30.4	49.9	50.1	**<u>61.1</u>**	54.9
Time (s/frame)	N/A	N/A	N/A	6.9	4	**<u>0.35</u>**	

The performances of our method are competitive, obtaining state-of-the-art results for three classes, against both supervised and unsupervised methods. Compared to the unsupervised methods, we obtain state-of-the-art results for seven classes. On average, our method performs better than all the others, and also in terms of execution time (also see Sect. 5.5.3). The fact that, on average, our algorithm outperforms other methods proves that it generalizes better for different classes of objects and different types of videos. Our solution performs poorly on the "horse" class, as many sequences contain multiple horses, and our method is not able to correctly separate the instances. Another class with low performance is the "cow" class, where we deal with the same problems as in the case of "horse" class, and where objects are usually still, being hard to segment in our system.

5.5.2 Tests on SegTrack V2 Dataset

Dataset: The SegTrack dataset was originally introduced by Tsai et al. [34], for evaluating tracking algorithms. Further, it was adapted for the task of video object segmentation [4]. We work with the second version of the dataset (SegTrack v2), which contains 14 videos ($\approx$1000 frames), with pixel-level ground truth annotations for the object of interest, in every frame. The dataset is difficult as the included objects can be easily confused with the background, appear in different sizes, and display complex deformations. There are eight videos with one primary object and six with multiple objects, from eight different categories (bird, cheetah, human, worm, monkey, dog, frog, parachute).

Metric: For the evaluation on the SegTrack, we have adopted the average intersection over union metric. We specify that for the purpose of this evaluation, we use GrabCut for refinement of the soft segmentation masks.

Results: We compare our method against [3, 4, 8, 9, 35]. We considered their results as originally reported by Wang et al. [35]. The comparison is presented in Table 5.4. Again, we compare our method against both supervised and unsupervised methods, and, in the table, we have marked state-of-the-art results for unsupervised methods (bold), and overall state-of-the-art results (underlined). The execution times are also introduced, to highlight that our method outperforms other approaches in terms of speed (see Sect. 5.5.3).

The performance of our method is competitive, while being an unsupervised method. Also, we prove that our method is one order of magnitude faster than the previous state-of-the-art [3] (see Sect. 5.5.3).

Table 5.4 The average IoU scores of our method and five other state-of-the-art methods, on the SegTrack v2 dataset. Our reported time also includes the computational time required for GrabCut

Method supervised?	Lee et al. [9] Y	Zhang et al. [8] Y	Wang et al. [35] Y	Papazoglou and Ferrari [3] N	Li et al. [4] N	Ours N
Bird of paradise	92	–	95	66	**94**	93
Birdfall	49	71	70	59	**63**	58
Frog	75	74	83	**77**	72	58
Girl	88	82	91	73	**89**	69
Monkey	79	62	90	65	**85**	69
Parachute	96	94	92	91	93	**94**
Soldier	67	60	85	69	**84**	60
Worm	84	60	80	74	83	**84**
Avg.	79	72	86	72	**83**	73
Time (s/frame)	>120	>120	N/A	4	242	**0.73**

5.5.3   Computation Time

One of the main advantages of our method is the reduced computational time. Note that all per-pixel classifications can be efficiently implemented by linear filtering routines, as all our classifiers are linear. It takes only **0.35** s/frame for generating the soft segmentation masks (initial object cues: 0.05 s/frame, object proposals refinement: 0.03 s/frame, patch-based regression model: 0.25 s/frame, motion estimation: 0.02 s/frame (Table 5.2)). The method was implemented in Matlab, with no special optimizations. All timing measurements were performed using a computer with an Intel Core i7 2.60 GHz CPU. The method of Papazoglou and Ferrari [3] report a time of 3.5 s/frame for the initial optical flow computation, on top of which they run their method, which requires 0.5 s/frame, leading to a total time of 4 s/frame. The method introduced in Stretcu and Leordeanu [10] has a total of 6.9 s/frame. For other methods, like the one introduced in Zhang et al. [8], Lee et al. [9], it takes up to 120 s/frame only for generating the initial object proposals using the method of Endres and Hoiem [13]. We have no information regarding computational time of other considered methods, but due to their complexity we expect them to be orders of magnitude slower than ours.

5.6 Conclusions and Future Work

We presented an efficient fully unsupervised method for object discovery in video that is both fast and accurate. It achieves state-of-the-art results on a challenging benchmark for bounding box object discovery and very competitive performance on a video object segmentation dataset. At the same time, our method is fast, being at least an order of magnitude faster than competition. We achieve an excellent combination of speed and performance by exploiting the contrasting properties between objects and their scenes, in terms of appearance and motion, which makes it possible to select positive feature samples with a very high precision. We show, theoretically and practically, that high precision is sufficient for reliable unsupervised learning (since positives are generally less frequent than negatives), which we perform both at the level of single pixels and at the higher level of groups of pixels, which capture higher order statistics about objects appearance, texture, and shape. The top speed and accuracy of our method, combined with theoretical guarantees that hold in practice under mild conditions, make our approach unique and valuable in the quest for solving the unsupervised learning problem in video.

In this chapter, we made an important step towards solving unsupervised learning. We introduced the concept of learning with HPP features, which generalizes the ideas we have discussed so far and relates to the general overview and principles given in Chap. 1. In the future chapters, we show how to further exploit this concept and use to train in an unsupervised way several generations of classifiers, in a novel paradigm that uses the previous generation as teacher for next-generation students (Chap. 7). The highly probable positive features are the ones that can be reliably used as positively labeled signal. In any attempt to perform true unsupervised learning, we need some sort of prior knowledge and, as shown in this chapter, the HPP features are ideal to play that role. We also prove that, under some mild assumptions, learning with HPP features provides some theoretical guarantees. Therefore, we could expect this concept to find its way deeper into the general idea of unsupervised learning and also suspect that there is probably no better HPP signal than the co-occurrences and agreements of classifier responses in neighborhoods of space and time that accidentally would otherwise be very unlikely. And that is the main theme of this book, seen here again, at work, as in the previous and next chapters.

References

1. Sivic J, Russell BC, Efros AA, Zisserman A, Freeman WT (2005) Discovering objects and their location in images. In: Tenth IEEE international conference on computer vision, ICCV 2005, vol 1. IEEE, pp 370–377
2. Leordeanu M, Collins R, Hebert M (2005) Unsupervised learning of object features from video sequences. In: IEEE Computer society conference on computer vision and pattern recognition. IEEE computer society, vol 1, p 1142
3. Papazoglou A, Ferrari V (2013) Fast object segmentation in unconstrained video. In: Proceedings of the IEEE international conference on computer vision, pp 1777–1784

4. Li F, Kim T, Humayun A, Tsai D, Rehg J (2013) Video segmentation by tracking many figure-ground segments. In: International conference on computer vision
5. Jain AK, Murty MN, Flynn PJ (1999) Data clustering: a review. ACM Comput Surv (CSUR) 31(3):264–323
6. Jun Koh Y, Jang WD, Kim CS (2016) POD: discovering primary objects in videos based on evolutionary refinement of object recurrence, background, and primary object models. In: Proceedings of the IEEE conference on computer vision and pattern recognition, pp 1068–1076
7. Zhang Y, Chen X, Li J, Wang C, Xia C (2015) Semantic object segmentation via detection in weakly labeled video. In: Proceedings of the IEEE conference on computer vision and pattern recognition, pp 3641–3649
8. Zhang D, Javed O, Shah M (2013) Video object segmentation through spatially accurate and temporally dense extraction of primary object regions. In: Proceedings of the IEEE conference on computer vision and pattern recognition, pp 628–635
9. Lee YJ, Kim J, Grauman K (2011) Key-segments for video object segmentation. In: 2011 IEEE international conference on computer vision (ICCV). IEEE, pp 1995–2002
10. Stretcu O, Leordeanu M (2015) Multiple frames matching for object discovery in video. In: BMVC, pp 186.1–186.12
11. Alexe B, Deselaers T, Ferrari V (2012) Measuring the objectness of image windows. IEEE Trans Pattern Anal Mach Intell 34(11):2189–2202
12. Felzenszwalb PF, Girshick RB, McAllester D, Ramanan D (2010) Object detection with discriminatively trained part-based models. IEEE Trans Pattern Anal Mach Intell 32(9):1627–1645
13. Endres I, Hoiem D (2010) Category independent object proposals. Comput Vis-ECCV 2010:575–588
14. Jain SD, Grauman K (2014) Supervoxel-consistent foreground propagation in video. In: European conference on computer vision. Springer, pp 656–671
15. Rother C, Kolmogorov V, Blake A (2004) Grabcut: interactive foreground extraction using iterated graph cuts. In: ACM transactions on graphics (TOG), vol 23. ACM, pp 309–314
16. Fulkerson B, Vedaldi A, Soatto S (2009) Class segmentation and object localization with superpixel neighborhoods. In: 2009 IEEE 12th international conference on computer vision. IEEE, pp 670–677
17. Levinshtein A, Stere A, Kutulakos KN, Fleet DJ, Dickinson SJ, Siddiqi K (2009) Turbopixels: fast superpixels using geometric flows. IEEE Trans Pattern Anal Mach Intell 31(12):2290–2297
18. Carreira J, Sminchisescu C (2012) CPMC: automatic object segmentation using constrained parametric min-cuts. IEEE Trans Pattern Anal Mach Intell 34(7):1312–1328
19. Li F, Carreira J, Lebanon G, Sminchisescu C (2013) Composite statistical inference for semantic segmentation. In: Proceedings of the IEEE conference on computer vision and pattern recognition, pp 3302–3309
20. Leordeanu MD (2010) Spectral graph matching, learning, and inference for computer vision. PhD Thesis, Carnegie Mellon University
21. Haller E, Leordeanu M (2017) Unsupervised object segmentation in video by efficient selection of highly probable positive features. In: The IEEE international conference on computer vision (ICCV)
22. Rother C, Kolmogorov V, Blake A (2004) Grabcut: interactive foreground extraction using iterated graph cuts. In: SIGGRAPH
23. Li Y, Sun J, Shum CTH (2004) Lazy snapping. In: SIGGRAPH
24. Collins R, Liu Y, Leordeanu M (2005) Online selection of discriminative tracking features. PAMI 27(10):1631–1643
25. Felzenszwalb P, McAllester D, Ramanan D (2008) A discriminatively trained, multiscale, deformable part model. In: CVPR
26. Borji A, Sihite D, Itti L (2012) Salient object detection: a benchmark. In: ECCV
27. Cheng M, Mitra N, Huang X, Torr P, Hu S (2015) Global contrast based salient region detection. PAMI 37(3)
28. Hou X, Zhang L (2007) Saliency detection: a spectral residual approach. In: 2007 IEEE conference on computer vision and pattern recognition, CVPR'07. IEEE, pp 1–8

29. Jiang H, Wang J, Yuan Z, Wu Y, Zheng N, Li S (2013) Salient object detection: a discriminative regional feature integration approach. In: Proceedings of the IEEE conference on computer vision and pattern recognition, pp 2083–2090
30. Leordeanu M, Hebert M (2008) Smoothing-based optimization. In: CVPR
31. Zitnick CL, Dollár P (2014) Edge boxes: locating object proposals from edges. In: European conference on computer vision. Springer, pp 391–405
32. Leordeanu M, Radu A, Baluja S, Sukthankar R (2015) Labeling the features not the samples: efficient video classification with minimal supervision. arXiv:151200517
33. Prest A, Leistner C, Civera J, Schmid C, Ferrari V (2012) Learning object class detectors from weakly annotated video. In: CVPR
34. Tsai D, Flagg M, Nakazawa A, Rehg JM (2012) Motion coherent tracking using multi-label MRF optimization. Int J Comput Vis 100(2):190–202
35. Wang L, Hua G, Sukthankar R, Xue J, Niu Z, Zheng N (2016) Video object discovery and co-segmentation with extremely weak supervision. IEEE Trans Pattern Anal Mach Intell

Chapter 6
Coupling Appearance and Motion: Unsupervised Clustering for Object Segmentation Through Space and Time

6.1 Introduction

Discovering objects in videos without human supervision, as they move and change appearance in space and time, is one of the most interesting and still unsolved problems in artificial intelligence. The task could have a strong impact on the way we learn about objects and how we process large amounts of video data that is widely available at low cost. One of our main goals is to understand how much could be learned from a video about the main object of interest, without any human supervision. We are interested in how object properties relate in both space and time and how we could exploit these consistencies in order to discover the objects in a fast and accurate manner. In a more general sense, how could the unsupervised learning principles proposed in Chap. 1 be applied on the problem tackled here, in the current chapter.

While human segmentation annotations of a given video are not always consistent, people usually agree on which is the main object shown in a video sequence. There are many questions to answer here: which are the specific features that make a group of pixels stand out as a single, primary object in a given video shot? Is it the pattern of motion different from the surrounding background? Is it the contrast in appearance, the symmetry, or good form of an object that makes it distinguish itself from the background? Or is it a combination of such factors and cues and maybe others, remaining to be discovered? What we know is the fact that humans can easily spot and segment the main, foreground object and this fact has been recognized and studied since the beginning of the twentieth century, when the Gestalt School of psychology was established [2].

We make two core assumptions, which constitute the basis of our approach to the problem of foreground object discovery in video:

1. Pixels that belong to the same object are highly likely to be connected through long-range optical flow chains (Sect. 6.2.1).

The material presented in this chapter is based in large part on the following paper:
Haller, Emanuela, Adina Magda Florea, and Marius Leordeanu. "Spacetime Graph Optimization for Video Object Segmentation." arXiv preprint arXiv:1907.03326 (2019).

© Springer Nature Switzerland AG 2020
M. Leordeanu, *Unsupervised Learning in Space and Time*,
Advances in Computer Vision and Pattern Recognition,
https://doi.org/10.1007/978-3-030-42128-1_6

2. Pixels of the same object often have similar motion and distinctive appearance patterns in space and time. In other words, what looks alike and moves together is likely to belong together. These ideas are not new, but next we will show that they can have a precise and elegant mathematical formulation. We first define a specific Feature-Motion matrix (Sect. 6.3.2), which captures motion and appearance properties which make the primary object form a strong subgraphs (cluster). This strong cluster is efficiently found optimally by a spectral clustering solution, based on computing the leading eigenvector of the Feature-Motion matrix.

Key result presented in the chapter: We present a unique graph structure in space and time, with nodes at the dense pixel level such that each pixel is connected to other pixels through short and medium range optical flow chains, which attempt to model the trajectories of the same physical object point through space and time. The graph combines such motion patterns with appearance information in a single space-time graph defined by the Feature-Motion matrix (Sect. 6.3.2). We formulate segmentation as a spectral clustering problem (Sect. 6.2.2) and propose an efficient algorithm that computes the global optimum as the principal eigenvector of the Feature-Motion matrix (Sect. 6.2.3). One of our main implementation tricks is that the matrix is never built, explicitly, but the algorithm is guaranteed to converge to its principal eigenvector. The idea is validated experimentally, as we obtain top results on three challenging benchmarks in the literature.

6.1.1   Relation to Principles of Unsupervised Learning

The ideas and algorithms discussed next are related to several principles of unsupervised learning that were introduced in the first chapter. That is somewhat expected as most principles are related to each other and follow logically one from another. However, the idea which is most related to this chapter comes from Principle 4 (in Chap. 1), which states the following:

Principle 4
Objects form strong clusters of motion trajectories and appearance patterns in their space-time neighborhood.

That is precisely the main idea in this chapter: discover objects as clusters in space and time. These clusters are formed by the strong links based on appearance and motion patterns between space-time points, which make the objects stand out, as strong structures within their space-time neighborhoods.

It is intuitive that if we add a fourth, temporal dimension to space, then the physical parts of an object would appear tightly interconnected in the 3D (space) + 1D (time) world. These tight connections between the parts of an object, due to physical laws and forces, should also become apparent in the 2D (image space) + 1D (time) world of temporally aligned video frame sequences. That is because each frame is a projection of what is in the 3D world. The same natural laws of physics, acting in space and time, make the parts of objects stay connected, move together, and look similar in images as they are and behave in the world.

Please note the immediate relation between the ideas above to the first three principles: such similar appearance and motion patterns, which make an object stand out against the background (Principle 1), become strong HPP features (Principle 3), which could be used reliably to pick, with high precision, data samples that belong to the main object of interest (Principle 2). These ideas make intuitive sense. Could we capture them within a single beautiful mathematical formulation? Could we define in the precise and elegant language of mathematics an object as a cluster in space and time and then come up with efficient algorithm to find it? This goal is the precise focus of this chapter. During the following sections, we will move from these initial intuitive ideas to the desired mathematical formulations and algorithms. We will find objects in video as principal eigenvectors of a special **Feature-Motion Matrix**.

6.1.2 Scientific Context

An important aspect that differentiates between different methods for foreground object segmentation in video is the level of human annotation during training. That could vary from complete absence [3–7] to using models heavily pretrained using human annotations [8–19]. We propose an approach that has the ability to accommodate supervised and unsupervised cases. In this book, we are mainly focused on the unsupervised approach, for which no human supervision was used during training, but we also exploit the supervised scenario, as shown in Sect. 6.4.2.

In order to define the structure of the space-time graph at the dense pixel level, we use optical flow. There are several important works based on optical flow or graph representations for problems related to video object segmentation (VOS). One such method [20] introduces a motion clustering approach that simultaneously estimates optical flow and object segmentation masks. It operates on pairs of images and minimizes an objective energy function, which incorporates classical optical flow constrains and a descriptor matching term. However, it has no mechanism for selecting the main object. There are other similar algorithms that take the same direction [21]. Another approach [22] creates salient motion masks, which are next combined with "objectness" segmentations in order to generate the final mask. Other works [23, 24] introduce approaches based on averaging saliency masks over connections defined by optical flow. Another method [5] based on motion segmentation uses elements of the scene that form clusters of motion patterns. Again, there is no special focus on the main object. They introduce a minimum cost multicut formulation over a graph with nodes that are defined by motion trajectories. Unlike our approach, none of these methods use a graph representation in space and time with a node for every pixel in the video.

The space-time graph, at the dense pixel level, which we introduce, considers direct connections between all video pixels. That is in contrast to super-pixel-level nodes or trajectory nodes. Different from previous approaches [5, 20], which do motion segmentation, we discover the strongest cluster in space and time by taking into consideration long-range motion links as well as local feature patterns (computed from motion and appearance). Another element of novelty is that we give a spectral clustering formulation, for which a global, optimum solution could be found fast by power iteration—as the principal eigenvector of a very large Feature-Motion matrix (of size $n \times n$, with n being the total number of pixels in the video), which, being very sparse, is never explicitly computed. We thus ensure a dense pixel-level consistency in space and time at the level of the whole video sequence and that is very different from previously published methods.

6.2   Our Spectral Approach to Segmentation

From a sequence of m consecutive video frames, we aim to extract a set of m soft object masks, one per frame, which should highlight the foreground, primary object of the video sequence. We represent the whole video as a graph with one node per pixel and an edge structure defined by optical flow chains (Sect. 6.2.1). In Sect. 6.2.2, we formulate segmentation as a clustering in the space-time graph. Then, the mask pertaining to a given frame is nothing else but a 2D slice of the 3D space-time cluster at the corresponding time step. Then, by an efficient optimization algorithm (Sect. 6.2.3), we compute the optimal mask as cluster, without accessing the whole graph simultaneously. By the efficient implementation of edges as adjacency pointers we avoid building the full adjacency matrix.

6.2.1   Creating the Space-Time Graph

Graph of pixels in space and time: We define the space-time graph $G = (V, E)$: each node $i \in V$ corresponds to a pixel of the video sequence ($|V| = n; n = m \cdot h \cdot w$; m is the number of frames; and (h, w) is the frame size).

Chains along optical flow: We start from a given pixel and follow the optical flow displacement vectors from one frame to the next, both forward and backward (for K frames in each time directions). The flow vectors form a path, one per direction, which starts from a given pixel. We term these paths, *optical flow chains*. Multiple optical flow chains could pass through a given pixel, at minimum one moving forward and another moving backward. While for a certain incoming direction, a pixel could have none, one, or several incoming flow chains, it will always have one and only one outgoing chain. Then, the flow chains define the graph structure: there is an undirected edge between nodes i and j if they are connected by a chain in at least

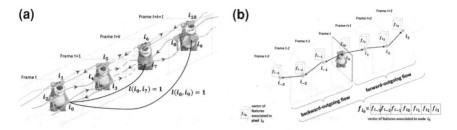

Fig. 6.1 Visual representation of how the space-time graph is created. **a** i_0 to i_{10} are graph nodes, corresponding to pixels from different frames. **Colored lines** are optical flow chains, formed by following the flow, forward or backward in time, sequentially from one frame to the next (or the previous one, respectively). Thus, two nodes are connected if there exists a chain of optical flow vectors that link them through intermediate frames. From a given node, there will always be two flows going out, one in each time direction (forward and backward). Note that there could be none or multiple chains coming in (e.g., node i_7). **Black lines** correspond to graph edges, present between nodes that are linked by at least one flow chain. **b** The flow chains, which are also used to create node features. For a given node i_0, the node features represented as a feature vector $\mathbf{f}_{i_0}$ are collected along the outgoing flow chains (forward and backward in time). They represent motion and appearance patterns along the chains

one direction. In Fig. 6.1a, we show how such long-range edges are formed through optical flow chains.

Graph adjacency matrix: $\mathbf{M} \in \mathbb{R}^{n \times n}$, with $\mathbf{M}_{i,j} = l(i, j) \cdot k(i, j)$, $k(i, j)$ is a Gaussian kernel, function of the temporal difference between nodes i and j and $l(i, j) = 1$ if there is an edge between nodes i and j and $l(i, j) = 0$ otherwise. Thus, $\mathbf{M}_{i,j} = k(i, j)$ if i and j are linked and zero otherwise. It follows that $\mathbf{M}$ is symmetric, semi-positive definite, has non-negative elements, and is expected to be very sparse. In fact, $\mathbf{M}$ is a Mercer kernel, as the pairwise terms $\mathbf{M}_{i,j}$ obey the Mercer's condition.

Appearance features at nodes: Each node i is described by node-level feature vectors $\mathbf{f}_i \in \mathbb{R}^{1 \times d}$, collected along the outgoing chains starting in i, one per outgoing temporal direction (Fig. 6.1b). Then we stack all features in the feature matrix $\mathbf{F} \in \mathbb{R}^{n \times d}$. In practice, we could consider any type feature, pretrained in a supervised manner or not (as presented in our experiments in Sect. 6.4.2).

Node segmentation labels: Each node i has an associated (soft) segmentation label $x_i \in [0, 1]$, which represents our confidence that the node is part of the object of interest. Thus, we could represent a segmentation solution, in both space and time, over the whole video, as a vector $\mathbf{x} \in \mathbb{R}^{n \times 1}$, with a certain label $\mathbf{x}_i$, for each i-th video pixel (graph node).

6.2.2 Segmentation as Spectral Clustering

We consider the primary object of interest to be the strongest cluster with respect to both space-time flow chain connections and appearance features. On one hand, object nodes that belong to the main object are strongly connected through long-range motions. On the other hand, which nodes belong to the objects should also be predictable from the object's appearance features. The (video segmentation) 3D cluster, in 2D image space and 1D time, is defined by the labels indicator vector $\mathbf{x}$, with $\mathbf{x}_i = 1$ if node i belongs to the cluster and $\mathbf{x}_i = 0$, otherwise.

Then we define the intra-cluster score to be $S_C = \mathbf{x}^T \mathbf{Mx}$, with the constraint that labels $\mathbf{x}$ live in the features subspace, as explained shortly. We relax $\mathbf{x}$ to be continuous and enforce that it is unit L2-norm, as only its relative node values really matter for segmentation. Solving segmentation then becomes a problem of maximizing the inter-cluster score $\mathbf{x}^T \mathbf{Mx}$, given the constraints $\|\mathbf{x}\|_2 = 1$ and $\mathbf{x}$ is a linear combination of the columns of feature matrix $\mathbf{F}$. Thus, node labels are a linear combination of the node features, which, in other words, means that the labels $\mathbf{x}$ could be predicted from the feature values by linear regression. The way the problem and the labels' constraints are set up directly implies that the object pixels form a strong motion cluster in both space and time, while also forming a strong cluster based on appearance features.

Next we see that solving the optimal segmentation $\mathbf{x}$ reduces to finding the principal cluster of a Feature-Motion matrix, which couples together the motion $\mathbf{M}$ with the features $\mathbf{F}$. Then, the constraint that $\mathbf{x}$ is a linear combination of the columns of $\mathbf{F}$ can be enforced by requiring that $\mathbf{x} = \mathbf{Px}$, where $\mathbf{P}$ is the feature projection matrix $\mathbf{P} = \mathbf{P^T} = \mathbf{F}(\mathbf{F}^T \mathbf{F})^{-1} \mathbf{F}^T$. Matrix $\mathbf{P}$ projects any vector onto the subspace spanned by the columns of the feature matrix $\mathbf{F}$.

We immediately obtain the following result.

Proposition 6.1 $\mathbf{x}^*$ *that maximizes* $\mathbf{x}^T \mathbf{Mx}$ *under constraints* $\mathbf{x} = \mathbf{Px}$ *and* $\|\mathbf{x}\|_2 = 1$, *also maximizes* $\mathbf{x}^T \mathbf{PMPx}$ *under constraint* $\|\mathbf{x}\|_2$.

Proof Since $\mathbf{x}^*$ maximizes $\mathbf{x}^T \mathbf{Mx}$ under constraints $\mathbf{x} = \mathbf{Px}$ and $\|\mathbf{x}\|_2 = 1$, it also maximizes $\mathbf{x}^T \mathbf{PMPx}$, under the same constraints, since $\mathbf{P} = \mathbf{P^T}$. Moreover, since $\mathbf{P}$ is a projection on the column space of $\mathbf{F}$, the optimum with fixed, unit norm $\mathbf{x}^*$ must live in that column space. Consequently, $\mathbf{x}^* = \mathbf{Px}^*$ and the result follows. ☐

Based on Proposition 6.1, we now define the final optimization problem as

$$\mathbf{x}^* = \arg \max_{\mathbf{x}} \mathbf{x}^T \mathbf{PMPx} \quad \text{s.t.} \quad \|\mathbf{x}\|_2 = 1. \tag{6.1}$$

Feature-Motion Matrix: Throughout the chapter we refer to $\mathbf{A} = \mathbf{PMP}$ as the Feature-Motion matrix, which couples the motion through space-time of an object cluster and its projection $\mathbf{P}$ on the column space of its features $\mathbf{F}$. This matrix is the central element of the segmentation problem we aim to solve (Problem 6.1).

6.2.3 Optimization by Power Iteration Method

Problem (Eq. 6.1) is solved optimally by the principal eigenvector $\mathbf{x}^*$ of $\mathbf{PMP}$. That is a case of classical spectral clustering [25], also directly related to our spectral approach to graph matching [26] Chap. 2. Note that $\mathbf{x}^*$ must have non-negative values by Perron-Frobenius theorem, as $\mathbf{PMP}$ has non-negative elements.

The efficient algorithm we propose in (Algorithm 6.1), converging to $\mathbf{x}^*$ by power iteration, cleverly exploits the space-time video graph, without explicitly computing $\mathbf{PMP}$. The steps of GO-VOS alternate between propagating the labels $\mathbf{x}$ according to their motion links (defined by optical flow chains in $\mathbf{M}$) and projecting $\mathbf{x}$ through $\mathbf{P}$ onto the feature columns' space of $\mathbf{F}$, while also normalizing $\|\mathbf{x}\|_2 = 1$.

Algorithm 6.1 GO-VOS: object discovery in video by power iteration on the Feature-Motion Matrix

The algorithm steps are presented using their equivalent mathematical formulations, but we highlight that our implementation avoids explicitly computing the adjacency matrix $\mathbf{M}$. $\mathbf{x}^{(it)}$ are node labels from iteration it

1: **Propagation:** $\mathbf{x}^{(it+1)} \leftarrow \mathbf{M}\mathbf{x}^{(it)}$
2: **Projection:** $\mathbf{x}^{(it+1)} \leftarrow \mathbf{P}\mathbf{x}^{(it+1)}$
3: **Normalization:** $\mathbf{x}^{(it+1)} \leftarrow \mathbf{x}^{(it+1)}/\|\mathbf{x}^{(it+1)}\|_2$

Propagation: It can be expressed for a node i as $\mathbf{x}_i^{(it+1)} = \sum_j \mathbf{M}_{i,j}\mathbf{x}_j^{(it)}$, s.t. the label of one node is an update function of the other nodes' labels. Thus, each node i transmits its current label, weighted by pairwise terms in $\mathbf{M}$, to all its neighbors in the graph. This is efficiently implemented as a propagation step, where starting from a node i, we move forward and backward on the time axis, along the optical flow chains, and cast node i's information, as a weighted vote according to $\mathbf{M}_{i,j}$, to all nodes j met along the chain: $\mathbf{x}_j \leftarrow \mathbf{x}_j + \mathbf{M}_{i,j}\mathbf{x}_i$. Since all our edges are undirected, the segmentation value of i will also change accordingly: $\mathbf{x}_i \leftarrow \mathbf{x}_i + \mathbf{M}_{i,j}\mathbf{x}_j$. When doing a propagation step, we propagate jointly information from all the nodes in one frame to all neighboring frames, backward and forward in time. As $\mathbf{M}_{i,j} = k(i, j)$ decreases rapidly towards zero with the distance in time between nodes (i, j), we only cast weighted votes between frames that are within a radius of K time steps.

Projection: It forces the solution $\mathbf{x}$ to be in the column space of feature matrix $\mathbf{F}$. At each iteration, we estimate the optimal set of weights, in the least squares sense, $\mathbf{w}^* = (\mathbf{F}^T\mathbf{F})^{(-1)}\mathbf{F}^T\mathbf{x}^{(it+1)}$, which best approximates current segmentation values $\mathbf{x}^{(it+1)}$, resulted from the previous propagation step at the current iteration. In essence, $\mathbf{w}^*$ becomes a linear regression predictor of video nodes' segmentation labels based on nodes' feature values. Then we set $\mathbf{x}^{(it+1)} \leftarrow \mathbf{F}\mathbf{w}^* = \mathbf{P}\mathbf{x}^{(it+1)}$, which does the required projection on the columns' feature space.

Normalization: It enforces the L2-norm constraint on the segmentation solution **x**. Only relative segmentation values of labels are important in practice for the final, binary segmentation problem. Therefore, as long as we do not change the direction of **x** we can impose the unit normalization.

6.3   Theoretical Properties

Next (Sect. 6.3.1) we show that our algorithm is guaranteed to converge to the principal eigenvector of the Feature-Motion matrix $\mathbf{A} = \mathbf{PMP}$ and that this solution is also the global optimum of our optimization problem defined in Eq. 6.1. Afterwards (Sect. 6.3.2), we discuss the core properties of the Feature-Motion matrix **A**.

6.3.1   Convergence Analysis

We have just introduced an iterative algorithm (Algorithm 6.1), which efficiently solves the clustering problem (Eq. 6.1). That problem effectively defines the segmentation tasks over space and time. Our method avoids computing **M** and $\mathbf{A} = \mathbf{PMP}$ explicitly. It solves the problem fast and with minimum computation cost. We theoretically prove that GO-VOS algorithm is guaranteed to converge to the leading eigenvector of the Feature-Motion matrix, which is the global optimum of the relaxed clustering problem with unit norm constraint.

Proposition 6.2 *Algorithm 6.1 converges to the principal eigenvector of Feature-Motion Matrix* $\mathbf{A} = \mathbf{PMP}$, *which is the global optimum of the optimization problem in Eq. 6.1.*

Proof The three main steps of our algorithm can be summarized into a single updated step as $\mathbf{x}^{(it+1)} = \frac{\mathbf{PMx}^{(it)}}{\|\mathbf{PMx}^{(it)}\|_2}$. This is the recurrence relation that describes the power iteration method, from which it follows that our method converges to the optimal solution $\mathbf{x}^*$, which is the leading eigenvector of matrix **PM**. It follows that $\mathbf{x}^*$ maximizes the Rayleigh quotient $R(\mathbf{PM}, \mathbf{x}) = \frac{\mathbf{x}^\mathsf{T}\mathbf{PMx}}{\mathbf{x}^\mathsf{T}\mathbf{x}}$. According to the projection and normalization steps of Algorithm 1.1, $\|\mathbf{x}^*\|_2 = 1$ and $\mathbf{x}^*$ is already projected into the feature space, so $\mathbf{x}^* = \mathbf{Px}^*$. It immediately follows that $\mathbf{x}^*$ also maximizes our objective $\mathbf{x}^\mathsf{T}\mathbf{PMPx}$, under constraint $\|\mathbf{x}\|_2 = 1$. Thus, the convergence point of our algorithm is also the global optimum of the defined optimization problem (Eq. 6.1), the leading eigenvector of the Feature-Motion matrix $\mathbf{A} = \mathbf{PMP}$. $\qquad\square$

We provided theoretical guarantees that the GO-VOS algorithm always converges to the global optimum of our clustering problem Eq. 6.1. Our algorithm will always find the foreground object in a video sequence as the strongest cluster in the defined space-time graph. Intuitively, that 3D space-time cluster should correspond to the most noticeable, distinct object, different from the surrounding background and strongly interconnected in terms of motion and appearance patterns.

Video-level projection versus frame-level projection: In practice, it is simpler and even more accurate to project the segmentation of each frame t ($\mathbf{x}_t \in \mathbb{R}^{hw \times 1}$) on its own frame-level feature space, defined by feature matrix $\mathbf{F}_t \in \mathbb{R}^{hw \times d}$ limited to nodes from the frame at time step t. This per-frame projection would give a richer representation power especially in the unsupervised scenario, where we deal with low level, less powerful features. In this case, the feature vector $\mathbf{f}_i$, associated to node i, is still formed by collecting features along the outgoing optical flow chains, which exploit the space-time consistency. It is relatively easy to show that, when considering frame-level projection, Algorithm 6.1 is still guaranteed to converge, but instead to the principal eigenvector of a slightly modified Feature-Motion matrix $\mathbf{A}^f = \mathbf{P}^f \mathbf{M} \mathbf{P}^f$, where $\mathbf{P}^f$ is the projection matrix corresponding to a block diagonal feature matrix $\mathbf{F}^f$ of the form: $\mathbf{F}^f = \begin{pmatrix} \mathbf{F}_0 & \mathbf{0} & \dots & \mathbf{0} \\ \mathbf{0} & \mathbf{F}_1 & \dots & \mathbf{0} \\ \vdots & \vdots & \ddots & \vdots \\ \mathbf{0} & \mathbf{0} & \dots & \mathbf{F}_m \end{pmatrix} \in \mathbb{R}^{n \times dm}$.

6.3.2 Feature-Motion Matrix

The Feature-Motion matrix $\mathbf{A} = \mathbf{PMP}$ is one of the key elements of our approach, defined by the adjacency matrix $\mathbf{M}$ and the feature projection matrix $\mathbf{P} = \mathbf{F}(\mathbf{F}^T \mathbf{F})^{-1} \mathbf{F}^T$. We have just provided theoretical guarantees that the algorithm converges to the leading eigenvector of this specific matrix $\mathbf{A} = \mathbf{PMP}$. At convergence, we could think of the final segmentation as an equilibrium state between its motion structure in space-time and its consistency with the features.

We formulate segmentation as a spectral clustering problem based on the assumption that pixels belonging to the primary foreground object form a strong cluster in the defined space-time graph. In order to validate experimentally the existence of a strong cluster in our space-time graph, we analyzed the Feature-Motion matrix $\mathbf{PMP}$, considering only unsupervised features (Sect. 6.4.2). The analysis is performed by considering only five consecutive frames, in order to keep everything in manageable dimensions.

In Fig. 6.2a, we represent the Feature-Motion matrix after reordering its rows and columns according to the decreasing order of the corresponding eigenvector elements. As it can be observed, there is a strong cluster in the space-time graph, expected to correspond to the primary object of the video sequence. In Fig. 6.2b, we present the first six eigenvalues of the Feature-Motion matrix, highlighting the large spectral gap between the first eigenvalue and the rest, which is an indication of the stability of the spectral solution. The larger the eigengap the more representative the main cluster is and the closer to the correct solution we could expect the principal

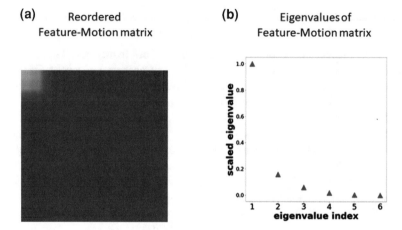

Fig. 6.2 **a** Feature-Motion matrix **PMP** reordered according to the leading eigenvector. **b** First six eigenvalues of Feature-Motion matrix **PMP**, normalized in range [0, 1]. To keep a manageable size, everything is computed for only five consecutive frames of a video sequence ("car-roundabout"— DAVIS dataset)

eigenvector to be. For this particular example, the spectral eigengap is 1521.1 with a spectral radius of 1811.45.

Based on perturbation theory [27, 28] this fact indicates that our leading eigenvector is stable and small perturbations will not affect its direction. Assuming that the Feature-Motion matrix $\mathbf{A}$ is a slightly perturbed version of the ideal matrix $\mathbf{A}^* = \mathbf{A} - \mathbf{E}$ (where $\mathbf{E}$ is the noise matrix, containing a small noise per element), the upper bound for any alterations of the leading eigenvector is [29]

$$\varepsilon \approx 8\frac{\parallel \mathbf{E} \parallel_F}{\parallel \mathbf{A}^* \parallel_F}, \tag{6.2}$$

where $\parallel \mathbf{E} \parallel_F$ is the Frobenius norm of the perturbation matrix $\mathbf{E}$. Equation 6.2 quantifies the stability of the eigenvector as a function of the quantity of perturbation. Thus, for a relatively small perturbation, our solution is stable and the space-time graph contains a strong cluster, well reflected on the principal eigenvector and expected to correspond, in practice, to the primary object of the video sequence.

6.4 Experimental Analysis

Next we do an in-depth analysis of our algorithm, from an experimental point of view (Sects. 6.4.1–6.4.3), addressing different aspects, such as the initialization point of the segmentation solution, the set of node features used, and the influence of the optical flow. In Sect. 6.4.4, we analyze the complexity of our algorithm, while in Sect. 6.4.5,

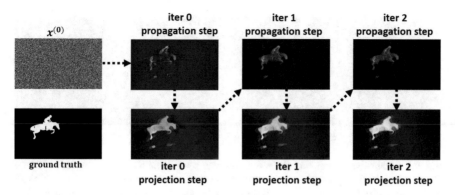

Fig. 6.3 Qualitative results of our method, over three iterations, when initialized with a random soft segmentation mask and using only unsupervised features (i.e., color a motion along flow chains—Sect. 6.4.2). Note that the main object emerges with each iteration of our algorithm

we compare the proposed algorithm against other state-of-the-art solutions for the problem of primary video object segmentation, on three challenging datasets.

6.4.1 The Role of Segmentation Initialization

In the iterative process defined in Algorithm 6.1, we need to establish the initial segmentation labels $\mathbf{x}^{(0)}$ associated to graph nodes. As proved in Sect. 6.3.1, the algorithm is equivalent to the power iteration method and computes the leading eigenvector of the Feature-Motion matrix $\mathbf{A}$. Thus, irrespective of the initial values of node labels, the method should always converge to the same solution, completely defined by the Feature-Motion matrix. In Fig. 6.3, we present an example, showing the evolution of the soft segmentation masks (namely, the principal eigenvector of the Feature-Motion matrix) over three iterations of our algorithm, when we start from a random mask. We observe that the main object of interest emerges from this initial random mask, as its soft segmentation mask is visibly improved after each iteration.

In Figs. 6.7 and 6.4, we present more examples regarding the evolution of our soft segmentation masks over several iterations.

In practice, we also observed that the method approaches the same point of convergence, regardless of the initialization—thus confirming the theory. We have considered different choices for $\mathbf{x}^{(0)}$, ranging from uninformative masks such as isotropic Gaussian soft mask placed in the center of each frame with varied standard deviations, a randomly initialized mask or a uniform full white mask, to masks given by state-of-the-art methods, such as ELM [3] and PDB [17]. In Fig. 6.5, we present quantitative results, which confirm the theoretical results. The performance evolves in terms of Jaccard index—J Mean, over 11 iterations of our algorithm, towards the

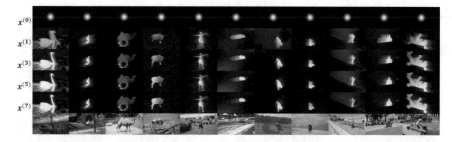

Fig. 6.4 Qualitative evolution of the soft masks over seven iterations of the proposed GO-VOS algorithm, in the fully unsupervised scenario: the segmentation is initialized (first row—$\mathbf{x}^{(0)}$) with a non-informative Gaussian, while the features in $\mathbf{F}$ are only motion directions along optical flow chains. Note how the object of interest emerges in very few iterations, even though no supervised information or features are used in the process (images from DAVIS 2016)

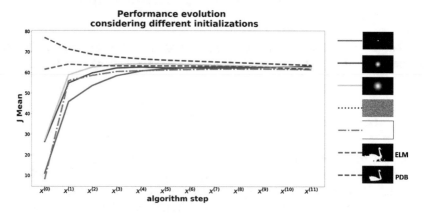

Fig. 6.5 Performance evolution of our method, over several iterations, considering different $\mathbf{x}^{(0)}$, but using the same unsupervised features (color and motion along flow chains—Sect. 6.4.2). The legend presents, for each experiment, a sample of the considered initialization: soft segmentation central Gaussians, with diverse standard deviations, random initial soft mask, uniform white mask, and two supervised state-of-the-art solutions PDB [17] and ELM [3]. Note that, regardless of the initialization, the final metric converges towards the same value, as theoretically proved in Sect. 6.3.1. Tests are performed on full DAVIS validation set

same common segmentation. Note that, as expected, convergence is faster for cases when we start closer to the convergence point.

To conclude, irrespective of the initialization, GO-VOS converges towards the same, unique solution, $\mathbf{x}^*$, that depends only on the Feature-Motion matrix $\mathbf{A}$. We have now proved this fact, both theoretically and experimentally.

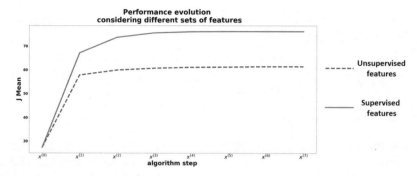

Fig. 6.6 Performance evolution over seven iterations, considering different sets of features. Tests are performed on full DAVIS validation set

6.4.2 The Role of Node Features

Most of our experiments are performed using **unsupervised features**, generated without requiring human-level annotations, namely, color and motion information. For color features we collect RGB pixel values, while for motion we consider optical flow displacements between consecutive frames. All features are collected along the outgoing flow chains, as discussed previously and illustrated in Fig. 6.1b. Our solution proves its effectiveness, having state-of-the-art results in the unsupervised scenario (Tables 6.2, 6.3, and 6.4).

We also considered adding **supervised features** to the set of unsupervised features, collected along the same outgoing flow chains. We tested two scenarios: **(i)** having a backbone trained for primary object segmentation in static images: U-Net [30] like architecture, with a ResNet34 [31] encoder; the encoder is pretrained on ImageNet [32], while the full network is trained on the training set of DAVIS [33] **(ii)** having a backbone trained for semantic segmentation: DeepLabv3 [34] model, with ResNet101, pretrained on COCO dataset [35]. As expected, the inclusion of stronger features significantly improves performance (Tables 6.2 and 6.3) and is a strong property of our algorithm that it has state-of-the-art results in the unsupervised scenario, while being able to incorporate informative high-level features and improve its performance accordingly. We emphasize that backbone information is only used for building the feature matrix **F**, while our algorithm is still initialized with non-informative masks.

In Fig. 6.6, we present the performance evolution (on average over the DAVIS validation video dataset) over seven iterations of the segmentation, depending on two types of features: unsupervised versus supervised. Note the strength of the node features used in **F** is an important factor that influences the quality of the final segmentation.

The GO-VOS algorithm can be used on top of other video object segmentation methods, considering their final masks as feature maps and concatenate them to our unsupervised features. These features will also be collected along the outgoing

Table 6.1   Quantitative results of our method with two different optical flow solutions. Experiments are on validation set of DAVIS 2016

Method	J Mean
GO-VOS with EpicFlow	61.0
GO-VOS with FlowNet2.0	65.0

optical flow chains, as described in Fig. 6.1b. Our solution incorporates those features, changing the structure of the Feature-Motion matrix, where the strongest cluster emerges. As the masks predicted by the considered methods are expected to be closer to the global optimum, we also use them to initialize our labels $\mathbf{x}^{(0)}$, to speed up convergence. We tested this scenario, applying GO-VOS over several state-of-the-art methods and proved experimentally that in only two iterations our solution is able to significantly improve the results (Table 6.2).

6.4.3   The Role of Optical Flow

The Feature-Motion matrix is strongly influenced by the graph structure (through matrix $\mathbf{M}$). While in the experiments conducted so far we have only considered altering the initialization and the node features, now we study the impact of the chosen optical flow solution. The adjacency matrix $\mathbf{M}$ is constructed using the optical flow provided by FlowNet2.0 [36], which is pretrained on synthetic data (FlyingThings3D—[37] and FlyingChairs—[38]), requiring no human annotations. Here, we consider replacing it with a more classical approach, such as EpicFlow [39], which is not based on deep learning. In Table 6.1, we compare the results of our method when considering the two optical flow solutions. Since EpicFlow is less accurate than FlowNet2.0, its effect, as expected is to lightly degrade the final segmentation. The fact is reflected by the evaluation of the algorithm with the two different optical flow modules. However, in both cases, we obtain competitive results.

6.4.4   Complexity Analysis and Computational Cost

Considering our formulation, as a graph with a node per each video pixel and long-range connections, one would expect it to be memory expensive and slow. However, as we have an efficient implementation of our algorithm steps, without explicitly constructing the adjacency matrix, computational complexity is only $O(n)$, where n is the number of video pixels.

In particular, it takes 0.04 s/frame for computing the optical flow and 0.17 s/frame for computing information related to matrices $\mathbf{M}$ and $\mathbf{F}$. Further, the power iteration method requires 0.4 s/frame, resulting in a total of 0.61 s/frame for the full algorithm. GO-VOS is implemented in PyTorch and the runtime analysis is performed on a computer with the following specifications: Intel(R) Xeon(R) CPU E5-2697A v4 @ 2.60 GHz, GPU GeForce GTX 1080.

We further provide a more detailed complexity analysis of the algorithm.

Complexity of the propagation step: The first step of Algorithm 1.1, $\mathbf{x}^{(it+1)} = \mathbf{Mx}^{(it)}$, could be expressed for a node i as $x_i \leftarrow \sum_j \mathbf{M}_{ij} x_j$. We implement the first step efficiently by propagating the soft labels x_i, weighted by the pairwise terms $\mathbf{M}_{ij}$, to all the nodes from the other frames to which i is connected in the graph. We consider only a window of k frames around the frame associated to node i, as $\mathbf{M}_{ij}$ decreases rapidly towards zero with the distance in time between i and j. Thus, the complexity of the propagation step is $O(nk)$ (with $k = 11$ in our experiments).

Complexity of the projection step: Step 3 of the algorithm (Algorithm 1.1) computes $\mathbf{x}^{(it+1)} = \mathbf{F}(\mathbf{F}^T\mathbf{F})^{-1}\mathbf{F}^T\mathbf{x}^{(it+1)}$. The complexity of computing $\mathbf{F}^T\mathbf{F}$ is $O(d^2n)$, the complexity of computing $\mathbf{F}^T\mathbf{x}^{(it+1)}$ is $O(dn)$, the complexity of computing the inverse $(\mathbf{F}^T\mathbf{F})^{-1}$ is $O(d^3)$, and the complexity of $\mathbf{F}y$ where $y \in \mathbb{R}^{n \times 1}$ is $O(dn)$. As n is the number of video pixels, $n \gg d$ and $O(d^2n)$ asymptotically dominates $O(dn)$ and $O(d^3)$. In consequence, the overall complexity of Step 3 is $O(d^2n)$.

We emphasize that in all our unsupervised experiments $d \leq 56$ and it takes only 0.009 s/frame to compute the inverse $(\mathbf{F}^T\mathbf{F})^{-1}$. In the supervised scenario, we have ≈ 100 features, which adds an insignificant computation time. It would require only 0.1 s/frame in case we raise d to consider thousands of features (e.g., 2048). Another important aspect is that $\mathbf{P} = \mathbf{F}(\mathbf{F}^T\mathbf{F})^{-1}\mathbf{F}^T$ should be computed only once, and further used in each iteration of GO-VOS.

6.4.5 Results

We compare our proposed approach, GO-VOS, against state-of-the-art solutions for video object segmentation, on three challenging datasets DAVIS 2016, SegTrackv2, and YouTube-Objects. We present some qualitative results in Fig. 6.7, in comparison to other methods on the DAVIS 2016 dataset.

6.4.5.1 Tests on DAVIS 2016 Dataset

DAVIS 2016 dataset [33] is composed of 50 videos, each accompanied by accurate, per pixel annotations of the main object of interest. For this version of the dataset, the main object can be composed of multiple strongly connected objects, considered as a single object. DAVIS is a challenging dataset as it contains many difficult cases such as appearance changes, occlusions, and motion blur.

Metric: We use J Mean (intersection over union) and F Mean (boundary F-Measure), along with their average (Avg).

Fig. 6.7 Qualitative results of the proposed GO-VOS algorithm, in the fully unsupervised case. We also present results of four other approaches: PDB [17], ARP [40], ELM [3], and FST [4]

Results: In Table 6.2, we compare our algorithm against supervised and unsupervised methods on the problem of unsupervised single object segmentation on DAVIS validation set (20 videos). We considered both unsupervised and supervised formulations of GO-VOS, with non-informative initial segmentation, along with the refinement scenario described in Sect. 6.4.2. The unsupervised version of GO-VOS outperforms the unsupervised methods while being much faster (GO-VOS 0.61 s/frame, ELM 20 s/frame, FST 4 s/frame, CUT 1.7 s/frame, and NLC 12 s/frame). When using supervised features the performance of GO-VOS improves with the strength of the considered node appearance features. In the refinement scenario, GO-VOS brings a substantial improvement to existing VOS solutions with up to 5% improvements in J Mean.

6.4.5.2   Tests on SegTrackv2 Dataset

SegTrackv2 [46] contains 14 videos with pixel-level annotations for the main object of interest, which could be composed of multiple strongly connected objects. SegTrack contains deformable and dynamic objects, with videos of a relative poor resolution—making it a very challenging dataset for video object segmentation. In Fig. 6.8, we present some visual example results, which are representative for Seg-Track dataset. While this dataset is relatively small, it contains very difficult cases, which explains the relatively low performance on SegTrackv2 of top current methods.

Metric: For evaluation, we used the average intersection over union score.

Results: In Table 6.3, we present quantitative results of our method, in both supervised and unsupervised scenarios (Sect. 6.4.2), with non-informative initialization. In the unsupervised scenario, our solution is surpassed only by NLC, which was initially published with results on SegTrack. Note that we significantly outperform NLC on DAVIS dataset. As expected, the addition of supervised features brings a performance boost. We observe that for this particular dataset, the DeepLabv3 backbone performs poorly by itself and is unable to identify objects such as worm or a birdfall. Still, GO-VOS efficiently exploits the space-time graph, its unsupervised

Table 6.2 Quantitative results of our method, compared with state-of-the-art solutions on DAVIS 2016 dataset. For each method, we report whether it makes use of features pretrained for video object segmentation, requiring human annotation (supervised). (black bold—best supervised; blue bold—best unsupervised)

Method	Supervised	J Mean	F Mean	Avg
COSNet [41]	✓	80.5	79.5	80.0
AGS [42]	✓	79.7	77.4	78.6
MotAdapt [43]	✓	77.2	77.4	77.3
LSMO [44]	✓	78.2	75.9	77.1
PDB [17]	✓	77.2	74.5	75.9
ARP [40]	✓	76.2	70.6	73.4
LVO [18]	✓	75.9	72.1	74.0
FSEG [19]	✓	70.7	65.3	68.0
LMP [45]	✓	70.0	65.9	68.0
U-Net backbone	✓	64.1	59.2	61.7
DeepLabv3 backbone	✓	65.9	65.6	65.8
GO-VOS with U-Net features	✓	70.8 (+6.7)	66.3 (+7.1)	68.6 (+6.9)
GO-VOS with DeepLabv3 features	✓	74.7 (+8.8)	71.1 (+5.5)	73.0 (+7.2)
GO-VOS with COSNet features	✓	**82.4** (+1.9)	80.9 (+1.4)	81.7 (+1.7)
GO-VOS with AGS features	✓	81.9 (+2.2)	79.3 (+1.9)	80.6 (+2.0)
GO-VOS with MotAdapt features	✓	82.5 (+5.3)	**81.0** (+3.6)	**81.8** (+4.5)
GO-VOS with LSMO features	✓	79.5 (+1.3)	76.6 (+0.7)	78.1 (+1.0)
GO-VOS with PDB features	✓	79.9 (+2.7)	78.1 (+3.6)	79.0 (+3.1)
GO-VOS with ARP features	✓	78.7 (+2.5)	73.1 (+2.5)	76.0 (+2.6)
GO-VOS with LVO features	✓	77.0 (+1.1)	73.7 (+1.6)	75.4 (+1.4)
GO-VOS with FSEG features	✓	74.1 (+3.4)	69.9 (+4.6)	72.0 (+4.0)
GO-VOS with LMP features	✓	73.7 (+3.7)	69.2 (+3.3)	71.5 (+3.5)
ELM [3]	X	61.8	**61.2**	61.5
FST [4]	X	55.8	51.1	53.5
CUT [5]	X	55.2	55.2	55.2
NLC [6]	X	55.1	52.3	53.7
GO-VOS	X	**65.0**	61.1	**63.1**

features, and any other useful information that may be extracted from the DeepLabv3 backbone, achieving a significant performance gain of 19%.

6.4.5.3 Tests on YouTube-Objects Dataset

YouTube-Objects (YTO) [49] is a very challenging dataset, containing 2511 video shots ($\approx$ 720k frames). Annotations are provided as object bounding boxes. Although

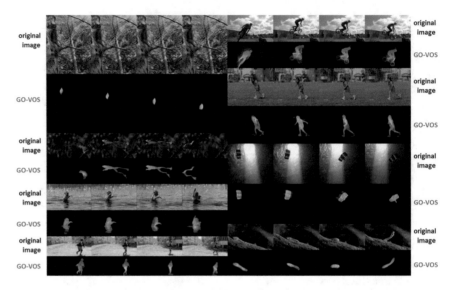

Fig. 6.8 Qualitative results on SegTrack dataset

Table 6.3 Quantitative results of our method, compared to state-of-the-art solutions on SegTrackv2 dataset. For each method, we report whether it makes use of features pretrained for video object segmentation, requiring human annotation (supervised). (black bold—best supervised; blue bold—best unsupervised)

Method	Supervised	IoU
KEY [47]	✓	57.3
FSEG [19]	✓	61.4
LVO [18]	✓	57.3
Li et al. [48]	✓	59.3
U-Net backbone	✓	67.1
DeepLabv3 backbone	✓	48.4
GO-VOS with U-Net features	✓	**70.2** (+3.1)
GO-VOS with DeepLabv3 features	✓	67.5 (+19.1)
NLC [6]	X	**67.2**
FST [4]	X	54.3
CUT [5]	X	47.8
HPP [7]	X	50.1
GO-VOS	X	62.2

Table 6.4 Quantitative results of our method (CorLoc metric), compared with state-of-the-art solutions on YTO v1.0 and v2.2. (blue bold—best solution for v2.2, black bold—best solution for v1.0)

Method	Aero	Bird	Boat	Car	Cat	Cow	Dog	Horse	Moto	Train	Avg	Ver
Croitoru et al. [50]	75.7	56.0	52.7	57.3	46.9	**57.0**	48.9	44.0	27.2	56.2	52.2	v2.2
Haller and Leordeanu [7]	76.3	68.5	**54.5**	50.4	**59.8**	42.4	53.5	30.0	**53.5**	**60.7**	54.9	v2.2
GO-VOS	**79.8**	**73.5**	38.9	**69.6**	54.9	53.6	**56.6**	**45.6**	52.2	56.2	**58.1**	v2.2
Prest et al. [49]	51.7	17.5	34.4	34.7	22.3	17.9	13.5	26.7	41.2	25.0	28.5	v1.0
Papazoglou and Ferrari [4]	65.4	67.3	38.9	65.2	46.3	40.2	65.3	48.4	39.0	25.0	50.1	v1.0
Zhang et al. [51]	75.8	60.8	43.7	71.1	46.5	54.6	55.5	54.9	42.4	35.8	54.1	v1.0
Jun Koh et al. [52]	64.3	63.2	73.3	68.9	44.4	62.5	71.4	52.3	78.6	23.1	60.2	v1.0
Haller and Leordeanu [7]	76.3	71.4	65.0	58.9	68.0	55.9	70.6	33.3	69.7	42.4	61.1	v1.0
Croitoru et al. [50]	77.0	67.5	**77.2**	68.4	54.5	**68.3**	72.0	**56.7**	44.1	34.9	62.1	v1.0
GO-VOS	**88.2**	**82.5**	62.7	**76.7**	**70.9**	50.0	**81.9**	51.8	**86.2**	**55.8**	**70.7**	v1.0

there are no pixel-level segmentation annotations, YTO tests are relevant considering the large number of videos and the wide diversity of testing cases covered. Following the methodology of published works, we test our solution on the train set, which contains videos with only one annotated object.

Metric: We used the CorLoc metric, computing the percentage of correctly localized object bounding boxes, according to PASCAL-criterion (IoU ≥ 0.5).

Results: In Table 6.4, we present the results on both versions of the dataset: v1.0 and v2.2, considering only our unsupervised formulation. For v1.0, we obtain top average score, while outperforming the other methods on 7 out of 10 object classes. For version v2.2 of the dataset, we also obtain the best average score and outperform other solutions on 5 out of 10 object classes. In Fig. 6.9, we show some representative visual results of GO-VOS on various YouTube-Objects videos.

6.5   Concluding Remarks

We present a novel graph representation at the dense pixel level for the problem of foreground object segmentation. We provide a spectral clustering formulation, in which the optimal solution is computed fast, by power iteration, as the principal eigenvector of a novel Feature-Motion matrix. The matrix couples local information at the level of pixels as well as long-range connections between pixels through optical flow chains. In this view, objects become strong, principal clusters of motion

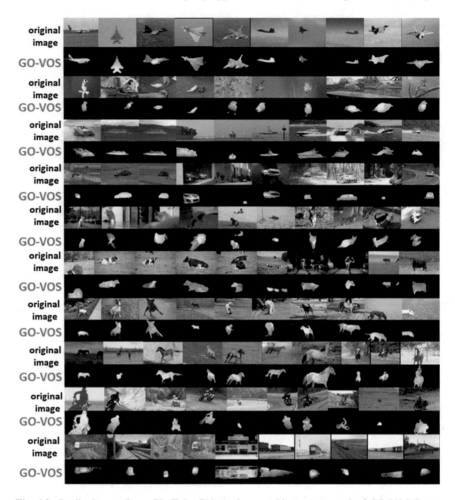

Fig. 6.9 Qualitative results on YouTube-Objects dataset with our unsupervised GO-VOS formulation

and appearance patterns in their immediate space-time neighborhood. Thus, the two "forces" in space and time, expressed through motion and appearance, are brought together into a single power iteration formulation that reaches the global optimum in a few steps. In extensive experiments, we show that the proposed algorithm, GO-VOS, is fast and obtains state-of-the-art results on three challenging benchmarks used in current literature.

Our method follows the same unsupervised learning principles that have been a main theme in this book. The agreements in the pattern of motions of different pixels as well as their agreements at the level of appearance provide the HPP signal that can trigger the unsupervised discovery process. These features in space and time are, in fact, much stronger than the simple color co-occurrences or patch-based classifiers that are used in Chap. 5. They are powerful enough, as demonstrated in our experiments, to often completely describe the object with both high precision and recall. However, the appearance features used as well as the motion ones are still not as strong as they could be. For example, in terms of features, we could incorporate in the motion matrix pairwise relationships between different nodes based on higher level, even pretrained deep features. That would make the cluster much stronger, through stronger and more discriminative connections in **M**. Such high-level features could also be incorporated in **F**, which would further improve the quality of the clustering. Another aspect is that of single versus multiple objects: the main eigenvector is always the same regardless of initialization, so the solution returned will be always showing the strongest cluster. How could we then discover multiple objects? One simple idea would be to remove the main object once it is found and repeat the process. However, that kind of greedy approach does not work that well in computer vision practice. We would rather design an approach that finds local strong clustering and that should be an important subject for future research. Another important property of our approach, which is not immediately visible, is that appearance and motion provide supervisory signals to each other. Motion by propagating the object information through motion flows constrains the prediction based on features only, while that prediction becomes the new object information to be propagated by motion at the next iteration. Motion and appearance take turns in being teacher and student to each other and that immediately relates our method to the approach presented in Chap. 7 and the more general idea of a visual story model presented in the last chapter of the book.

References

1. Nicolicioiu A, Duta I, Leordeanu M (2019) Recurrent spacetime graph neural networks. In: Neural information processing systems
2. Koffka K (2013) Principles of Gestalt psychology. Routledge
3. Lao D, Sundaramoorthi G (2018) Extending layered models to 3d motion. In: Proceedings of the European conference on computer vision (ECCV), pp 435–451
4. Papazoglou A, Ferrari V (2013) Fast object segmentation in unconstrained video. In: Proceedings of the IEEE international conference on computer vision, pp 1777–1784
5. Keuper M, Andres B, Brox T (2015) Motion trajectory segmentation via minimum cost multi-cuts. In: Proceedings of the IEEE international conference on computer vision, pp 3271–3279
6. Faktor A, Irani M (2014) Video segmentation by non-local consensus voting. In: BMVC, vol 2, p 8
7. Haller E, Leordeanu M (2017) Unsupervised object segmentation in video by efficient selection of highly probable positive features. In: Proceedings of the IEEE international conference on computer vision, pp 5085–5093
8. Luiten J, Voigtlaender P, Leibe B (2018) PReMVOS: proposal-generation, refinement and merging for the DAVIS challenge on video object segmentation 2018. In: The 2018 DAVIS challenge on video object segmentation-CVPR workshops
9. Maninis KK, Caelles S, Chen Y, Pont-Tuset J, Leal-Taixé L, Cremers D, Van Gool L (2017) Video object segmentation without temporal information. arXiv:170906031

10. Voigtlaender P, Leibe B (2017) Online adaptation of convolutional neural networks for the 2017 DAVIS challenge on video object segmentation. In: The 2017 DAVIS challenge on video object segmentation-CVPR workshops, vol 5
11. Bao L, Wu B, Liu W (2018) CNN in MRF: video object segmentation via inference in a CNN-based higher-order spatio-temporal MRF. In: Proceedings of the IEEE conference on computer vision and pattern recognition, pp 5977–5986
12. Wug Oh S, Lee J Y, Sunkavalli K, Joo Kim S (2018) Fast video object segmentation by reference-guided mask propagation. In: Proceedings of the IEEE conference on computer vision and pattern recognition, pp 7376–7385
13. Cheng J, Tsai YH, Hung WC, Wang S, Yang MH (2018) Fast and accurate online video object segmentation via tracking parts. In: Proceedings of the IEEE conference on computer vision and pattern recognition, pp 7415–7424
14. Caelles S, Maninis KK, Pont-Tuset J, Leal-Taixé L, Cremers D, Van Gool L (2017) One-shot video object segmentation. In: Proceedings of the IEEE conference on computer vision and pattern recognition, pp 221–230
15. Perazzi F, Khoreva A, Benenson R, Schiele B, Sorkine-Hornung A (2017) Learning video object segmentation from static images. In: Proceedings of the IEEE conference on computer vision and pattern recognition, pp 2663–2672
16. Chen Y, Pont-Tuset J, Montes A, Van Gool L (2018) Blazingly fast video object segmentation with pixel-wise metric learning. In: Proceedings of the IEEE conference on computer vision and pattern recognition, pp 1189–1198
17. Song H, Wang W, Zhao S, Shen J, Lam KM (2018) Pyramid dilated deeper ConvLSTM for video salient object detection. In: Proceedings of the European conference on computer vision (ECCV), pp 715–731
18. Tokmakov P, Alahari K, Schmid C (2017) Learning video object segmentation with visual memory. arXiv:170405737
19. Jain SD, Xiong B, Grauman K (2017) FusionSeg: learning to combine motion and appearance for fully automatic segmentation of generic objects in videos. arXiv:170105384 2(3):6
20. Brox T, Malik J (2010) Object segmentation by long term analysis of point trajectories. In: European conference on computer vision. Springer, pp 282–295
21. Tsai YH, Yang MH, Black MJ (2016) Video segmentation via object flow. In: Proceedings of the IEEE conference on computer vision and pattern recognition, pp 3899–3908
22. Zhuo T, Cheng Z, Zhang P, Wong Y, Kankanhalli M (2018) Unsupervised online video object segmentation with motion property understanding. arXiv:181003783
23. Li J, Zheng A, Chen X, Zhou B (2017) Primary video object segmentation via complementary CNNs and neighborhood reversible flow. In: Proceedings of the IEEE international conference on computer vision, pp 1417–1425
24. Wang W, Shen J, Porikli F (2015) Saliency-aware geodesic video object segmentation. In: Proceedings of the IEEE conference on computer vision and pattern recognition, pp 3395–3402
25. Meila M, Shi J (2001) A random walks view of spectral segmentation. In: AISTATS
26. Leordeanu M, Sukthankar R, Hebert M (2012) Unsupervised learning for graph matching. Int J Comput Vis 96:28–45
27. Stewart GW (1990) Matrix perturbation theory
28. Ng AY, Zheng AX, Jordan MI (2001) Link analysis, eigenvectors and stability. In: International joint conference on artificial intelligence, vol 17. Lawrence Erlbaum Associates Ltd., pp 903–910
29. Leordeanu M, Hebert M (2005) A spectral technique for correspondence problems using pairwise constraints. In: Tenth IEEE international conference on computer vision (ICCV'05), vols 1 and 2. IEEE, pp 1482–1489
30. Ronneberger O, Fischer P, Brox T (2015) U-net: convolutional networks for biomedical image segmentation. In: International conference on medical image computing and computer-assisted intervention. Springer, pp 234–241

31. He K, Zhang X, Ren S, Sun J (2016) Deep residual learning for image recognition. In: 2016 IEEE conference on computer vision and pattern recognition (CVPR), pp 770–778
32. Deng J, Dong W, Socher R, Li LJ, Li K, Fei-Fei L (2009) Imagenet: a large-scale hierarchical image database. In: 2009 IEEE conference on computer vision and pattern recognition. IEEE, pp 248–255
33. Perazzi F, Pont-Tuset J, McWilliams B, Van Gool L, Gross M, Sorkine-Hornung A (2016) A benchmark dataset and evaluation methodology for video object segmentation. In: Computer vision and pattern recognition
34. Chen LC, Papandreou G, Schroff F, Adam H (2017) Rethinking atrous convolution for semantic image segmentation. arXiv:170605587
35. Lin TY, Maire M, Belongie S, Hays J, Perona P, Ramanan D, Dollár P, Zitnick L (2014) Microsoft coco: common objects in context. arXiv:14050312
36. Ilg E, Mayer N, Saikia T, Keuper M, Dosovitskiy A, Brox T (2017) Flownet 2.0: evolution of optical flow estimation with deep networks. In: Proceedings of the IEEE conference on computer vision and pattern recognition, pp 2462–2470
37. Mayer N, Ilg E, Hausser P, Fischer P, Cremers D, Dosovitskiy A, Brox T (2016) A large dataset to train convolutional networks for disparity, optical flow, and scene flow estimation. In: Proceedings of the IEEE conference on computer vision and pattern recognition, pp 4040–4048
38. Dosovitskiy A, Fischer P, Ilg E, Hausser P, Hazirbas C, Golkov V, Van Der Smagt P, Cremers D, Brox T (2015) Flownet: learning optical flow with convolutional networks. In: Proceedings of the IEEE international conference on computer vision, pp 2758–2766
39. Revaud J, Weinzaepfel P, Harchaoui Z, Schmid C (2015) Epicflow: edge-preserving interpolation of correspondences for optical flow. In: Proceedings of the IEEE conference on computer vision and pattern recognition, pp 1164–1172
40. Koh YJ, Kim CS (2017) Primary object segmentation in videos based on region augmentation and reduction. In: Proceedings of the IEEE conference on computer vision and pattern recognition, vol 1, p 6
41. Lu X, Wang W, Ma C, Shen J, Shao L, Porikli F (2019) See more, know more: unsupervised video object segmentation with co-attention Siamese networks. In: Proceedings of the IEEE conference on computer vision and pattern recognition, pp 3623–3632
42. Wang W, Song H, Zhao S, Shen J, Zhao S, Hoi SC, Ling H (2019) Learning unsupervised video object segmentation through visual attention. In: Proceedings of the IEEE conference on computer vision and pattern recognition, pp 3064–3074
43. Siam M, Jiang C, Lu S, Petrich L, Gamal M, Elhoseiny M, Jagersand M (2019) Video object segmentation using teacher-student adaptation in a human robot interaction (HRI) setting. In: 2019 international conference on robotics and automation (ICRA). IEEE, pp 50–56
44. Tokmakov P, Schmid C, Alahari K (2019) Learning to segment moving objects. Int J Comput Vis 127(3):282–301
45. Tokmakov P, Alahari K, Schmid C (2017) Learning motion patterns in videos. In: 2017 IEEE conference on computer vision and pattern recognition (CVPR). IEEE, pp 531–539
46. Li F, Kim T, Humayun A, Tsai D, Rehg J (2013) Video segmentation by tracking many figure-ground segments. In: International conference on computer vision
47. Lee YJ, Kim J, Grauman K (2011) Key-segments for video object segmentation. In: 2011 IEEE international conference on computer vision (ICCV). IEEE, pp 1995–2002
48. Li S, Seybold B, Vorobyov A, Fathi A, Huang Q, Jay Kuo CC (2018) Instance embedding transfer to unsupervised video object segmentation. In: Proceedings of the IEEE conference on computer vision and pattern recognition, pp 6526–6535
49. Prest A, Leistner C, Civera J, Schmid C, Ferrari V (2012) Learning object class detectors from weakly annotated video. In: CVPR
50. Croitoru I, Bogolin SV, Leordeanu M (2017) Unsupervised learning from video to detect foreground objects in single images. In: 2017 IEEE international conference on computer vision (ICCV). IEEE, pp 4345–4353

51. Zhang Y, Chen X, Li J, Wang C, Xia C (2015) Semantic object segmentation via detection in weakly labeled video. In: Proceedings of the IEEE conference on computer vision and pattern recognition, pp 3641–3649
52. Jun Koh Y, Jang WD, Kim CS (2016) POD: discovering primary objects in videos based on evolutionary refinement of object recurrence, background, and primary object models. In: Proceedings of the IEEE conference on computer vision and pattern recognition, pp 1068–1076

Chapter 7
Unsupervised Learning in Space and Time over Several Generations of Teacher and Student Networks

7.1 Introduction

Unsupervised learning is one of the most difficult and interesting problems in computer vision and machine learning today. Many researchers believe that learning from large collections of unlabeled videos could help decode hard questions regarding the nature of intelligence and learning. Moreover, as unlabeled images and videos are easy to collect at relatively low cost, unsupervised learning could be of real practical value in many computer vision and robotic applications. In this article, we propose a novel approach to unsupervised learning that successfully tackles many of the challenges associated with this problem. We present a system that is composed of two main pathways, one that does unsupervised object discovery in videos or large image collections—the teacher branch, and the other—the student branch, which learns from the teacher to segment foreground objects in single images. The unsupervised learning process could continue over several generations of students and teachers. In Algorithm 7.1, we present the high-level description of our method. We will use throughout the chapter the terms "generation" and "iteration" of Algorithm 7.1 interchangeably. The key aspects of our approach, which ensure improvement in performance from one generation to the next are (1) the existence of an unsupervised selection module that is able to pick up good quality masks generated by the teacher and pass them for training to the next-generation students; (2) training of multiple students with different architectures, able through their diversity to help train a better selection module for the next iteration and form together with the selection a more powerful teacher pathway at the next iteration; and (3) access to larger quantities of, and potentially more complex, unlabeled data, which becomes more useful as the generations become stronger.

Our approach is general in the sense that the student or teacher pathways do not depend on a specific neural network architecture or implementation. Through many experiments and comparisons to state-of-the-art methods, we also show that it is applicable to different problems in computer vision, such as object discovery in

The material presented in this chapter is based in large part on the following paper:
Croitoru, Ioana, Simion-Vlad Bogolin, and Marius Leordeanu. "Unsupervised Learning of Foreground Object Segmentation." International Journal of Computer Vision (IJCV) 2019: 1–24.

211
M. Leordeanu, *Unsupervised Learning in Space and Time*,
Advances in Computer Vision and Pattern Recognition,
https://doi.org/10.1007/978-3-030-42128-1_7

video, unsupervised image segmentation, saliency detection, and transfer learning. The same approach, as the one presented here, could be applied to any visual learning task for which an initial teacher exists, since the main steps of the algorithm are based on ideas that are very general and do not depend on a specific problem, such as it is that of foreground object segmentation.

In Fig. 7.1, we present a graphic overview of our full system. In the unsupervised training stage, the student network (module A) learns, frame by frame, from an unsupervised teacher pathway (modules B and C) to produce similar object masks in single images. Module B discovers objects in images or videos, while module C selects which masks produced by module B are sufficiently good to be passed to module A for training. Thus, the student branch tries to imitate the output of module B for the frames selected by module C, having as input only a single image—the current frame, while the teacher can have access to an entire video sequence.

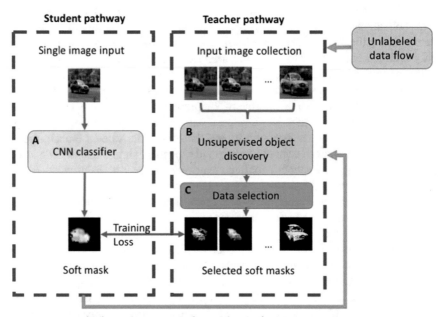

Multiple students create the teacher in the next generation

Fig. 7.1 The dual student-teacher system proposed for unsupervised learning to segment foreground objects in images, functioning as presented in Algorithm 7.1. It has two pathways: along the teacher branch, an object discoverer in videos or large image collections (module B) detects foreground objects. The resulting soft masks are then filtered based on an unsupervised data selection procedure (module C). The resulting final set of pairs—input image (or video frame) and soft mask for that particular frame (which acts as an unsupervised label)—is used to train the student pathway (module A). The whole process can be repeated over several generations. At each generation, several student CNNs are trained, and then they collectively contribute to train a more powerful selection module C (modeled by a deep neural network, Sect. 7.4.3) and form an overall more powerful teacher pathway at the next iteration of the overall algorithm

The strength of the trained student (module A) depends on the performance of module B. However, as we see in experiments, the power of the selection module C contributes to the fact that the new student will outperform its initial teacher module B. Therefore, throughout the chapter, we refer to B as the initial "teacher" and to both B and C together as the full "teacher pathway". The method presented in Algorithm 7.1 follows the main steps of the system as it learns from one iteration (generation) to the next. The steps are discussed in more detail in Sect. 7.3.

During the first iteration of Algorithm 7.1, the unsupervised teacher (module B) has access to information over time—a video. In contrast, the student is deeper in structure, but it has access only to a single image—the current video frame. Thus, the information discovered by the teacher in time is captured by the student in added depth, over neural layers of abstraction. Several student nets with different architectures are trained at the first iteration. In order to use as supervisory signal only good quality masks, an unsupervised mask selection procedure (very simple at Iteration 1) is applied (module C), as explained in Sect. 7.4. Once several student nets are trained, they can form (in various ways, as explained in Sects. 7.4.1 and 7.5.1) the teacher pathway at the next iteration, along with a stronger unsupervised selection module C, represented by a deep neural network, EvalSeg-Net, trained as explained in detail in Sect. 7.4.3. In short, EvalSeg-Net learns to predict the output masks agreement among the generally diverse students, which statistically takes place when the masks are of good quality. Thus, EvalSeg-Net could be used as an unsupervised mask evaluation procedure and a strong selection module. Then, we run, at the next generation, the newly formed teacher pathway (modules B and C) on a larger set of unlabeled videos or collections of images, to produce supervisory signal for the next-generation students. In experiments, we show that the improvement of both modules B and C at the next iterations, together with the increase in the amount of data, are all important, while not all necessary, for increasing accuracy at the next generation.

Note that, while at the first iteration the teacher pathway is required to receive video sequences as input, from the second generation onwards, it could receive as input large image collections, as well. Due to the very high computational and storage costs, required during training time, we limit our experiments to learning over two generations, but our algorithm is general and could run over many iterations. We show in extensive experiments that even two generations are sufficient to outperform the current state of the art on object discovery in videos and images. We also demonstrate experimentally a solid improvement from one generation to the next for each component involved, the individual students (module A), the teacher (module B) as well as the selection module C.

Now we enumerate below the main results of the work presented in this chapter and originally published in Croitoru et al. [1, 2]:

1. We present an original and highly effective approach to unsupervised learning to segment foreground objects in images. The overview of our system and algorithm is presented in Fig. 7.1 and Algorithm 7.1. The system has two main pathways— one that acts as a teacher (module B) and discovers objects in videos or large collections of images followed by an unsupervised selection module C that filters

out low-quality masks, and the other that acts as student and learns from the teacher pathway to detect the foreground objects in single input images. We provide a general algorithm for unsupervised learning over several generations of students and teachers. We also show how to learn an unsupervised mask selection deep network (EvalSeg-Net, see Sect. 7.4.3), which is important in improving the teacher pathway at the next iteration, over all cases tested: when the teacher (module B) is formed by a single student network, by all students combined into an ensemble or by all students taken separately. The whole unsupervised training at the second generation is a novelty over the conference work, with significantly improved experimental results (see Sect. 7.5).

2. At the higher level, our proposed algorithm is sufficiently general to accommo-date different implementations, various neural network architectures, and multi-ple visual learning tasks. Here we provide a specific implementation which we describe in detail. We demonstrate its performance on three related unsupervised learning problems, namely, video object discovery tested on YouTube-Objects [3], unsupervised foreground segmentation in images tested on Object Discovery in Internet Images [4], and saliency detection tested on Pascal-S [5], on which we obtain state-of-the-art results. We further apply our approach to a well-known transfer learning setup and obtain competitive results when compared to top trans-fer learning methods in the field. We also compare experimentally to the work most related to ours [6], on both foreground segmentation and transfer learning problems and show that our method obtains better results on foreground object segmentation, while theirs is more effective for transfer learning. To the best of our knowledge, at the time of its initial publication [1], our method along with the one proposed in Pathak et al. [6] was the first to present a system that learns to detect and segment foreground objects in images in unsupervised fashion, with no pretrained features given or manual labeling, while requiring only a single image at test time.

7.1.1　Relation to Unsupervised Learning Principles

Principle 6
When several weak independent classifiers consistently fire together, they sig-nal the presence of a higher level class for which a stronger classifier can be learned.

The work in this chapter is directly related to the sixth principle of unsupervised learning proposed in Chap. 1. It is highly unlikely for many independent classifiers to fire together unless they fire at the same thing, potentially unknown class at a higher level of semantics. In this chapter, we propose a teacher-student system, which trains itself, in an unsupervised and also self-supervised way, such that an entire generation of students trained at iteration k together form the teacher at the next iteration $k + 1$.

Their collective consensus becomes very strong HPP supervisory signal (directly related to Principle 3) such that they are collectively becoming capable to "see" what not a single one of them can.

The students together can now signal the presence of a "higher level" class, such that the segmentation obtained from their combined agreement is of higher quality, refinement, or of higher kind than the segmentation obtained by each, separately. For example, each student at some iteration could be able to produce a blob that is roughly correlated with the location, size, and shape of the main object of interest. However, after several generations, the students (which are the training product of agreements from previous generation students) can produce object masks that are of different and fundamentally better quality, such that the fine shape details of objects could now be observed.

In Sect. 7.3, we introduce the different components of our approach, from a general overview of the overall system, to more detailed descriptions of each of its modules. It is important to note that the entire teacher pathway (see Fig. 7.1) is, in fact, based on the idea expressed by Principle 6.

We soon realize the real need for learning over several generations—if we want the system to transcend to learning fundamentally new classes or better levels of recognition and segmentation. This fact relates our approach to the next proposed principle of unsupervised learning.

Principle 7
To improve or learn new classes, we need to increase the quantity and level of difficulty of the training data as well as the power and diversity of the classifiers. In this way, we could learn in an unsupervised way over several generations, by using the agreements between the existing classifiers as a teacher supervisory signal for the new ones.

Note that such *new classes* could consist of different categories the students do not know yet about. They could also consist of previously seen classes, but at a fundamentally different level of segmentation quality, dramatically better than the previous one (e.g., a new level of refinement, specialization, or attention to details).

However, in order to learn to segment such "new classes", at superior levels of quality, we can only use the classifiers that we already have. Therefore, we need to rely on their consistent (but rare) co-firing in order to spot space-time regions where we might catch the presence of a new class, object, or object part (or detail), which has been unseen until now.

Learning to "see" a new class requires the co-firing of the existing classifiers along the teacher pathway (as discussed in relation to Principle 6), but it also requires access to the right amount of data that contains sufficient evidence of that new class. Then, unsupervised learning of object segmentation has to be done in stages, step by step, going from one level of difficulty to the next, by sequentially increasing the quantity of data and also by training multiple generations of classifiers. The idea is directly related to the concept of curriculum learning [7], by going during the training process from simple to complex cases. The approach has been validated in all the experiments we performed (Sect. 7.5): the proposed unsupervised teacher-student system is able to train multiple generations of increasingly better teachers and students using increasingly larger, more varied, and complex unlabeled training sets.

7.2   Scientific Context

The literature on unsupervised learning follows two main directions: (1) One is to learn powerful features in an unsupervised way and then use them for transfer learning, within a supervised scheme and in combination with different classifiers, such as SVMs or CNNs [8–10]. (2) The second direction is to discover, at test time, common patterns in unlabeled data, using clustering, feature matching, or data mining formulations [11–13].

Belonging to the first category and closely related to our work, the approach in Pathak et al. [6] proposes a system in which a deep neural network learns to produce soft object masks from an unsupervised module that uses optical flow cues in video. The deep features learned in this manner are then applied to several transfer learning problems. Their work, together with ours, is probably the first two that show ways to learn in an unsupervised fashion to segment objects in single images. While the two approaches are clearly different at the technical and algorithmic level, we also provide some interesting comparisons in the experiments in Sect. 7.5.3, on both tasks of transfer learning and foreground object segmentation. Our results reveal that while their approach is better on transfer learning tasks, ours is more effective on unsupervised segmentation as tested on several datasets.

Recently, researchers have started to use the natural, spatial, and temporal structure in images and videos as supervisory signals in unsupervised learning approaches that are considered to follow a *self-supervised learning* paradigm [14–16]. Methods that fall into this category include those that learn to estimate the relative patch positions in images [17], predict color channels [18], solve jigsaw puzzles [19], and inpaint [20]. One trend is to use as supervisory signal, spatial and appearance information collected from raw single images. In such single-image cases, the amount of information that can be learned is limited to a single moment in time, as opposed to the case of learning from video sequences. Using unlabeled videos as input is closely related to our work and includes learning to predict the temporal order of frames [15], generate the future frame [21–23], or learn from optical flow [24].

For most of these papers, the unsupervised learning scheme is only an intermediate step to train features that are eventually used on classic supervised learning tasks, such as object classification, object detection, or action recognition. Such pretrained features do better than randomly initialized ones, as they contain valuable information implicit in the natural structure of the world used as supervisory signal. While the unsupervised features might not contain semantic, class-specific information [25], it is clear that they capture general objectness properties, useful for problems such as segmenting the main objects in the scene or transfer learning to specific supervised classification problems. In our work, we focus mostly on specific unsupervised tasks on which we perform extensive evaluations, but we also show some results on transfer learning experiments.

The second main approach to unsupervised learning includes methods for image co-segmentation [4, 26–32] and weakly supervised localization [33–35]. Earlier methods are based on local features' matching and detection of their co-occurrence patterns [13, 36–39], while more recent ones [3, 40, 41] discover object tubes by linking candidate bounding boxes between frames with or without refining their location. Traditionally, the task of unsupervised learning from image sequences has been formulated as a feature matching or data clustering optimization problem, which is computationally very expensive due to its combinatorial nature.

There are also other papers [42–45] that tackle unsupervised learning tasks but are not fully unsupervised, using powerful features that are pretrained in supervised fashion on large datasets, such as ImageNet [46] or VOC 2012 [47]. Such works take advantage of the rich source of supervised information learned from other datasets, through features trained to respond to general object properties over tens or hundreds of object categories. In another paper some amount of supervision is necessary, as in Tokmakov et al. [48] where a system is proposed having a motion-CNN that learns from weakly annotated videos and optical flow cues to segment objects in video frames. One key difference from our work is that their approach requires the class labels of the training video frames.

With respect to the end goal, our work is more related to the second research direction, on unsupervised discovery in video. However, unlike that research, we do not discover objects at test time, but during the unsupervised training process, when the student pathway learns to detect foreground objects. Therefore, from the learning perspective, our work is more related to the first research direction based on self-supervised training. While there are published methods that leverage spatiotemporal information in video, our method at the second iteration is able to learn even from collections of unrelated images. Related to our idea of improving segmentations over several iterations, there is the method proposed in Khoreva et al. [49], which is not unsupervised as it requires the ground truth bounding box information in order to improve the segmentations over several iterations, in conjunction with a modified version of GrabCut [50].

7.3 Learning over Multiple Teacher-Student Generations

We propose a genuine unsupervised learning algorithm (see Algorithm 7.1) for foreground object segmentation that offers the possibility to improve over several iterations. Our method combines in complementary ways multiple modules that are well suited for this problem.

It starts with a teacher (module B, Fig. 7.1) that discovers objects in unlabeled videos and produces a soft mask of the foreground object in each frame. There are several available methods for video discovery in the literature, with good performance [51–53]. We choose the VideoPCA algorithm introduced as part of the system in Stretcu and Leordeanu [36] because it is very fast (50–100 fps), uses very simple features (individual pixel colors), and it is completely unsupervised, with no usage of supervised pretrained features. It learns how to separate the foreground from the background and it exploits the spatiotemporal consistency in appearance, shape, movement, and location of objects, common in video shots, along with the contrasting properties, in size, shape, motion, and location, between the main object and the background scene. Note that it would be much harder, at this first stage, to discover objects in collections of unrelated images, where there is no smooth variation in shape, appearance, and location over time. Only at the second iteration of the algorithm, the simpler VideoPCA is replaced by a more powerful teacher which is able to discover objects in collections of images as well.

The resulting soft masks of lower quality are then filtered out automatically (module C, Fig. 7.1), using at the first iteration a very simple automatic procedure. Next, the remaining ones are passed to a student ConvNet, which learns to predict object masks in single images. When several student nets of different architectures are learned, they give the possibility of learning a stronger selection network (module C, Fig. 7.1) and form a more powerful teacher pathway (modules B and C, Fig. 7.1) for the next generation. Then, the whole process is repeated. As discussed in Sect. 7.1, three key aspects contribute to improvement at the second iteration: learning of a more powerful teacher (module B), which could be formed by a single student model or an entire ensemble, learning a stronger selection module C (modeled by a EvalSeg-Net, Sect. 7.4.3) and last but not least, increasing the amount of unlabeled data. As shown in the experiments in Sect. 7.5.1, bringing in more data helps only at the second generation when both the teacher and the selection module are improved.

In Algorithm 7.1, we enumerate concisely the main steps of our approach.

Algorithm 7.1 Unsupervised learning of foreground object segmentation

---

Step 1: perform unsupervised object discovery in unlabeled videos (or image collections, at later iterations), along the teacher pathway (module B in Fig. 7.1).

Step 2: automatically filter out poor soft masks produced at the previous step (module C in Fig. 7.1).

Step 3: use the remaining masks as supervisory signal for training one or more student nets, along the student pathway (module A in Fig. 7.1).

Step 4: use as new teacher one or several student nets from the current generation (a new module B) and learn a more powerful soft mask selector (a new module C), for the next iteration.

Step 5: extend the unlabeled video or image dataset and return to Step 1 to train the next generation (Note: from the first iteration forward, the training dataset can also be extended with collections of unlabeled images, not just videos).

7.4   Our Teacher-Student System Architecture

We detail the architecture and training process of our system, module by module, as seen in Fig. 7.1. We first present the student pathway (module A in Fig. 7.1), which takes as input an individual image (e.g., current frame in the video) and learn to predict foreground soft masks from an unsupervised teacher. The teacher (represented by module B) and the selection module C are explained in Sects. 7.4.2 and 7.4.3.

7.4.1 Student Pathway: Single-Image Segmentation

The student pathway (module A in Fig. 7.1) consists of a deep convolutional network. We test different network architectures, some of which are commonly used in the recent literature on semantic image segmentation. We create a small pool of relatively diverse architectures, presented next.

The first convolutional network architecture for semantic segmentation that we test is based on a more traditional CNN design. We term it LowRes-Net (see Fig. 7.2) due to its low-resolution soft mask output. It has ten layers (seven convolutional, two pooling, and one fully connected) and skip connections. Skip connections have proved to offer a boost in performance, as shown in the literature [54, 55]. We also observed a similar improvement in our experiments when using skip connections. The LowRes-Net takes as input a 128 × 128 RGB image (along with its hue, saturation, and derivatives w.r.t. x and y) and produces a 32 × 32 soft segmentation of the main objects present in the image. Because LowRes-Net has a fully connected layer at the

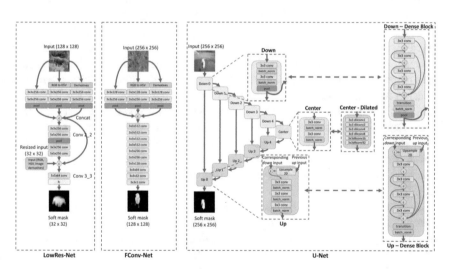

Fig. 7.2 Different architectures for the "student" networks, each processing a single image. They are trained to predict the unsupervised label masks given by the teacher pathway, frame by frame. The architectures vary from the more classical baseline LowRes-Net (left), with low-resolution output, to more recent architectures, such as the fully convolutional one (middle) and different types of U-Nets (right). For the U-Net architecture, the blocks denoted with double arrows can be interchanged to obtain a new architecture. We noticed that on the task of bounding box fitting the simpler low-resolution network did very well, while being outperformed by the U-Nets on fine object segmentation

top, we reduced the output resolution of the soft segmentation mask to limit memory cost. While the derivatives w.r.t x and y are not needed, in principle (as they could be learned by appropriate filters during training), in our tests explicitly providing the derivatives along with HSV and by using skip connections boosted the accuracy by over 1%. The LowRes-Net has a total of 78M parameters, most of them being in the last, fully connected layer.

The second CNN architecture tested, termed FConv-Net, is fully convolutional [56], as also presented in Fig. 7.2. It has a higher resolution output of 128×128, with input size 256×256. Its main structure is derived from the basic LowRes-Net model. Different from LowRes-Net, it is missing the fully connected layer at the end and has more parameters in the convolutional layers, for a total of 13M parameters.

We also tested three different nets based on the U-Net [57] architecture, which proved very effective in the semantic segmentation literature. Our U-net networks are (1) BasicU-Net, (2) DilateU-Net—similar to BasicU-Net but using atrous (dilated) convolutions [58] in the *center* module, and (3) DenseU-Net—with dense connections in the *down* and *up* modules [59].

The BasicU-Net has five *down* modules with two convolutional layers each with 32, 64, 128, 256, and 512 features maps, respectively. In the *center* module, the BasicU-Net has two convolutional layers with 1024 feature maps each. The *up* modules have three convolutional layers and the same number of feature maps as the corresponding *down* modules. The only difference between BasicU-Net and DilateU-Net is that the former has a different *center* module with six atrous convolutions and 512 feature maps each. Then, DenseU-Net has four *down* modules with four corresponding *up* modules. Each *down* and *up* module has four convolutions with skip connections (as presented in Fig. 7.2). The modules have 12, 24, 48, and 64 features maps, respectively. The *transition* represents a convolution, having the role of reducing the output number of feature maps from each module. The BasicU-Net has 34M parameters, while the DilateU-Net has 18M parameters. DenseU-Net has only 3M parameters, but uses skip connections inside the *up* and *down* blocks in order to make up for the difference in the number of parameters. All three U-Nets have 256×256 input and same resolution output. All networks use ReLU activation functions. Please see Fig. 7.2 for more specific details regarding the architectures of the different models.

Given the current setup, the student nets do not learn to identify specific object classes. They will learn to softly segment the main foreground objects present, regardless of their particular category. The main difference in their performance is in their ability to produce fine object segmentations. While the LowRes-Net tends to provide a good support for estimating the object's bounding box due to its simpler output, the other ConvNets (especially the U-Nets), with higher resolution, are better at finely segmenting objects. The different student architectures bring diversity to their outputs. Due to the different ways in which the particular models make mistakes, they are stronger when forming an ensemble and can also be used, as seen in Sect. 7.4.3, to train a network for segmentation evaluation, used as the new selection module C. As explained later EvalSeg-Net will learn to predict the output masks' agreement among the students, which statistically takes place when the masks are of good quality. In experiments, we also show that the student nets outperform their teacher and are able to detect objects from categories that were not seen during training.

Combining several student nets: The student networks with different architectures produce varied results that differ qualitatively. While the bounding boxes computed from their soft masks have similar accuracy, the actual soft segmentation output looks different. They have different strengths, while making different kinds of mistakes. Their diversity will be the basis for creating the teacher pathway at the next generation (Sects. 7.4.2 and 7.4.3).

We experimented with the idea of using several student networks, by combining them to form an ensemble or by letting them produce separate independent segmentations for each image. In our final system, we preferred the latter approach, which is more practical, easier to implement, and gives the freedom of having the students run independently, in parallel with no need to synchronize their outputs. As shown in Sects. 7.4.2 and 7.4.3, together with the EvalSeg-Net used for selection, independent individual students from Iteration 1 will form the teacher pathway at the next generation. However, note that even a single student net along with the new EvalSeg-Net selector can be effectively used as next teacher pathway (See experimental Sect. 7.5.1, Table 7.6, and Fig. 7.7).

When forming an actual ensemble, which we term Multi-Net, the final output is the one obtained by multiplying pixelwise the soft masks produced by each individual student net. Thus, only positive pixels, on which all nets agree, survive to the final segmentation. As somehow expected, Multi-Net offers robust masks of higher precision than each individual network. However, it might lose details around the border of objects having a lower recall (see Fig. 7.5). We provide results of the Multi-Net ensemble only for comparison purposes. Please note, however, that in our final system the output of the ensemble was not used to train the students at the next generation. The students at the second iteration are all trained directly on outputs from individual students at the first iteration, filtered with EvalSeg-Net. As explained in more detail later in this section, Multi-Net is used only to train the unsupervised selection network, EvalSeg-Net.

Technical details: training the students: We treat foreground object segmentation as a multidimensional regression problem, where the soft mask given by the unsupervised video segmentation system acts as the desired output. Let $\mathbf{I}$ be the input RGB image (a video frame) and $\mathbf{Y}$ be the corresponding 0–255 valued soft segmentation given by the unsupervised teacher for that particular frame. The goal of our network is to predict a soft segmentation mask $\hat{\mathbf{Y}}$ of width W and height H (where $W = H = 32$ for the basic architecture, $W = H = 128$ for fully convolutional architecture, and $W = H = 256$ for U-Net architectures) that approximates as well as possible the mask $\mathbf{Y}$. For each pixel in the output image, we predict a 0–255 value, so that the total difference between $\mathbf{Y}$ and $\hat{\mathbf{Y}}$ is minimized. Thus, given a set of N training examples, let $\mathbf{I}^{(n)}$ be the input image (a video frame), $\hat{\mathbf{Y}}^{(n)}$ be the predicted output mask for $\mathbf{I}^{(n)}$, $\mathbf{Y}^{(n)}$ be the soft segmentation mask (corresponding to $\mathbf{I}^{(n)}$), and $\mathbf{w}$ be the network parameters. $\mathbf{Y}^{(n)}$ is produced by the video discoverer after processing the video that $\mathbf{I}^{(n)}$ belongs to. Then, our loss is

$$L(\mathbf{w}) = \frac{1}{N} \sum_{n=1}^{N} \sum_{p=1}^{W \times H} (\mathbf{Y}_{p}^{(n)} - \hat{\mathbf{Y}}_{p}^{(n)}(\mathbf{w}, \mathbf{I}^{(n)}))^{2}, \qquad (7.1)$$

where $\mathbf{Y}_{p}^{(n)}$ and $\hat{\mathbf{Y}}_{p}^{(n)}$ denote the p-th pixel from $\mathbf{Y}^{(n)}$, respectively $\hat{\mathbf{Y}}^{(n)}$.

We observed that in our tests, the L2 loss performed better than the cross-entropy loss, due to the fact that the soft masks used as labels have real values, not discrete ones. Also, they are not perfect, so the idea of thresholding them for training does not do as well as directly predicting their real values. We train our network using the Tensorflow [60] framework with the Adam optimizer [61]. All models are trained end to end using a fixed learning rate of 0.001 for 10 epochs. The training time for any given model is about 3–5 days on a Nvidia GeForce GTX 1080 GPU, for the first iteration, and about 2 weeks, for the second iteration.

Post-processing: The student CNN outputs a $W \times H$ soft mask. In order to fairly compare our models with other methods, we have two different post-processing steps: (1) bounding box fitting and (2) segmentation refinement. For fitting a box around the soft mask, we first up-sample the $W \times H$ output to the original size of the image, then threshold the mask (validated on a small subset), determine the connected components, and fit a tight box around each of the components. We do segmentation refinement (point 2) in a single case, on the Object Discovery in Internet Images Dataset as also specified in the experiments section. For that, we use the OpenCV implementation of GrabCut [50] to refine our soft mask, up-sampled to the original size. In all other tests, we use the original output of the networks.

7.4.2 Teacher Pathway: Unsupervised Object Discovery

There are several methods available for discovering objects and salient regions in images and videos [51–53, 62–64] with reasonably good performance. More recent methods for foreground objects discovery such as Papazoglou and Ferrari [65] are both relatively fast and accurate, with runtime around 4 s/frame. However, that runtime is still long and prohibitive for training the student CNN that requires millions of images. For that reason we used at the first generation (Iteration 1 of Algorithm 7.1) for module B in Fig. 7.1, the VideoPCA algorithm, which is a part of the whole system introduced in Stretcu and Leordeanu [36]. It has lower accuracy than the full system, but it is much faster, running at 50–100 fps. At this speed, we can produce one million unsupervised soft segmentations in a reasonable time of about 5–6 h.

VideoPCA: The main idea behind VideoPCA (also presented in Chap. 5) is to model the background in video frames with Principal Component Analysis. It finds initial foreground regions as parts of the frames that are not reconstructed well with the PCA model. Foreground objects are smaller than the background and have contrasting appearance and more complex movements. They could be seen as outliers, within the larger background scene. That makes them less likely to be captured well by the first PCA components. Thus, for each frame, an initial soft mask is produced from an error image, which is the difference between the original image and the PCA reconstruction. These error images are first smoothed with a large Gaussian filter and then thresholded. The binary masks obtained are used to learn color models of foreground and background, based on which individual pixels are classified as belonging to foreground or not. The object masks obtained are further multiplied with a large centered Gaussian, based on the assumption that foreground objects are often closer to the image center. These are the final masks produced by VideoPCA. For more technical details, the reader is invited to consult [36]. In this work, we use the method exactly as found online[1] without any parameter tuning.

Teacher at the next generation: At the next iteration of Algorithm 7.1, VideoPCA (in module B) is replaced by student nets trained at the previous generation. We tested with three different ideas: one is to use a single student network and combine it with the more powerful selection module to form a stronger full teacher pathway (modules B and C). While this approach is very effective and prove the relevance of selection, it is not the most competitive. Using all student nets is always more powerful and this can be done in two ways, as discussed in the previous section. One possibility is to create Multi-Net ensemble by multiplying their outputs and the other, equally powerful but easier to implement is to use all student nets independently and let each image have several output masks, as separate (input image, soft mask) pairs for training the next generation. We prefer the latter approach which, in combination with the EvalSeg-Net network, will constitute the full teacher pathway at the second iteration. Next, we present in detail how we do mask selection and how we train EvalSeg-Net.

7.4.3 Unsupervised Soft Masks Selection

The performance of the student net is influenced by the quality of the soft masks provided as labels by the teacher branch. The cleaner the masks, the more chances the student has to learn to segment well objects in images. VideoPCA tends to produce good results if the object present in the video stands out well against the background scene, in terms of motion and appearance. However, if the object is occluded at some point, it does not move w.r.t the scene or it has a similar appearance to its background, then the resulting soft masks might be poor. In the first generation, we used a simple measure of masks' quality to select only the good soft masks for training the student pathway, based on the following observation: when VideoPCA masks are close to the ground truth, the average of their non-zero values is usually high. Thus, when the discoverer is confident, it is more likely to be right. The average value of non-zero pixels in the soft mask is then used as a score indicator for each segmented frame. Only masks of certain quality according to this indicator are selected and used for training the student nets. This represents module C in Fig. 7.1 at the first generation

---

[1] https://sites.google.com/site/multipleframesmatching/.

of Algorithm 7.1. While being effective at iteration 1, the simple average value over all pixels cannot capture the goodness of a segmentation at the higher level of overall shape. At the next iterations, we, therefore, explore new ways to improve it.

Training EvalSeg-Net: At the next iterations, we propose an unsupervised way for learning the EvalSeg-Net to estimate segmentation quality. As mentioned previously, Multi-Net provides masks of higher quality as it cancels errors from individual student nets. Thus, we use the cosine similarity between a given individual segmentation and the ensemble Multi-Net mask, as a cost for "goodness" of segmentation. Having this unsupervised segmentation cost we train the EvalSeg-Net deep neural net to predict it. As previously mentioned, this net acts as an automatic mask evaluation procedure, which in subsequent iterations becomes module C in Fig. 7.1, replacing the simple mask average value used at Iteration 1. Only masks that pass a certain threshold are used for training the student path. As it turns out in experiments, EvalSeg-Net becomes an effective selection procedure (module C) that improves the teacher pathway regardless of the teacher module B used.

The architecture of EvalSeg-Net is similar to LowRes-Net (Fig. 7.2), with the difference that the input channel containing image derivatives is replaced by the actual soft segmentation that requires evaluation and it does not have skip connections. After the last fully connected layer (size 512) we add a last one-neuron layer to predict the segmentation quality score, which is a single real-valued number.

Let $\mathbf{I}$ be an input RGB image, $\mathbf{S}$ an input soft mask, $\hat{\mathbf{Y}} = \prod_{i=1}^{5} \hat{\mathbf{Y}}_{N_i}$ be the output of our Multi-Net where $\hat{\mathbf{Y}}_{N_i}$ denotes the output of network N_i. We treat the segmentation "goodness" evaluation task as a regression problem where we want to predict the cosine similarity between $\mathbf{S}$ and $\hat{\mathbf{Y}}$. So, our loss for EvalSeg-Net is defined as follows:

$$ L(\mathbf{w}) = \frac{1}{K} \sum_{k=1}^{K} \left(\hat{o}^{(k)}(\mathbf{w}, \mathbf{I}^{(k)}, \mathbf{S}^{(k)}) - \frac{\mathbf{S}^{(k)} \cdot \hat{\mathbf{Y}}^{(k)}}{\mathbf{S}^{(k)} \hat{\mathbf{Y}}^{(k)}} \right)^2, \quad (7.2) $$

where K represents the number of training examples and $\hat{o}^{(k)}(\mathbf{w}, \mathbf{I}^{(k)}, \mathbf{S}^{(k)})$ represents the output of EvalSeg-Net for image $\mathbf{I}^{(k)}$ and soft mask $\mathbf{S}^{(k)}$.

Given a certain metric for segmentation evaluation (depending on the learning iteration), we keep only the soft masks above a threshold for each dataset (e.g., VID [46], YTO [3], YouTube Bounding Boxes [66]). In the first iteration, this threshold was obtained by sorting the VideoPCA soft masks based on their score and keeping only the top 10 percentile, while on the second iteration we validate a threshold ($= 0.8$) on a small dataset and select each mask independently by using this threshold on the single value output of EvalSeg-Net.

Mask selection evaluation: In Fig. 7.3, we present the dependency of segmentation performance w.r.t ground truth object boxes (used only for evaluation) versus the percentile p of masks kept after the automatic selection, for each generation. We notice the strong correlation between the percentage of frames kept and the quality of segmentations. It is also evident that the EvalSeg-Net is vastly superior to the

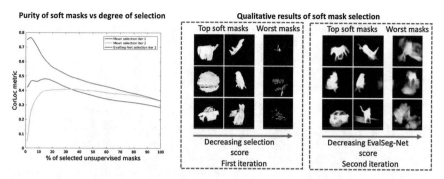

Fig. 7.3 Quality of soft masks versus degree of selection at module C. When selectivity increases, the true quality of the training frames that pass through selection improves. At the first iteration of Algorithm 7.1, we select masks using a simple selection procedure based on the mean value of non-zero mask pixels. At the second iteration, we select masks using the much more powerful EvalSeg-Net. The plots are computed using results from the VID dataset, where there is an annotation for each input frame. Note the superior quality of masks selected at the second iteration (red vs. blue lines, in the left plot). We have also compared the simple "mean"-based selection procedure used at iteration 1 (yellow line) with EvalSeg-Net used at iteration 2 (red line), on the same soft masks from iteration 2. The EvalSeg-Net is clearly more powerful, which justifies its use at the second iteration when it replaces the very simple "mean"-based procedure

simpler procedure used at iteration 1. EvalSeg-Net is able to correctly evaluate soft segmentations even in more complex cases (see Fig. 7.4).

Even though we can expect to improve the quality of the unsupervised masks by drastically pruning them (e.g., keeping a smaller percentage), the fewer we are left with, the less training data we get, increasing the chance to overfit. We make up for the losses in training data by augmenting the set of training masks and by also enlarging the actual unlabeled training set at the second generation. There is a trade-off between level of selectivity and training data size: the more selective we are about what masks we accept for training, the more videos we need to collect and process through the teacher pathway, to obtain the sufficient training data size.

Data augmentation: A drawback of the teacher at the first learning iteration (VideoPCA) is that it can only detect the main object if it is close to the center of the image. The assumption that the foreground is close to the center is often true and indeed helps that method, which has no deep learned knowledge, to produce soft masks with a relatively high precision. Not surprisingly, it often fails when the object is not in the center, and therefore its recall is relatively low. Our data augmentation procedure addresses this limitation and can be concisely described as follows: randomly crop patches of the input image, covering 80% of the original image and scale up the patch to the expected input size. This produces slightly larger objects at locations that cover the whole image area, not just the center. As experiments show, the student net is able to see objects at different locations in the image, unlike its raw teacher (VideoPCA at iteration 1), which is strongly biased towards the image center.

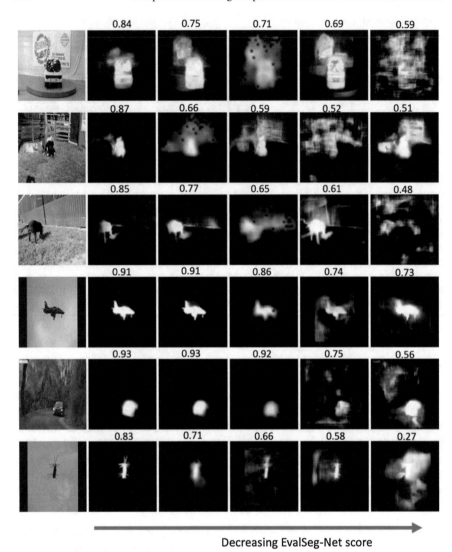

Decreasing EvalSeg-Net score

Fig. 7.4 Qualitative results of the unsupervised EvalSeg-Net used for measuring segmentation "goodness" and filtering bad masks (Module C, iteration 2). For each input image, we present five soft mask candidates (from first iteration students) along with their "goodness" scores given by EvalSeg-Net, in decreasing order of scores. Note the effectiveness of EvalSeg-Net at ranking soft segmentations

At the second generation, the teacher branch is superior at detecting objects at various locations and scales in the image. Therefore, while artificial data augmentation remains useful (as it is usually the case in deep learning), its importance diminishes at the second iteration of learning (Algorithm 7.1). Adding more unlabeled data helps at both generations up to a point. If more difficult training cases are added, they improve learning only at the second generation, as discussed in the experimental section (Table 7.5).

7.4.4 Implementation Pipeline

Now that we have presented in technical detail all major components of our system, we concisely present the actual steps taken in our experiments, in sequential order, and show how they relate to our general Algorithm 7.1 for unsupervised learning to detect foreground objects in images.

1. Run VideoPCA on input images from VID and YouTube-Objects datasets (Algorithm 7.1, Iteration 1, Step 1).
2. Select VideoPCA masks using first-generation selection procedure (Algorithm 7.1, Iteration 1, Step 2).
3. Train first-generation student ConvNets on the selected masks, namely, LowRes-Net, FConv-Net, BasicU-Net, DilateU-Net, and DenseU-Net (Algorithm 7.1, Iteration 1, Step 3).
4. Create first-generation student ensemble Multi-Net by multiplying the outputs of all students and train EvalSeg-Net to predict the similarity between a particular mask and the mask of Multi-Net (Algorithm 7.1, Iteration 1, Step 4).
5. Add new data from YouTube Bounding Boxes. (Algorithm 7.1, Iteration 1, Step 5).
6. Return to Step 1, the teacher pathway: predict multiple soft masks per input image on the enlarged unlabeled video set, using the student nets from Iteration 1 (Module B, Iteration 2), which will be then selected with EvalSeg-Net at Module C (Algorithm 7.1, Iteration 2, Step 1).
7. Select only sufficiently good masks evaluated with EvalSeg-Net (Algorithm 7.1, Iteration 2, Step 2).
8. Train the second-generation students on the newly selected masks. We use the same architectures as in Iteration 1 (Algorithm 7.1, Iteration 2, Step 3).

The method presented in the introduction sections (Algorithm 7.1) is a general algorithm for unsupervised learning from video to detect objects in single images. It presents a sequence of high-level steps followed by different modules for an unsupervised learning system. The modules are complementary to each other and function in tandem, each focusing on a specific aspect of the unsupervised learning process. Thus, we have a module for generating data, where soft masks are produced. There is a module that selects good quality masks. Then, we have a module for training the next-generation classifiers. While our concept is first presented in high-level terms, we also present a specific implementation that represents the first two iterations of the algorithm. While our implementation is costly during training, in terms of storage and computation time, at test time it is very fast.

Computation and storage costs: During training, the computation time for passing through the teacher pathway during the first iteration of Algorithm 7.1 is about 2–3 days: it requires processing data from VID and YTO datasets, including running the VideoPCA module. Afterwards, training the first iteration students, with access to six GPUs, takes about 5 days: six GPUs are needed for training the five different student architectures, since training FConv-Net requires two GPUs in parallel. Next, training the EvalSeg-Net requires four additional days on one GPU. At the second iteration, processing the data through the teacher pathway takes about 1 week on six GPUs in parallel—it is more costly due to the larger training set from which only a small percent (about 10%) is kept after selection with EvalSeg-Net in order to have in the end 1M data for training. Finally, training the second-generation students takes 2 additional weeks. In conclusion, the total computation time required for training, with full access to six GPUs is about 5 weeks, when everything is optimized. The total storage cost is about 4TB. At test time the student nets are fast, taking approximately 0.02 s/image, while the ensemble nets take around 0.15 s/image.

7.5   Experimental Analysis

In the first set of experiments, we evaluate the impact of the different components of our system. We experimentally verify that at each iteration the students perform better than their teachers. Then, we test the ability of the system to improve from one generation to the next. We also test the effects of data selection and increasing training data size. Then, we compare the performances of each individual network and their combined ensembles.

In Sect. 7.5.2, we compare our algorithm to state-of-the-art methods on object discovery in videos and images. We do tests on three datasets: YouTube-Objects [3], Object Discovery in Internet images [4], and Pascal-S [5]. In Sect. 7.5.3, we verify that our unsupervised deep features are also useful on a well-known transfer learning for object detection on the Pascal VOC 2012 dataset [67].

Datasets: Unsupervised learning requires large quantities of unlabeled video data. We have chosen for training data, videos from three large datasets: ImageNet VID dataset [46], YouTube-Objects (YTO) [3], and YouTube Bounding Boxes (YouTubeBB) [66]. VID is one of the largest video datasets publicly available, being fully annotated with ground truth bounding boxes. The dataset consists of about 4000 videos, having a total of about 1.2M frames. The videos contain objects that belong to 30 different classes. Each frame could have zero, one, or multiple objects annotated. The benchmark challenge associated with this dataset focuses on the supervised object detection and recognition problem, which is different from the one that we tackle here. Our system is not trained to identify different object categories, so we do not report results compared to the state of the art on object class recognition and detection, on this dataset.

YouTube-Objects (YTO) is a challenging video dataset with objects undergoing strong changes in appearance, scale, and shape, going in and out of occlusion against a varying, often cluttered background. YTO is at its second version now and consists of about 2500 videos, having a total of about 700K frames. It is specifically created for unsupervised object discovery, so we perform comparisons to state of the art on this dataset.

YouTube Bounding Boxes (YTBB or YouTubeBB) is a large-scale video dataset, having approximately 240k videos with single object bounding box annotation. We use a subset of the large number of videos to augment our existing video database. In this dataset, there are 23 types of object categories often undergoing strong changes in appearance, scale, and shape, making it the most difficult dataset we use in our foreground object segmentation setup.

For unsupervised training of our system, we used approximately 200k frames (after selection) from videos chosen from each dataset (120k from VID and 80k from YTO), at learning iteration 1—those frames which survived after the data selection module. At the second learning iteration, besides improving the classifier, it is important to have access to larger quantities of new unlabeled data. Therefore, for training the second generation of classifiers, we enlarge our training dataset to 1 million soft masks as follows: 600k frames from VID + YTO and 400k from the YouTubeBB dataset—those frames which survived after filtering with the EvalSeg-Net data selection module. For experiments presenting results without selection, the frames were randomly chosen from each set, VID, YTO, or YouTubeBB, until the total of 1M was reached. We did not add more frames due to heavy computation and storage limitations.

Evaluation metrics: We use different kinds of metrics in our experiments, which depend on the specific problem that requires either bounding box fitting or fine segmentation:

- *CorLoc*—for evaluating the detection of bounding boxes the most commonly used metric is CorLoc. It is defined as the percentage of images correctly localized according to the PASCAL criterion: $\frac{B_p \cap B_{GT}}{B_p \cup B_{GT}} \geq 0.5$, where B_P is the predicted bounding box and B_{GT} is the ground truth bounding box.
- F-$\beta = \frac{(1-\beta^2)precision \times recall}{\beta^2 \times precision + recall}$ for evaluating the segmentation score on Pascal-S dataset. We use the official evaluation code when reporting results. As in all previous works, we set $\beta^2 = 0.3$.
- *P-J metric*—P refers to the precision per pixel, while J is the Jaccard similarity (the intersection over union between the output mask and the ground truth segmentations). We use this metric only on Object Discovery in Internet images. For computing the reported results we use the official evaluation code.
- *MAE*—Mean Absolute Error is defined as the average pixelwise difference between the predicted mask and the ground truth. Different from the other metrics, for this metric a lower value is better.
- *mean IoU* score is defined as $\frac{|G \cap Y|}{|G \cup Y|}$ where G represents the ground truth and Y represents the predicted mask.

- *mAP* represents the mean average precision. It is used when reporting results for the transfer learning experiments on the Pascal VOC 2012 dataset.

7.5.1 Ablation Study

Student versus teacher: In Fig. 7.8, we present qualitative results on VID dataset as compared to VideoPCA and between iterations. We can see that the masks produced by VideoPCA are of lower quality, often having holes, non-smooth boundaries, and strange shapes. In contrast, the students (at both iterations) learn more general shape and appearance characteristics of objects in images, reminding of the grouping principles governing the basis of visual perception as studied by the Gestalt psychologists [68] and the more recent work on the concept of "objectness" [69]. The object masks produced by the students are simpler, with very few holes, have nicer and smoother shapes, and capture well the foreground-background contrast and organization. Another interesting observation is that the students are sometimes able to detect multiple objects, a feature that is less commonly achieved by the teacher.

A key fact in our experiments with learning over two generations is that every single module becomes better from one iteration to the next: all individual models and the selector (Module C), all improve and each contributes, in a complementary way, along with the addition of extra unlabeled data, to the overall improvement at the next iteration. The result suggests that we can repeat the process over several iterations and continue to improve. It is also encouraging that the individual nets, which see a single image, are able to generalize and detect objects better than what the initial VideoPCA teacher discovers in videos.

As seen in Tables 7.1, 7.2 and 7.3 and Fig. 7.5 at the second generation, we obtain a clear gain over the first, on all experiments and datasets. In Fig. 7.3, left plot shows the significant improvement of the unsupervised selection network at iteration 2 (EvalSeg-Net) versus the simple selection procedure (based on the mean value of white mask pixels) used at iteration 1.

Our proposed algorithm starts from a completely unsupervised object discoverer in video (VideoPCA) and is able to train neural nets for foreground object segmentation, while improving their accuracy over two generations. It uses the students from iteration 1 as teachers at iteration 2. At the second iteration, it also uses more unlabeled training data and it is better at automatically filtering out poor quality segmentations.

Training data size versus learning iteration: Next, we consider the influence of increasing the data size from one iteration to the next versus learning from a more powerful teacher pathway. In order to better understand the importance of each, we have tested our models at each iteration with two training datasets: a smaller set consisting of 200k images (only from VID + YTO datasets) and a larger dataset formed by increasing the dataset size to 1M frames by adding frames from VID, YTO, and also from YouTubeBB. Each generation of student nets is trained using

Table 7.1 Results of our networks and ensembles on YouTube-Objects v1 [3] dataset (CorLoc metric) at both iterations (generations). We present the average of CorLoc metric of all 10 classes from YTO dataset for each model and the ensemble, as well as the average of all single models. As it can be seen, at the second generation, there is a clear increase in performance for all models. Note that, at the second generation, a single model is able to outperform all the methods (single or ensemble) from the first generation. Also, we want to emphasize that the students of the second generation were not trained on the output of Multi-Net, but on the output of individual generation 1 students (Module B), selected with the unsupervised EvalSeg-Net (Module C)

	LowRes-Net	FConv-Net	DenseU-Net	BasicU-Net	DilateU-Net	Avg	Multi-Net
Iteration 1	62.1	57.6	54.6	59.1	61.8	59.0	65.3
Iteration 2	65.7	64.9	59.5	65.5	66.4	64.4	67.1
Gain	3.6 ↑	7.3 ↑	4.9 ↑	6.4 ↑	4.6 ↑	5.4 ↑	1.8 ↑

Table 7.2 Results of our networks and ensemble on Object Discovery in Internet Images [4] dataset (CorLoc metric) at both iterations. We presented the CorLoc metric of single models and the ensemble per class as well as the overall average. Note that at the second generation there is a clear increase in performance for all methods. The students of the second generation were not trained on the output of the Multi-Net ensemble, but on the output of individual generation 1 students (Module B), selected with the unsupervised EvalSeg-Net (Module C)

	LowRes-Net	FConv-Net	DenseU-Net	BasicU-Net	DilateU-Net	Avg	Multi-Net
Iteration 1	85.8	79.8	83.3	86.8	85.6	84.3	85.8
Iteration 2	87.5	84.2	87.9	87.9	88.3	87.2	89.1
Gain	1.7 ↑	4.4 ↑	4.6 ↑	1.1 ↑	2.7 ↑	2.9 ↑	3.3 ↑

Table 7.3 Results of our networks and ensemble on Pascal-S [5] dataset (F-β metric), for all of our methods for first and second generations as well as their average performance. Note that in this case, since we evaluate actual segmentations and not bounding box fitting, nets with higher resolution output do better (DenseU-Net, BasicU-Net, and DilateU-Net). We mention that the students of the second generation were not trained on the output of Multi-Net, but on the output of individual generation 1 students (Module B), selected with the unsupervised EvalSeg-Net (Module C). Again, the ensemble outperforms single models and the second iteration brings a clear gain in every case

	LowRes-Net	FConv-Net	DenseU-Net	BasicU-Net	DilateU-Net	Avg	Multi-Net
Iteration 1	64.6	51.5	65.2	65.4	65.8	62.5	67.8
Iteration 2	66.8	58.9	69.1	67.9	68.2	66.2	69.6
Gain	2.2 ↑	7.4 ↑	3.9 ↑	2.5 ↑	2.4 ↑	3.7 ↑	1.3 ↑

the teacher and selection method corresponding to that particular iteration. We present results in Table 7.4 (mean CorLoc and standard deviation over five students).

The results are interesting: at Iteration 2, as expected we obtain better accuracy when adding more data (by 1.1%). However, at Iteration 1 adding more data helps initially (as seen in Table 7.5 when using the LowRes-Net model), but as data becomes more difficult it could hurt performance. We have a similar change in performance on tests on image segmentation on the Object Discovery dataset, using the same three

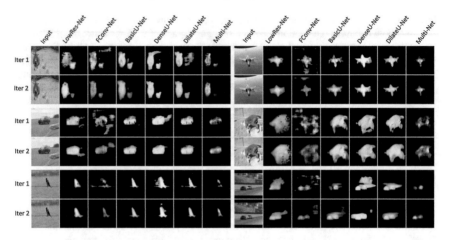

Fig. 7.5 Visual comparison between models at each iteration (generation). The Multi-Net, shown for comparison, represents the pixelwise multiplication between the five models. Note the superior masks at the second-generation students, with better shapes, fewer holes, and sharper edges. Also note the relatively poorer recall of the ensemble Multi-Net, which produces smaller, eroded masks

Table 7.4 Influence of adding more unlabeled data, tested on YTO dataset. The results represent the average of all five students for each iteration trained with the specified number of frames obtained after selection. As it can be seen adding more data increases the performance in iteration 2 when the teacher and the selection module are better

	Training data	No fr	Avg CorLoc
Iteration 1	VID + YTO	200k	59.0 ± 3.1
Iteration 1	VID + YTO + YTBB	1M	58.6 ± 1.2
Iteration 2	VID + YTO	200k	63.3 ± 2.6
Iteration 2	VID + YTO + YTBB	1M	64.4 ± 2.7

training sets on Iteration 1 as in Table 7.5, where the LowRes-Net model initially improves results (meanP from 87.7 to 88.4% and meanJ from 61.2 to 62.3%), and then it starts losing accuracy as data increases from 200k to 1M frames (meanP goes down to 86.8% and meanJ goes down to 60.7%).

This phenomenon, related to observations in Tokmakov et al. [48] when working with more difficult images, is probably due to the weaker teacher path at iteration 1. Images in YouTubeBB are significantly more difficult than the ones from the initial 200k frame sets. Neither VideoPCA nor the very simple selection method used along the teacher pathway, at Iteration 1, are powerful enough to cope with these images and produce good training masks. Therefore, even though we have more masks to train on, their quality is poorer and the overall result degrades.

On the other hand, the second iteration, with a stronger teacher, which is able to produce and select good masks on the more difficult frames from YouTubeBB set is able to take advantage of the extra large amounts of unlabeled data. It is important to increase the data from one generation to the next in order to avoid simply imitating the teacher of the previous generation. The idea of increasing the data size and

Table 7.5 Influence of adding more unlabeled training data at iteration 1 and iteration 2 for the LowRes-Net student net, evaluated on YTO with the CorLoc metric. The number of frames represents the actual number of training frames—that remain after selection. Note that the performance initially increases as data is added for both iterations. When data becomes more difficult only the second iteration LowRes-Net student benefits

	Training data	No fr	CorLoc
LowRes-Net iter1	VID	120k	56.1
LowRes-Net iter1	VID + YTO	200k	62.2
LowRes-Net iter1	VID + YTO + YTBB	1M	58.8
LowRes-Net iter2	VID + YTO	200k	63.7
LowRes-Net iter2	VID + YTO + YTBB	1M	65.7

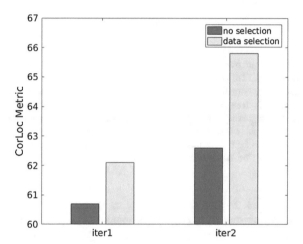

Fig. 7.6 Impact of data selection for both iterations. Data selection (module C) strongly affects the results at each iteration. The results from iteration 2 with no selection are only slightly better than the ones from iteration 1 with selection. Note that the students trained with selection at iteration 1 become the teacher at the second iteration (module B), without selection. The slight improvement is due to the increase in the training data size. The results represent the average over 10 classes on YouTube-Objects using CorLoc percentage metric

complexity in stages, from fewer easy cases to many and more complex ones, is also related to insights from curriculum learning [7].

Data selection versus teacher: Data selection is important (see Figs. 7.6, 7.3, and 7.7 and Table 7.6). The more selective we are, when accepting or rejecting soft masks used for training, the better the end result. Also note that being more selective means decreasing the training set. There is a trade-off between selectivity and training data size.

We study the impact of data selection (Module C) along the teacher pathway w.r.t to the masks produced by the teacher (Module B), which could be a group of students or a single student net learned at the previous iteration. We want to better

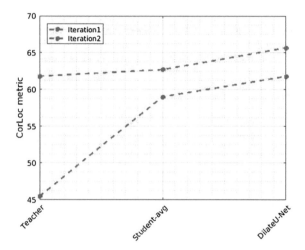

Fig. 7.7 Comparison across two generations (blue line—first iteration; red line—second iteration) when the individual model DilateU-Net trained at Iteration 1 becomes the teacher for the second generation. DilateU-Net is helped along the teacher pathway at the second iteration, by the EvalSeg-Net selection module, which explains the improvement from one iteration to the next. Note that, in this case, DilateU-Net improves while being trained, from scratch, on its own good masks allowed to pass by EvalSeg-Net. Also note that individual students (for which we report average values) outperform the teacher on both iterations. The plots are computed over results on the YouTube-Objects dataset using the CorLoc metric (percentage)

understand the roles of the two modules in learning and how they can work best in combination. We did the following experiments: (1) we trained all our student models at iteration 2 with soft segmentations extracted from Multi-Net created from students trained at iteration one (active module B), but no data selection applied (no module C); (2) then we performed the same experiment as above, but with data selection using EvalSeg-Net (active module C), such that only the masks that passed through selection were used for training; (3) we trained all our models with soft segmentation masks obtained from a single student, DilateU-Net (active module B) and selected using EvalSeg-Net (active module C); and (4) we used as teacher all student models acting independently, with EvalSeg-Net active, such that for each input image we could have several masks, for a unique (image, mask) pair. As stated before, this setup is our choice in the final system.

For these experiments, we used a small set of 200k images. We report average CorLoc on YTO dataset for all five students trained at Iteration 2 with different choices of teacher pathways (Table 7.6). The results indicate the power of data selection, which could overcome the advantage brought by an ensemble. The ensemble is generally stronger than each individual, as it outputs the mask that represents the multiplication of each student soft segmentation. While its output is more robust to noises, it does not guarantee agreement between student models nor quality. In fact, the final mask obtained by multiplication could be destroyed in the process. For example, when a good mask existed among the students that would be lost through

Table 7.6 Different results, averaged over all student nets after being trained at Iteration 2 with different teachers, with or without data selection by EvalSeg-Net. In the cases when data selection was used, we present three results, when the selection is applied to the output of the Multi-Net ensemble (second row), to the output of a single model (third row) and the output of our proposed approach (fourth row), where we applied selection to all masks from all students without multiplying them as in Multi-Net (so we obtain five times more masks from single models). We compared these cases "with selection" with the case of learning from an ensemble without data selection (first row). Note that data selection clearly brings an improvement. Even when a single model is used with data selection as a teacher, we could outperform the case of learning from an ensemble without selection

Teacher	Avg CorLoc
Multi-Net without selection	62.49
Multi-Net with selection	63.32
One model (DilateU-Net) with selection	62.69
All models with selection	63.34

multiplication. This is a limitation of an ensemble which could be overcome by our approach in which all students are allowed to speak, independently and separately. The EvalSeg-Net, which is a mask selection network trained to predict the agreement among the student models, brings in novel, complementary information and whose output is strongly correlated with the goodness of segmentation (Figs. 7.6 and 7.3). Such a network could be used to select only good masks. Thus, any teacher, being it a single model or an ensemble, in combination with the selection module is more powerful than without.

The performance of each trained student is boosted through the selection process by 0.8% on average, when the Multi-Net is used as teacher. The relevance of selection could also be seen in the fact that even a simpler teacher with a single model and no ensemble (third row) can be more effective than the ensemble by itself (by 0.2%). The role of selection is again evident when we compare the average results of models at Iteration 1 and those at Iteration 2 when trained by a single model from Iteration 1 (DilateU-Net) with selection, with an increase by 3.7% (compare results in Tables 7.7 and 7.6).

Maybe the most conclusive result in favor of selection is when the student model (DilateU-Net) itself improves its own performance when trained (from scratch) on its own outputs from Iteration 1, used as teacher, with selection (third row in Table 7.6), by no less than 3.89%, increasing the CorLoc on YTO from 61.8% at Iteration 1 to 65.7% at Iteration 2. This improvement can also be seen in Fig. 7.7 where we presented the results having DilateU-Net acting as a teacher in the second iteration.

The fourth row presents the case when we do not use the Multi-Net ensemble (as teacher), and let all segmentations from all models pass through selection, as explained in Sect. 7.4.3. As we see, the performance of this approach is almost identical to that of using the ensemble with selection (compare rows 2 and 4 in Table 7.6). As previously mentioned, this approach is more effective: if for Multi-Net we need to wait for all five models to produce an output until we consider a mask for selection,

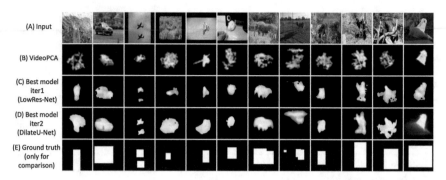

Fig. 7.8 Qualitative results on the VID dataset [46]. For each iteration, we show results of the best individual models, in terms of CorLoc metric. Note the superior quality of our models compared to the VideoPCA (iteration 1 teacher). We also present the ground truth bounding boxes. For more qualitative results, please visit our project page https://sites.google.com/view/unsupervisedlearningfromvideo

in this "All models" case we can pass masks through the selection process as they are produced, in parallel, without having to synchronize all five. For this reason, as discussed previously this is our first choice, generally being referred to as "our proposed approach" and tested in the following sections.

Analysis of different network architectures: As seen in Tables 7.1, 7.2, and 7.3, different network architectures yield different results, while the ensemble always outperforms individual models. Our experiments show that different architectures are better at different tasks. LowRes-Net, for example, does well on the task of box fitting since that does not require a fine sharp object mask. On the other hand, when evaluating the exact segmentation, nets with higher resolution output, such as the ones based on the U-Net design which are more specialized for this task, perform better. Among those, qualitatively, we observed that DenseU-Net produces masks with fewer "holes" when compared to DilateU-Net, which turns out to be the top model for segmentation. The quantitative differences between architectures are shown in Tables 7.1, 7.2, and 7.3, while the qualitative differences can be seen in Figs. 7.5 and 7.8.

7.5.2   Tests on Foreground Segmentation

Object discovery in video: We first did comparisons with methods specifically designed for object discovery in video. For that, we choose the YouTube-Objects dataset and compare it to the best methods on this dataset in the literature (Table 7.7). Evaluations are conducted on both versions of YouTube-Objects dataset, YTOv1 [3] and YTOv2.2 [72]. On YTOv1 we follow the same experimental setup as Jun Koh et al. [70], Prest et al. [3], by running experiments on all annotated frames from the training split. We have not included in Table 7.7 the results reported by Stretcu and Leordeanu [36] because they use a different setup, testing on all videos from YTOv1. It is important to stress out, again, the fact that while the methods presented here for comparison have access to whole video shots, ours only needs a single image at test time. Despite this limitation, our method outperforms the others on 7 out of 10 classes and has the best overall average performance. Note that even our baseline LowRes-Net at the first iteration achieves top performance. The feed-forward CNN processes each image in 0.02 s, being at least one to two orders of magnitude faster than all other methods (see Table 7.7). We also mention that in all our comparisons, while our system is faster at test time, it takes much longer during its training phase and requires large quantities of unsupervised training data.

Object discovery in images: We compare our system against other methods that perform object discovery in images. We use two different datasets for this comparison: Object Discovery in Internet Images and Pascal-S datasets. We report results using metrics that are commonly used for these tasks, as presented at the beginning of the experimental section.

Object Discovery in Internet Images is a representative benchmark for foreground object detection in single images. This set contains Internet images and it is annotated with high detail segmentation masks. In order to enable comparison with previous methods, we use 100 image subsets provided for each of the three categories: airplane, car, and horse. The methods evaluated on this dataset in the literature aim to either discover the bounding box of the main object in a given image or its fine segmentation mask. We evaluate our system on both. Note that different from other works, we do not need a collection of images during test time, since each image can be processed independently by our system. Therefore, unlike other methods, our performance is not affected by the structure of the image collection or the number of classes of interest being present in the collection.

In Table 7.8, we present the performance of our method as compared to other unsupervised object discovery methods in terms of CorLoc on the Object Discovery dataset. We compare our predicted box against the tight box fitted around the ground truth segmentation as done in Cho et al. [12], Tang et al. [73]. Our system can be considered in the mixed class category: it does not depend on the structure of the image collection. It treats each image independently. The performance of other algorithms degrades as the number of main categories increases in the collection (some are not even tested by their authors on the mixed class case), which is not the case with our approach.

We obtain state-of-the-art results on all classes, improving by 6% over the method of Cho et al. [12]. When the method in Cho et al. [12] is allowed to see a collection of images that are limited to a single majority class, its performance improves and it is equal with ours on one class. However, our method has no other information necessary besides the input image, at test time.

We also tested our method on the problem of fine foreground object segmentation and compared it to the best performers in the literature on the Object Discovery dataset in Table 7.9. For refining our soft masks, we apply the GrabCut method, as

Table 7.7 Results on Youtube-Objects dataset, versions v1 [3] and v2.2 [72]. We achieve state-of-the-art results on both versions. Please note that the baseline LowRes-Net already achieves top results on v1, while being close to the best on v2.2. We present results of the top individual models and the ensemble and also keep the baseline LowRes-Net at both iterations, for reference. Note that complete results on this dataset v1 for all models are also presented in Table 7.1. For each column, we highlight with bold the best model and in blue italic the cases where the ensemble is better or equal

Method	Aero	Bird	Boat	Car	Cat	Cow	Dog	Horse	Mbike	Train	Avg	Time	Version
Prest et al. [3]	51.7	17.5	34.4	34.7	22.3	17.9	13.5	26.7	41.2	25.0	28.5	N/A	v1
Papazoglou and Ferrari [65]	65.4	67.3	38.9	65.2	46.3	40.2	65.3	48.4	39.0	25.0	50.1	4 s	
Jun Koh et al. [70]	64.3	63.2	73.3	68.9	44.4	62.5	71.4	52.3	**78.6**	23.1	60.2	N/A	
Haller and Leordeanu [71]	76.3	**71.4**	65.0	58.9	**68.0**	55.9	70.6	33.3	69.7	42.4	61.1	0.35 s	
LowRes-Net_iter1	77.0	67.5	**77.2**	68.4	54.5	68.3	72.0	**56.7**	44.1	34.9	62.1	0.02 s	
LowRes-Net_iter2	**83.3**	**71.4**	74.3	69.6	57.4	**80.0**	77.3	**56.7**	50.0	37.2	65.7	0.02 s	
DilateU-Net_iter2	**83.3**	66.2	**77.2**	**70.9**	63.4	75.0	**80.0**	53.3	50.0	**44.2**	**66.4**	0.02 s	
Multi-Net_iter2 (ensemble)	*87.4*	*72.7*	*77.2*	*64.6*	*62.4*	*75.0*	*82.7*	*56.7*	*52.9*	*39.5*	*67.1*	*0.15 s*	
Haller and Leordeanu [71]	76.3	**68.5**	**54.5**	50.4	**59.8**	42.4	53.5	30.0	**53.5**	**60.7**	54.9	0.35 s	v2.2
LowRes-Net_iter1	75.7	56.0	52.7	57.3	46.9	57.0	48.9	44.0	27.2	56.2	52.2	0.02 s	
LowRes-Net_iter2	79.0	48.2	51.0	**62.1**	46.9	65.7	55.3	**50.6**	36.1	52.4	54.7	0.02 s	
DilateU-Net_iter2	**84.3**	49.9	52.7	61.4	50.3	**68.8**	**56.4**	47.1	36.1	56.7	**56.4**	0.02 s	
Multi-Net_iter2 (ensemble)	*83.1*	*53.2*	*54.3*	*63.7*	*50.6*	*69.2*	*61.0*	*51.1*	*37.2*	*48.7*	*57.2*	*0.15 s*	

Table 7.8 Results on the Object Discovery in Internet images [4] dataset (CorLoc metric). The results obtained in the first iteration are further improved in the second one. We present the best single models and ensemble along with the baseline LowRes-Net at both iterations. Among the single models, DilateU-Net is often the best when evaluating box fitting. For each column, we highlight with bold the best model and in blue italic the cases where the ensemble is better or equal

Method	Airplane	Car	Horse	Avg
Kim et al. [27]	21.95	0.00	16.13	12.69
Joulin et al. [26]	32.93	66.29	54.84	51.35
Joulin et al. [28]	57.32	64.04	52.69	58.02
Rubinstein et al. [4]	74.39	87.64	63.44	75.16
Tang et al. [73]	71.95	93.26	64.52	76.58
Cho et al. [12]	82.93	94.38	*75.27*	84.19
Cho et al. [12] mixed	81.71	94.38	70.97	82.35
LowRes-Net$_{iter1}$	87.80	**95.51**	74.19	85.83
LowRes-Net$_{iter2}$	93.90	93.25	**75.27**	87.47
DilateU-Net$_{iter2}$	**95.12**	**95.51**	74.19	**88.27**
Multi-Net$_{iter2}$ (ensemble)	*97.56*	*95.51*	74.19	*89.09*

it is available in OpenCV. We evaluate based on the same P, J evaluation metric as described by Rubinstein et al. [4]—the higher P and J, the better. In Figs. 7.9 and 7.10, we present some qualitative results for each class. As mentioned previously, these segmentation experiments on Object Discovery in Internet Images are the only ones on which we apply GrabCut as a post-processing step, as also used by all other methods presented in Table 7.9.

Another important dataset used for the evaluation of a related problem, that of salient object detection, is Pascal-S dataset, consisting of 850 images annotated with segmentation mask. As seen from Table 7.10, we achieve top results on all three metrics against methods that do not use any supervised pretrained features. Being a foreground object detection method, our approach is usually biased towards the main object in the image—even though it can also detect multiple ones. Images in Pascal-S usually have more objects, so we consider our results very encouraging being close to approaches that use features pretrained in a supervised manner. Also note that we did not use GrabCut for these experiments.

On single-image experiments, our system was trained, as discussed before on other, video datasets (VID, YTO, and YTBB). It has not previously seen any of the images in Pascal-S or Object Discovery datasets during training.

7.5.3 Tests on Transfer Learning

While our main focus here is unsupervised learning of foreground object segmentation, we also want to test the usefulness of our features in a transfer learning setup,

Table 7.9 Results on the Object Discovery in Internet images [4] dataset using (P, J metric) on segmentation evaluation. We present results of the top single model and the ensemble, along with LowRes-Net at both iterations. On the problem of fine object segmentation, the best individual model tends to be DenseU-Net as also mentioned in the text. Note that we applied GrabCut only on these experiments as a post-processing step, since all methods reported in this table also used it. For each column, we highlight with bold the best model and in blue italic the cases where the ensemble is better

	Airplane		Car		Horse	
	P	J	P	J	P	J
Kim et al. [27]	80.20	7.90	68.85	0.04	75.12	6.43
Joulin et al. [26]	49.25	15.36	58.70	37.15	63.84	30.16
Joulin et al. [28]	47.48	11.72	59.20	35.15	64.22	29.53
Rubinstein et al. [4]	88.04	55.81	85.38	64.42	82.81	51.65
Chen et al. [74]	90.25	40.33	87.65	64.86	86.16	33.39
LowRes-Net$_{iter1}$	**91.41**	61.37	86.59	70.52	87.07	55.09
LowRes-Net$_{iter2}$	90.70	63.15	**87.00**	73.24	**87.78**	**55.67**
DenseU-Net$_{iter2}$	90.27	**63.37**	86.08	**73.25**	87.40	55.49
Multi-Net$_{iter2}$ (ensemble)	91.39	*65.07*	86.61	73.09	*88.34*	55.53

Table 7.10 Results on the PASCAL-S dataset compared against other unsupervised methods. For MAE score lower is better, while for F_β and mean IoU higher is better. We reported max F_β, min MAE, and max mean IoU for every method. In bold, we presented the top results when no supervised pretrained features were used per single model and in blue italic the cases where the ensemble is better

Method	F_β	MAE	Mean IoU	Pretrained supervised features?
Wei et al. [75]	56.2	22.6	41.6	No
Li et al. [76]	56.8	19.2	42.4	No
Zhu et al. [77]	60.0	19.7	43.9	No
Yang et al. [78]	60.7	21.7	43.8	No
Zhang et al. [79]	60.8	20.2	44.3	No
Tu et al. [80]	60.9	19.4	45.3	No
Zhang et al. [81]	68.0	*14.1*	*54.9*	Init VGG
LowRes-Net$_{iter1}$	64.6	19.6	48.7	No
LowRes-Net$_{iter2}$	66.8	18.2	**51.7**	No
DenseU-Net$_{iter2}$	**69.1**	**17.6**	50.9	No
Multi-Net$_{iter2}$ (ens)	*69.6*	19.5	*52.3*	No

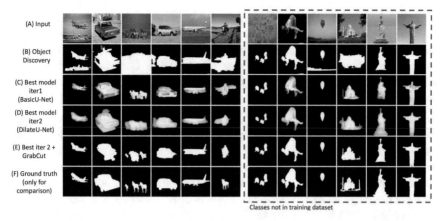

Fig. 7.9 Qualitative results on the Object Discovery dataset as compared to (B) [4]. For both iterations, we present the results of the top model (C, D), without using GrabCut. We also present results when GrabCut is used with the top model (E). Note that our models are able to segment objects from classes that were not present in the training set (examples on the right side). Also, note that the initial VideoPCA teacher cannot be applied on single images

Fig. 7.10 Qualitative results on the object discovery in Internet images [4] dataset. For each example, we show the input RGB image and immediately below our segmentation result, with GrabCut post-processing for obtaining a hard segmentation. Note that our method produces good quality segmentation results, even in images with cluttered background

namely, for the problem of multiple object detection. For this purpose, we follow the well-known experimental setup used in the recent transfer learning literature, in which an AlexNet-like [82] network, initialized in an unsupervised way, is fine-tuned on supervised object detection within the Fast R-CNN framework [83].[2] We closely follow the work of Pathak et al. [6], with code, documentation, and training

---

[2]https://github.com/rbgirshick/py-faster-rcnn.

data available online, which, as mentioned in the related work section, also starts by learning from videos to segment objects in single images in an unsupervised fashion. In these experiments, we adapted in the same way the last part of AlexNet in order to produce a soft segmentation of the image (instead of an output class). We used the same base architecture as the methods we compared against, to make sure that the results come from the learned features, not from the architecture we used.

Initial unsupervised training for object segmentation: We used the adapted AlexNet-based model (which we term AlexNet-Seg) described in this section as a student in our unsupervised learning framework at iteration 2. Thus, the AlexNet-Seg will be trained by the unsupervised teacher pathway at our Iteration 2—in this case, the teacher will be a single network, namely, DialtedU-Net at module B, combined with the EvalSeg-Net mask selector at module C. In order to see how the actual training data influences the final transfer learning outcome, we experimented with both our data and the data used by Pathak et al. [6] which is obtained from YFCC100m [84] dataset, having 1.6M frames. The results are presented in Table 7.13.

As it is, our unsupervised learning system prefers to segment main, foreground objects in images and it is less versatile on segmenting complex images containing many objects. Since the images in Pascal VOC 2012 contain complex scenes with multiple objects and the final transfer learning problem is of multiple object detection, we also tested the case when we adapted our system to better cope with multiple objects. For that, we divide each training image into five large patches (a grid with one image at each corner and one in the middle, each crop being about 60% of the original size for both dimensions) which we pass through the teacher pathway at Iteration 2. The results are combined into a single image, by superimposing the soft masks and taking the maximum over all, at each location in the original image. Thus, we obtain soft segmentations that better capture multiple objects in the input image. Note that the original image passed through the teacher pathway without the five-point grid division is referred to as the Single Object (SO) teacher in Tables 7.12 and 7.13 and Fig. 7.11, while the five-grid version just described is referred to as the Multiple Object (MO) teacher in the same tables and figures. We train the AlexNet-Seg student on these multiple object soft-segs. We thus transfer knowledge from our unsupervised student models to AlexNet-Seg and prepare it for the next task of supervised object detection.

Transferring to object detection: Similar to the other methods we compare to, we conduct transfer learning experiments on the Pascal VOC 2012 [67] dataset. We train on the *train* split of VOC 2012 and we report our results on the *validation* split. We also use multi-scale training and testing and remove difficult objects during training. We report the comparison results in Table 7.11. We see that the unsupervised knowledge learned by our approach is indeed useful for transfer learning, as our results are in the top three among current published methods. This is interesting, as our unsupervised learning algorithm is mainly designed for foreground object segmentation, not classification.

Foreground segmentation versus multiple object detection: The transfer learning problem, which we test our approach on, is both about localization (detection) and

Table 7.11 Comparison with state of the art on VOC 2012 [67] using Fast R-CNN initialized with different methods. The numbers represent the mAP for the validation split of VOC 2012. Each column represents to what extent the network was fine-tuned, so, for example, >c1 represents that all layers after the first convolution are fine-tuned, while "All" represents the case where all the layers are fine-tuned

Method	All	>c1	>c2	>c3	>c4	>c5
Agrawal et al. [85]	37.4	36.9	34.4	28.9	24.1	17.1
Pathak et al. [20]	39.1	36.4	34.1	29.4	24.8	13.4
Krähenbühl et al. [86]	42.8	42.2	40.3	37.1	32.4	26.0
Owens et al. [87]	42.9	42.3	40.6	37.1	32.0	26.5
Wang and Gupta [16]	43.5	44.6	44.6	44.2	41.5	35.7
Zhang et al. [88]	43.8	45.6	45.6	46.1	44.1	37.6
Zhang et al. [89]	44.5	44.9	44.7	44.4	42.6	38.0
Donahue et al. [90]	44.9	44.6	44.7	42.4	38.4	29.4
Pathak et al. [6]	48.6	48.2	**48.3**	**47.0**	**45.8**	**40.3**
Doersch et al. [17]	**49.9**	**48.8**	44.4	44.3	42.1	33.2
Ours—AlexNet-Seg (MO)	46.8	46.2	43.9	41.0	36.5	29.9

Table 7.12 Comparison with [6] on unsupervised object discovery. For the YTO and Obj-Disc datasets, we report the CorLoc metric and for Pascal-S the F_β metric. With SO we represent our AlexNet-Seg student trained from a Single Object (SO) teacher, while with MO, we represent our student trained from a Multiple Object (MO) teacher—as explained in the text. As it can be seen, our proposed multi-object scheme affects in a negative way the results on foreground object segmentation. This happens, because these datasets have mainly one single object

Method	YTO	Obj-Disc	Pascal-S	Training data
Pathak et al. [6]	39.4	82.1	64.7	YFCC100m
AlexNet-Seg (SO)	62.7	86.4	67.4	Our data
AlexNet-Seg (SO)	56.6	83.6	65.5	YFCC100m
AlexNet-Seg (MO)	54.7	86.0	65.9	Our data
AlexNet-Seg (MO)	58.3	84.4	65.2	YFCC100m
AlexNet-Seg (MO)	58.1	85.2	66.1	Both

classification. In this context, as already discussed in the introduction section, the work of Pathak et al. [6] is most related to ours in the sense that they also learn from video to segment objects in single images in an unsupervised manner. They do so by using a teacher that produces soft masks from optical flow in video. Beyond the theoretical connection between the two works, we wanted to better understand how the two approaches relate on actual foreground segmentation experiments. Their method generally produces masks that cover larger areas in the image than ours and are better suited for transfer learning experiments, as results show.

On the other hand, when tested on foreground segmentation tasks, our approach, in turn, seems to yield better results (see Table 7.12). The results we obtained are

Table 7.13 Comparison between different types of training images and training data for the AlexNet-Seg student we used. With SO we represent our AlexNet-Seg student trained from a single object (SO) teacher, while with MO, we represent our student trained from a Multiple Object (MO) teacher—as explained in the text. We show how the training data affects the transfer learning results. The numbers represent the mAP metric for the Pascal VOC 2012 validation split when fine-tuning the whole network. Please note that the proposed multiple object approach brings only a small improvement. Also, on these transfer learning tests, the larger training dataset—YFCC100m—tends to bring improved results, by a small margin

Model	VOC 2012 (mAP)	Training data
AlexNet-Seg (SO)	46.2	Our data
AlexNet-Seg (SO)	45.7	YFCC100m
AlexNet-Seg (MO)	46.1	Our data
AlexNet-Seg (MO)	46.8	YFCC100m
AlexNet-Seg (MO)	46.8	Both

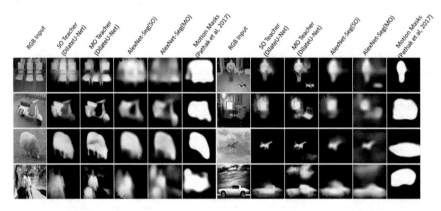

Fig. 7.11 Representative visual results on Pascal VOC 2012. We show the output of the unsupervised teacher pathway used in transfer learning for both the original case (SO—DilateU-Net) and for the multiple objects scenario (MO—combined DilateU-Net outputs combined on a five-grid), the trained students for both cases (AlexNet-Seg (SO) and AlexNet-Seg (MO)), as well as the output of Pathak et al. [6]

in agreement with observations made by the authors when testing their method on detecting main objects in single frames against human annotations (e.g., Precision: 29.9, Recall: 59.3, Mean IoU: 24.8). Their high recall and lower precision agree with our observations that their segmentation covers larger parts of the image, while ours provides sharper and smaller masks. This observation leads us towards extracting large crops on a five-point grid and combining the results (termed AlexNet-Seg MO). As seen in experiments, taking multiple outputs over the grid eventually brought a relatively small improvement of 0.6% (see Table 7.13). In Fig. 7.11, we present qualitative results of our DilatedU-Net teacher for the AlexNet-Seg student trained with a single foreground object detected (termed SO Teacher) and our five-grid multiple objects (termed MO Teacher) segmentation result. We also present the results of our AlexNet-Seg student on both cases as well as the outputs of Pathak et al. [6] for comparison. While our method has better results on the task of segmentation, their method is more suited for transfer learning experiments. We suspect that their larger masks (Fig. 7.11), with lower precision but relatively high recall and high confidence values, could be more flexible and less conservative for the final transfer learning stage where multiple objects need to be detected over the whole image. At the same time, ours is specialized in obtaining generally sharper and better quality foreground segmentation masks. Overall, the transfer learning experiments show that our approach is suited for such task, as we obtain a performance that is in the top three among the state-of-the-art methods using the same experimental setup.

7.5.4 Concluding Remarks on Experiments

One of the interesting conclusions in our experimental analysis is that the system is able to improve its performance from Iteration 1 to Iteration 2. There are several factors that are involved in this outcome, which we believe are related through the following relationship: (1) multiple students of diverse structures ensure diversity and somewhat independent mistakes; (2) in turn, point (1) makes possible the unsupervised training of a mask selection module that learns to predict agreements; (3) thus, the selection module at (2) becomes a good mask evaluation network; (4) once that evaluation network (from 3) is available, we can then add larger and potentially more complex data to select a larger set with good object masks of more interesting cases at the next iteration; (5) finally, (4) ensures the improvement at the next iteration and now we could return to point (1).

7.6 Concluding Discussion on Unsupervised Learning

The ultimate goal of unsupervised learning might not be about matching the performance of the supervised case but rather about reaching beyond the capabilities of the classical supervised scenario. An unsupervised system should be able to learn and recognize different object classes, such as animals, plants, and man-made objects, as they evolve and change over time, from the past and into the unknown future. It should also be able to learn about new classes that might be formed, in relation to others, maybe known ones. We see this case as fundamentally different from the supervised one in which the classifier is forced to learn from a distribution of samples that is fixed and limited to a specific period of time—that when the human labeling was done. Therefore, in the supervised learning paradigm, a car from the future should not be classified as car, because it is not a car according to the supervised distribution of cars given at present training time, when human annotations are collected. On the other hand, a system that learns by itself should be able to track how cars have been changing in time and recognize such objects as "cars"—with no step-by-step human intervention.

Current unsupervised learning methods might still not be able to learn profound semantic information [25], but the ability to learn to segment foreground objects in an unsupervised fashion constitutes evidence that we are moving in the right direction. In order to understand and learn about semantic classes, the system would need to learn by itself about how such objects interact with each other and what role they play within the larger spatiotemporal story. While our unsupervised methods are still far from reaching this level of interpretation, the ability to learn about and detect objects that constitute the foreground within their local spatial context could constitute an important building block. It is an element that could be used to further learn about more complex interactions and behavior in both space and time.

From the larger spatiotemporal perspective, unsupervised learning is about continuous learning and adaptation to huge quantities of data that are perpetually changing. Human annotation is extremely limited in an ocean of data and not able to provide the so-called "ground truth" information continuously. Therefore, unsupervised learning, and especially its weaker version—learning from large quantities of data with minimal human intervention—will soon become a core part, larger than the supervised one, in the future of artificial intelligence.

7.7 Overall Conclusions and Future Work

In this chapter, we present a novel and effective approach to learning from large collections of images and videos, in an unsupervised fashion, to segment foreground objects in single images. We present a relatively general algorithm for this task, which offers the possibility of learning several generations of students and teachers. We demonstrate in practice that the system improves its performance over the course of two generations. We also test the impact of the different system components on performance and show state-of-the-art results on three different datasets, while also showing top performance on challenging transfer learning experiments. Our system is one of the first in the literature that learns to detect and segment foreground objects in images in an unsupervised fashion, with no pretrained features given or manual labeling, while requiring only a single image at test time.

The convolutional networks trained along the student pathway are able to learn general "objectness" characteristics, which include good form, closure, smooth contours, as well as contrast with the background. What the simpler initial VideoPCA teacher discovers over time, the deep, complex student is able to learn across several layers of image features at different levels of abstraction. Our results on transfer learning experiments are also encouraging and show additional cases in which such a system could be useful. In future work, with extended computational and storage capabilities, it would be interesting to demonstrate the power of our unsupervised learning algorithm along many generations of student and teacher networks. The presented approach, tested here in extensive experiments, could bring a valuable contribution to computer vision research by giving the possibility of continuous unsupervised learning within the teacher-student paradigm, where the teacher is essentially the entire previous generation of students. This immediately suggests the next idea, which is presented conceptually in the next chapter.

7.7.1 Towards a Universal Visual Learning Machine

Let us imagine that we use an entire population of neural networks that address all known aspects of visual recognition and we link them through other connecting networks, which take the output of one and attempt to predict the output of another. Wouldn't it be possible, for any given visual learning problem and any given student network, to use all of the others collectively as a teacher in an unsupervised setting? What are the limitations of this idea, how much pretraining do we need and what constraints would we have with respect to the set of visual tasks considered and amount and type of unlabeled data used? These are the questions of our next and final chapter.

References

1. Croitoru I, Bogolin SV, Leordeanu M (2017) Unsupervised learning from video to detect foreground objects in single images. In: 2017 IEEE international conference on computer vision (ICCV). IEEE, pp 4345–4353
2. Croitoru I, Bogolin SV, Leordeanu M (2019) Unsupervised learning of foreground object segmentation. Int J Comput Vis 1–24
3. Prest A, Leistner C, Civera J, Schmid C, Ferrari V (2012) Learning object class detectors from weakly annotated video. In: CVPR
4. Rubinstein M, Joulin A, Kopf J, Liu C (2013) Unsupervised joint object discovery and segmentation in internet images. In: CVPR
5. Li Y, Hou X, Koch C, Rehg JM, Yuille AL (2014) The secrets of salient object segmentation. Georgia Institute of Technology
6. Pathak D, Girshick R, Dollar P, Darrell T, Hariharan B (2017) Learning features by watching objects move. In: CVPR
7. Bengio Y, Louradour J, Collobert R, Weston J (2009) Curriculum learning. In: Proceedings of the 26th annual international conference on machine learning. ACM, pp 41–48
8. Radenović F, Tolias G, Chum O (2016) CNN image retrieval learns from bow: unsupervised fine-tuning with hard examples. In: ECCV
9. Misra I, Zitnick CL, Hebert M (2016) Shuffle and learn: unsupervised learning using temporal order verification. In: ECCV
10. Li D, Hung WC, Huang JB, Wang S, Ahuja N, Yang MH (2016) Unsupervised visual representation learning by graph-based consistent constraints. In: ECCV
11. Jain AK, Murty MN, Flynn PJ (1999) Data clustering: a review. ACM Comput Surv (CSUR) 31(3):264–323
12. Cho M, Kwak S, Schmid C, Ponce J (2015) Unsupervised object discovery and localization in the wild: Part-based matching with bottom-up region proposals. In: Proceedings of the IEEE conference on computer vision and pattern recognition, pp 1201–1210

13. Sivic J, Russell B, Efros A, Zisserman A, Freeman W (2005) Discovering objects and their location in images. In: ICCV
14. Raina R, Battle A, Lee H, Packer B, Ng AY (2007) Self-taught learning: transfer learning from unlabeled data. In: Proceedings of the 24th international conference on Machine learning. ACM, pp 759–766
15. Lee HY, Huang JB, Singh M, Yang MH (2017) Unsupervised representation learning by sorting sequences. In: 2017 IEEE international conference on computer vision (ICCV). IEEE, pp 667–676
16. Wang X, Gupta A (2015) Unsupervised learning of visual representations using videos. arXiv:150500687
17. Doersch C, Gupta A, Efros AA (2015) Unsupervised visual representation learning by context prediction. In: Proceedings of the IEEE international conference on computer vision, pp 1422–1430
18. Larsson G, Maire M, Shakhnarovich G (2016) Learning representations for automatic colorization. In: European conference on computer vision. Springer, pp 577–593
19. Noroozi M, Favaro P (2016) Unsupervised learning of visual representations by solving Jigsaw puzzles. In: European conference on computer vision. Springer, pp 69–84
20. Pathak D, Krahenbuhl P, Donahue J, Darrell T, Efros AA (2016) Context encoders: feature learning by inpainting. In: Proceedings of the IEEE conference on computer vision and pattern recognition, pp 2536–2544
21. Finn C, Goodfellow I, Levine S (2016) Unsupervised learning for physical interaction through video prediction. In: Advances in neural information processing systems, pp 64–72
22. Xue T, Wu J, Bouman K, Freeman B (2016) Visual dynamics: probabilistic future frame synthesis via cross convolutional networks. In: Advances in neural information processing systems, pp 91–99
23. Goroshin R, Mathieu MF, LeCun Y (2015) Learning to linearize under uncertainty. In: Advances in neural information processing systems, pp 1234–1242
24. Wang X, Gupta A (2015) Unsupervised learning of visual representations using videos. In: The IEEE international conference on computer vision (ICCV)
25. Bau D, Zhou B, Khosla A, Oliva A, Torralba A (2017) Network dissection: quantifying interpretability of deep visual representations. In: International conference on computer vision and pattern recognition (CVPR)
26. Joulin A, Bach F, Ponce J (2010) Discriminative clustering for image co-segmentation. In: CVPR
27. Kim G, Xing E, Fei-Fei L, Kanade T (2011) Distributed cosegmentation via submodular optimization on anisotropic diffusion. In: ICCV
28. Joulin A, Bach F, Ponce J (2012) Multi-class cosegmentation. In: CVPR
29. Kuettel D, Guillaumin M, Ferrari V (2012) Segmentation propagation in ImageNet. In: ECCV
30. Vicente S, Rother C, Kolmogorov V (2011) Object cosegmentation. In: CVPR
31. Rubio J, Serrat J, López A (2012) Video co-segmentation. In: ACCV
32. Leordeanu M, Sukthankar R, Hebert M (2012) Unsupervised learning for graph matching. Int J Comput Vis 96:28–45
33. Deselaers T, Alexe B, Ferrari V (2012) Weakly supervised localization and learning with generic knowledge. IJCV 100(3)
34. Nguyen M, Torresani L, la Torre FD, Rother C (2009) Weakly supervised discriminative localization and classification: a joint learning process. In: CVPR
35. Siva P, Russell C, Xiang T, Agapito L (2013) Looking beyond the image: unsupervised learning for object saliency and detection. In: CVPR
36. Stretcu O, Leordeanu M (2015) Multiple frames matching for object discovery in video. In: BMVC, pp 186.1–186.12
37. Leordeanu M, Collins R, Hebert M (2005) Unsupervised learning of object features from video sequences. In: CVPR
38. Parikh D, Chen T (2007) Unsupervised identification of multiple objects of interest from multiple images: discover. In: Asian conference on computer vision

39. Liu D, Chen T (2007) A topic-motion model for unsupervised video object discovery. In: CVPR
40. Joulin A, Tang K, Fei-Fei L (2014) Efficient image and video co-localization with Frank-Wolfe algorithm. In: ECCV
41. Rochan M, Wang Y (2014) Efficient object localization and segmentation in weakly labeled videos. In: Advances in visual computing. Springer, pp 172–181
42. Lee YJ, Kim J, Grauman K (2011) Key-segments for video object segmentation. In: 2011 IEEE international conference on computer vision (ICCV). IEEE, pp 1995–2002
43. Cheng J, Tsai YH, Wang S, Yang MH (2017) Segflow: joint learning for video object segmentation and optical flow. In: The IEEE international conference on computer vision (ICCV)
44. Dutt Jain S, Xiong B, Grauman K (2017) Fusionseg: learning to combine motion and appearance for fully automatic segmentation of generic objects in videos. In: The IEEE conference on computer vision and pattern recognition (CVPR)
45. Tokmakov P, Alahari K, Schmid C (2017) Learning motion patterns in videos. In: The IEEE conference on computer vision and pattern recognition (CVPR)
46. Russakovsky O, Deng J, Su H, Krause J, Satheesh S, Ma S, Huang Z, Karpathy A, Khosla A, Bernstein M et al (2015) Imagenet large scale visual recognition challenge. IJCV 115(3)
47. Everingham M, Eslami SMA, Van Gool L, Williams CKI, Winn J, Zisserman A (2015) The pascal visual object classes challenge: a retrospective. Int J Comput Vis 111(1):98–136
48. Tokmakov P, Alahari K, Schmid C (2016) Learning semantic segmentation with weakly-annotated videos. In: ECCV, vol 1, p 6
49. Khoreva A, Benenson R, Hosang JH, Hein M, Schiele B (2017) Simple does it: weakly supervised instance and semantic segmentation. In: CVPR, vol 1, p 3
50. Rother C, Kolmogorov V, Blake A (2004) Grabcut: interactive foreground extraction using iterated graph cuts. In: ACM transactions on graphics (TOG), vol 23. ACM, pp 309–314
51. Borji A, Sihite D, Itti L (2012) Salient object detection: a benchmark. In: ECCV
52. Cheng M, Mitra N, Huang X, Torr P, Hu S (2015) Global contrast based salient region detection. PAMI 37(3)
53. Barnich O, Van Droogenbroeck M (2011) Vibe: a universal background subtraction algorithm for video sequences. IEEE Trans Image Process 20(6):1709–1724
54. Raiko T, Valpola H, LeCun Y (2012) Deep learning made easier by linear transformations in perceptrons. AISTATS 22:924–932
55. Pinheiro PO, Lin TY, Collobert R, Dollár P (2016) Learning to refine object segments. In: ECCV
56. Long J, Shelhamer E, Darrell T (2015) Fully convolutional networks for semantic segmentation. In: Proceedings of the IEEE conference on computer vision and pattern recognition, pp 3431–3440
57. Ronneberger O, Fischer P, Brox T (2015) U-net: convolutional networks for biomedical image segmentation. In: International conference on medical image computing and computer-assisted intervention. Springer, pp 234–241
58. Yu F, Koltun V (2015) Multi-scale context aggregation by dilated convolutions. arXiv:151107122
59. Jégou S, Drozdzal M, Vazquez D, Romero A, Bengio Y (2017) The one hundred layers tiramisu: fully convolutional densenets for semantic segmentation. In: 2017 IEEE conference on computer vision and pattern recognition workshops (CVPRW). IEEE, pp 1175–1183
60. Abadi M, Agarwal A, Barham P, Brevdo E, Chen Z, Citro C, Corrado GS, Davis A, Dean J, Devin M, Ghemawat S, Goodfellow I, Harp A, Irving G, Isard M, Jia Y, Jozefowicz R, Kaiser L, Kudlur M, Levenberg J, Mané D, Monga R, Moore S, Murray D, Olah C, Schuster M, Shlens J, Steiner B, Sutskever I, Talwar K, Tucker P, Vanhoucke V, Vasudevan V, Viégas F, Vinyals O, Warden P, Wattenberg M, Wicke M, Yu Y, Zheng X (2015) TensorFlow: large-scale machine learning on heterogeneous systems. Software available from http://www.tensorflow.org
61. Kingma D, Ba J (2014) Adam: a method for stochastic optimization. arXiv:14126980
62. Hou X, Zhang L (2007) Saliency detection: a spectral residual approach. In: 2007 IEEE conference on computer vision and pattern recognition CVPR'07. IEEE, pp 1–8

63. Jiang H, Wang J, Yuan Z, Wu Y, Zheng N, Li S (2013) Salient object detection: a discriminative regional feature integration approach. In: Proceedings of the IEEE conference on computer vision and pattern recognition, pp 2083–2090
64. Cucchiara R, Grana C, Piccardi M, Prati A (2003) Detecting moving objects, ghosts, and shadows in video streams. PAMI 25(10)
65. Papazoglou A, Ferrari V (2013) Fast object segmentation in unconstrained video. In: Proceedings of the IEEE international conference on computer vision, pp 1777–1784
66. Real E, Shlens J, Mazzocchi S, Pan X, Vanhoucke V (2017) Youtube-boundingboxes: a large high-precision human-annotated data set for object detection in video. In: 2017 IEEE conference on computer vision and pattern recognition (CVPR). IEEE, pp 7464–7473
67. Everingham M, Van Gool L, Williams CK, Winn J, Zisserman A (2010) The pascal visual object classes (VOC) challenge. Int J Comput Vis 88(2):303–338
68. Rock I, Palmer S (1990) Gestalt psychology. Sci Am 263:84–90
69. Alexe B, Deselaers T, Ferrari V (2010) What is an object? In: Computer vision and pattern recognition
70. Jun Koh Y, Jang WD, Kim CS (2016) POD: discovering primary objects in videos based on evolutionary refinement of object recurrence, background, and primary object models. In: Proceedings of the IEEE conference on computer vision and pattern recognition, pp 1068–1076
71. Haller E, Leordeanu M (2017) Unsupervised object segmentation in video by efficient selection of highly probable positive features. In: The IEEE international conference on computer vision (ICCV)
72. Kalogeiton V, Ferrari V, Schmid C (2016) Analysing domain shift factors between videos and images for object detection. PAMI 38(11)
73. Tang K, Joulin A, Li LJ, Fei-Fei L (2014) Co-localization in real-world images. In: CVPR
74. Chen X, Shrivastava A, Gupta A (2014) Enriching visual knowledge bases via object discovery and segmentation. In: CVPR
75. Wei Y, Wen F, Zhu W, Sun J (2012) Geodesic saliency using background priors. In: European conference on computer vision. Springer, pp 29–42
76. Li N, Sun B, Yu J (2015) A weighted sparse coding framework for saliency detection. In: Proceedings of the IEEE conference on computer vision and pattern recognition, pp 5216–5223
77. Zhu W, Liang S, Wei Y, Sun J (2014) Saliency optimization from robust background detection. In: Proceedings of the IEEE conference on computer vision and pattern recognition, pp 2814–2821
78. Yang C, Zhang L, Lu H, Ruan X, Yang MH (2013) Saliency detection via graph-based manifold ranking. In: 2013 IEEE conference on computer vision and pattern recognition (CVPR). IEEE, pp 3166–3173
79. Zhang J, Sclaroff S, Lin Z, Shen X, Price B, Mech R (2015) Minimum barrier salient object detection at 80 fps. In: Proceedings of the IEEE international conference on computer vision, pp 1404–1412
80. Tu WC, He S, Yang Q, Chien SY (2016) Real-time salient object detection with a minimum spanning tree. In: Proceedings of the IEEE conference on computer vision and pattern recognition, pp 2334–2342
81. Zhang D, Han J, Zhang Y (2017) Supervision by fusion: towards unsupervised learning of deep salient object detector. In: Proceedings of the IEEE conference on computer vision and pattern recognition, pp 4048–4056
82. Krizhevsky A, Sutskever I, Hinton GE (2012) Imagenet classification with deep convolutional neural networks. In: Advances in neural information processing systems, pp 1097–1105
83. Girshick R (2015) Fast R-CNN. In: Proceedings of the IEEE international conference on computer vision, pp 1440–1448
84. Thomee B, Shamma DA, Friedland G, Elizalde B, Ni K, Poland D, Borth D, Li LJ (2015) Yfcc100m: the new data in multimedia research. arXiv:150301817
85. Agrawal P, Carreira J, Malik J (2015) Learning to see by moving. In: Proceedings of the IEEE international conference on computer vision, pp 37–45

86. Krähenbühl P, Doersch C, Donahue J, Darrell T (2015) Data-dependent initializations of convolutional neural networks. arXiv:151106856
87. Owens A, Wu J, McDermott JH, Freeman WT, Torralba A (2016) Ambient sound provides supervision for visual learning. In: European conference on computer vision. Springer, pp 801–816
88. Zhang R, Isola P, Efros AA (2017) Split-brain autoencoders: unsupervised learning by cross-channel prediction. In: CVPR, vol 1, p 5
89. Zhang R, Isola P, Efros AA (2016) Colorful image colorization. In: European conference on computer vision. Springer, pp 649–666
90. Donahue J, Krähenbühl P, Darrell T (2016) Adversarial feature learning. arXiv:160509782

Chapter 8
Unsupervised Learning Towards the Future

8.1 Introduction

Visual data in space and time is available everywhere. While image-level recognition is better understood, visual learning in space and time is far from being solved. One of the main challenges remaining is how to model rich and complex interactions between lower level entities (such as pixels, depth, or motion) to objects and higher level concepts, within the large spatiotemporal context. At this point, we would want to put everything together into a single unified structure. We would like to use all the key ideas, models, and methods presented in the book and start working on a general model for unsupervised learning in the world of space and time. Before we start we should ask ourselves whether that general model is even possible. We believe it is, given the existence of biological brains who learn by themselves in the natural world to see and understand their environment [1]. In this chapter, we argue not only that this is possible, but we will also show how, in the immediate future, the starting ideas and paths taken in this work could converge towards such a unified model, which we name the Visual Story Graph Neural Network or, in short, Visual Story Network.

Before we present the general concept we first present a simpler yet very general and ambitious model, the recurrent graph neural network, which we have already built and obtained state-of-the-art results on the higher level task of recognizing complex activities. The recurrent graph neural network is general and can be applied to any task that involves processing of space-time data. It also brings together, into a single model, one of the two main computational models in vision today, that of graphical models with predefined nodes and edges and their iterative message passing schemes for optimization and that of neural networks, which learn features directly from data and have a single feed-forward pass at inference time.

The Recurrent Space-time Graph Neural Network model was first published in the following paper: Nicolicioiu, Andrei, Iulia Duta, and Marius Leordeanu. "Recurrent Space-time Graph Neural Networks." In Advances in Neural Information Processing Systems (NeurIPS) 2019.

© Springer Nature Switzerland AG 2020
M. Leordeanu, *Unsupervised Learning in Space and Time*,
Advances in Computer Vision and Pattern Recognition,
https://doi.org/10.1007/978-3-030-42128-1_8

8.2   Recurrent Graph Neural Networks in Space and Time

Learning in the space-time domain remains a very challenging problem in machine learning and computer vision. Current computational models for understanding spatiotemporal visual data are heavily rooted in the classical single-image-based paradigm [2, 3]. It is not yet well understood how to integrate information in space and time into a single, general model. We propose a neural graph model, recurrent in space and time, suitable for capturing both the local appearance and the complex higher level interactions of different entities and objects within the changing world scene. Nodes and edges in our graph have dedicated neural networks for processing information. Nodes operate over features extracted from local parts in space and time and previous memory states. Edges process messages between connected nodes at different locations and spatial scales or between past and present time. Messages are passed iteratively in order to transmit information globally and establish long-range interactions. Our model is general and could learn to recognize a variety of high-level spatiotemporal concepts and be applied to different learning tasks. We demonstrate, through extensive experiments and ablation studies, that our model outperforms strong baselines and top published methods on recognizing complex activities in video. Moreover, we obtain state-of-the-art performance on the challenging Something-Something human-object interaction dataset.

Often, for different learning tasks, different models are preferred, such that they capture the specific domain priors and biases of the problem [4]. Convolutional Neural Networks (CNNs) are preferred on tasks involving strong local and stationary assumptions about the data. More traditional graphical models are preferred when learning is more limited and search and optimization are required at test time. When data is temporal or sequential in nature recurrent models are chosen. Then, fully connected models could also be preferred when there is no known structure in the data. Our recurrent neural graph efficiently processes information in both space and time and can be applied to different learning tasks in video.

We present Recurrent Space-time Graph (RSTG) neural networks, in which each node receives features extracted from a specific region in space-time using a backbone deep neural network. Global processing is achieved through iterative message passing in space and time. Spatiotemporal processing is factorized, into a space-processing stage and a time-processing stage, which are alternated within each iteration. We aim to decouple, conceptually, the data from the computational machine that processes the data. Thus, our nodes are processing units that receive inputs from several sources: local regions in space at the present time, their neighbor spatial nodes as well as their past memory states (Fig. 8.1).

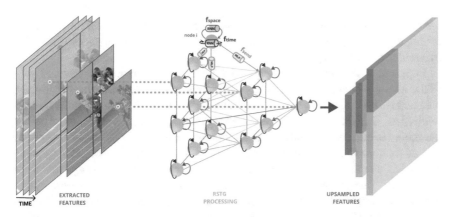

Fig. 8.1 The RSTG-to-map architecture: the input to RSTG is a feature volume, extracted by a backbone network, down-sampled according to each scale. Each node receives input from a cell, corresponding to a region of interest in space. The green links represent messages in space, the red ones are spatial updates, while the purple links represent messages in time. All the extracted (input to graph) and up-sampled features (output from graph) have the same spatial and temporal dimension $T \times H \times W \times C$ and are only represented at different scales for a better visualization

8.2.1 Scientific Context

Iterative graph-based methods have a long history in machine learning and are currently enjoying a fast-growing interest [4, 5]. Their main paradigm is the following: at each iteration, messages are passed between nodes, information is updated at each node, and the process continues until convergence or a stopping criterion is met. Such ideas trace back to work on image denoising, restoration, and labeling [6–9], with many inference methods, graphical models, and mathematical formulations being proposed over time for various tasks [10–16].

Current approaches combine the idea of message passing between graph nodes, from graphical models, with convolution operations. Thus, the idea of graph convolutions was born. Initial methods generalizing convNets to the case of graph structured data [17–19] learn in the spectral domain of the graph. They are approximated [20] by message passing based on linear operations [21] or MLPs [22]. Aggregation of messages needs permutation-invariant operators such as max or sum, the last one being proved superior in Xu et al. [23], with attention mechanism [24] as an alternative.

Recurrence in graph models has been proposed for sequential tasks [25, 26] or for iteratively processing the input [27, 28]. Recurrence is used in graph neural nets [25] to tackle symbolic tasks with single input and sequential language output. Different from them, we have two types of recurrent stages, with distinct functionality, one over space and the other over time.

The idea of modeling complex, higher order, and long-range spatial relationships by the spatial recurrence relates to more classical work using pictorial structures [29] to model object parts and their relationships and perform inference through iterative optimization algorithms. The idea of combining information at different scales also relates to classic approaches in object recognition, such as the well-known spatial pyramid model [30, 31].

Long-range dependencies in sequential language are captured in Vaswani et al. [32] with a self-attention model. It has a stack of attention layers, each with different parameters. It is improved in Dehghani et al. [27] by performing operations recurrently. This is similar to our recurrent spatial-processing stage. As mentioned before, our model is different by adding another complementary dimension—the temporal one. In Santoro et al. [28], new information is incorporated into the existing memory by self-attention using a temporary new node. Then each node is updated by an LSTM [33]. Their method is applied on program evaluation, simulated environments used in reinforcement learning and language modeling where they do not have a spatial dimension. Their nodes act as a set of memories. Different from them, we receive new information for each node and process them in multiple interleaved iterations of our two stages.

Initial node information could come from each local spatiotemporal point in convolutional feature maps [34, 35] or from features corresponding to detected objects [36]. Different from that work our nodes are not attached to specific volumes in time and space. Also, we do not need pretrained higher level object detectors. While the above methods need access to the whole video at test time, ours is recurrent and can function in an online, continuous manner in time. Also, the approach in Baradel et al. [37] is to extract objects and form relations between objects from pairs of time steps randomly chosen. In contrast, we treat space and time differently and prove the effectiveness of our choice in experiments. Thus, we do not connect all space-time positions in the input volume as in Wang et al. [35], Wang and Gupta [36], Chen et al. [38]. We could see our different handling of time and space as an efficient factorization into simpler mechanisms that function together along different dimensions. The work in Szegedy et al. [39], Chollet [40] confirms our hypothesis that features could be more efficiently processed by factorization into simpler operations. The models in Sun et al. [41], Xie et al. [42], Tran et al. [43] factorize 3D convolutions into 2D spatial and 1D temporal convolutions, but we are the first to use similar factorization in the domain of neural graph processing.

For spatiotemporal processing, some methods, which do not use explicit graph modeling, encode frames individually using 2D convolutions and aggregate them in different ways [2, 3, 44]; others form relations as functions (MLPs) over sets of frames [45] or use 3D convolution inflated from existing 2D convolutional networks [46]. Optical flow could be used as input to a separate branch of a 2D ConvNet [47] or used as part of the model to guide the kernel of 3D convolutions [48]. To cover both spatial and temporal dimensions simultaneously, convolutional LSTM [49] can be used, augmented with additional memory [50] or self-attention in order to update LSTM hidden states [51].

8.2.2 Recurrent Space-Time Graph Model

The Recurrent Space-time Graph (RSTG) model is designed to process data in both space and time, to capture both local and long-range spatiotemporal interactions (Fig. 8.1). RSTG takes into consideration local information by computing over features extracted from specific locations and scales at each moment in time. Then it integrates long-range spatial and temporal information by iterative message passing at the spatial level between connected nodes and by recurrence in time, respectively. The space and time message passing is coupled with the two stages succeeding one after another.

Our model takes a video and processes it using a backbone function into a feature volume $F \in \mathbb{R}^{T \times H \times W \times C}$, where T is the time dimension and H, W the spatial ones, and C the number of feature channels. The backbone function could be modeled by any deep neural network that operates over single frames or over space-time volumes. Thus, we extract local spatiotemporal information from the video volume and we process it using our graph, sequentially, time step after time step. This approach makes it possible for our graph to also process a continuous flow of spatiotemporal data and function in an online manner.

Instead of fully connecting all positions in time and space, which is costly, we establish long-range interactions through recurrent and complementary space- and time-processing stages. Thus, in the temporal-processing stage, each node receives a message from the previous time step. Then, at the spatial stage, the graph nodes, which now have information from both present and past, start exchanging information through message passing. Space and time are coupled and performed alternatively: after each space iteration *iter*, another time iteration follows, with a message coming from past memory associated with the same space iteration *iter*. The processing stages of our algorithm are succinctly presented in Algorithm 8.1 and Fig. 8.2. They are detailed below.

8.2.2.1 Graph Creation

We create N nodes connected in a graph structure and use them to process a feature volume $F \in \mathbb{R}^{T \times H \times W \times C}$. Each node receives input from a specific region (a window defined by a location and scale) of the feature volume at each time step t (Fig. 8.1). At each scale, we down-sample the $H \times W$ feature maps into $h \times w$ grids, each cell corresponding to one node. Two nodes are connected if they are neighbors in space or if their regions at different scales intersect.

8.2.2.2 Space Processing

Spatial interactions are established by exchanging messages between nodes. The process involves three steps: **send** messages between all connected nodes, **gather**

Fig. 8.2 Two
Space-Processing Stages
($K = 2$) from top to bottom,
each one preceded by a
temporal-processing stage

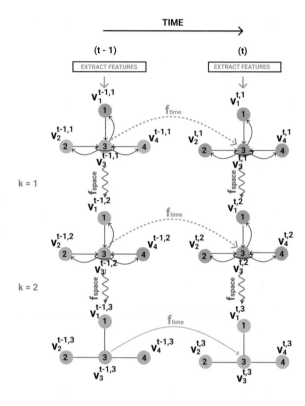

information at node level from the received messages, and **update** internal nodes representations. Each step has its own dedicated MLP. Message passing is iterated K times, with time processing steps followed by space-processing steps, at each iteration.

A given message between two nodes should represent relevant information about their pairwise interaction. Thus, the message is a function of both the source and destination nodes j and i, respectively. The **message sending function**, $f_{send}(\mathbf{v}_j, \mathbf{v}_i)$, is modeled as a Multilayer Perceptron (MLP) applied on the concatenation of the two node features:

$$f_{send}(\mathbf{v}_j, \mathbf{v}_i) = \mathrm{MLP}_s([\mathbf{v}_j | \mathbf{v}_i]) \in \mathbb{R}^D. \tag{8.1}$$

$$\mathrm{MLP}_a(\mathbf{x}) = \sigma(W_{a_2}\sigma(W_{a_1}(\mathbf{x}) + b_{a_1}) + b_{a_2}). \tag{8.2}$$

The pairwise interactions between nodes should have positional awareness—each node should be aware of the position of the neighbor that sends a particular message. Therefore, we include the position information as a (linearized) low-resolution 6×6 map in the **position-aware message** body sent with f_{send}, by concatenating the map to the rest of the message. The actual map is formed by putting ones for the cells corresponding to the region of interest of the sending nodes and zeros for the remaining cells, and then applying filtering with a Gaussian kernel.

Algorithm 8.1 Space-time processing in RSTG model

Input: Time-space features $F \in \mathbb{R}^{T \times H \times W \times C}$

repeat

$\quad \mathbf{v}_i \leftarrow \textit{extract\_features}(F_t, i)$ $\qquad\qquad\qquad\qquad\qquad\qquad\qquad\qquad \forall i$

$\quad$ **for** $k = 0$ **to** $K - 1$ **do**

$\qquad \mathbf{v}_i = \mathbf{h}_i^{t,k} = f_{time}(\mathbf{v}_i, \mathbf{h}_i^{t-1,k})$ $\qquad\qquad\qquad\qquad\qquad\qquad \forall i$

$\qquad \mathbf{m}_{j,i} = f_{send}(\mathbf{v}_j, \mathbf{v}_i)$ $\qquad\qquad\qquad\qquad\qquad \forall i, \forall j \in \mathcal{N}(i)$

$\qquad \mathbf{g}_i = f_{gather}(\mathbf{v}_i, \{\mathbf{m}_{j,i}\}_{j \in \mathcal{N}(i)})$ $\qquad\qquad\qquad\qquad\qquad \forall i$

$\qquad \mathbf{v}_i = f_{space}(\mathbf{v}_i, \mathbf{g}_i)$ $\qquad\qquad\qquad\qquad\qquad\qquad\qquad \forall i$

$\quad$ **end for**

$\quad \mathbf{h}_i^{t,K} = f_{time}(\mathbf{v}_i, \mathbf{h}_i^{t-1,K})$ $\qquad\qquad\qquad\qquad\qquad\qquad \forall i$

$\quad t = t + 1$

until end-of-video

$\mathbf{v}_{final} = f_{aggregate}(\{\mathbf{h}_i^{1:T,K}\}_{\forall i})$

Each node receives a message from each of its neighbors and aggregates them using the **message gathering function** f_{gather}, which could be a simple sum of all messages or an attention mechanism that gives a different weight to each message, according to its importance. In this way, a node could choose what information to receive. In our implementation, the attentional weight function α is computed as the dot product between features at the two nodes, measuring their similarity.

$$f_{gather}(\mathbf{v}_i) = \sum_{j \in \mathcal{N}(i)} \alpha(\mathbf{v}_j, \mathbf{v}_i) f_{send}(\mathbf{v}_j, \mathbf{v}_i) \in \mathbb{R}^D. \qquad (8.3)$$

$$\alpha(\mathbf{v}_j, \mathbf{v}_i) = (W_{\alpha_1} \mathbf{v}_j)^T (W_{\alpha_2} \mathbf{v}_i) \in \mathbb{R}. \qquad (8.4)$$

We update the representation of each node with the information gathered from its neighbors, using the **node update function** f_{space} modeled as a Multilayer Perceptron (MLP). We want each node to be capable of taking into consideration global information while also maintaining its local identity. The MLP is able to combine efficiently new information received from neighbors with the local information from the node's input features.

$$f_{space}(\mathbf{v}_i) = \text{MLP}_u([\mathbf{v}_i | f_{gather}(\mathbf{v}_i)]) \in \mathbb{R}^D. \qquad (8.5)$$

In general, the parameters W_u, b_u could be shared among all nodes at all scales or each set could be specific to the actual scale.

8.2.2.3 Time Processing

Each node updates its state in time by aggregating the current spatial representation $f_{space}(\mathbf{v}_i)$ with its time representation from the previous step using a recurrent function. In order to model more expressive spatiotemporal interactions and to give it the ability to reason about all the information in the scene, with knowledge about past states, we put a time-processing stage before each space-processing stage, at each iteration, and another time-processing stage after message passing ends, at each time step. Thus, messages are passed iteratively in both space and time, alternatively. The time-processing stage at iteration k updates each node's internal state $v_i^{t,k}$ with information from its corresponding state $v_i^{t-1,k}$, at iteration k, in the previous time $t-1$, resulting in features that take into account both spatial interactions and history (Fig. 8.2).

$$\mathbf{h}_{i,time}^{t,k} = f_{time}(\mathbf{v}_{i,space}^k, \mathbf{h}_{i,time}^{t-1,k}). \tag{8.6}$$

8.2.2.4 Information Aggregation

The aggregation $f_{aggregate}$ function could produce two types of final representations, a 1D vector or a 3D map. In the first case, denoted **RSTG-to-vec**, we obtain the vector encoding by summing the representation of all the nodes from the last time step. In the second case, denoted **RSTG-to-map**, we create the inverse operation of the node creation, by sending the processed information contained in each node back to the original region in the space-time volume as shown in Fig. 8.1. For each scale, we have $h * w$ nodes with C-channel features, which we arrange in a $h \times w$ grid resulting in a volume of size $h \times w \times C$. We up-sample the grid map for each scale into $H \times W \times C$ maps and sum all maps for all scales for the final 3D $H \times W \times C$ representation.

8.2.2.5 Computational Complexity

We analyze the computational complexity of the RSTG model. If N is the number of nodes in a frame and E the number of edges, we have $O(2E)$ messages per space-processing stage, as there are two different spatial messages in each edge direction. With a total of T time steps and K (=3) spatiotemporal message passing iterations, each of the K spatial message passing iterations is preceded by a temporal iteration, resulting in a total complexity of $O(T \times (2E) \times K + T \times N \times (K+1))$. Note that E is upper bounded by $N(N-1)/2$. Without the factorization, with messages between all the nodes in time and space, we would arrive at a complexity of $O(T^2 \times N^2 \times K)$ in the number of messages, which is quadratic in time and similar to Wang et al. [35], Wang and Gupta [36] when the non-local graphs are full. Note that our lower complexity is due to the recurrent nature of our model and the space-time factorization.

8.2.3 Experiments: Learning Patterns of Movements and Shapes

We perform experiments on two video classification tasks, which involve complex object interactions. We experiment on a video dataset that we create synthetically, containing complex patterns of movements and shapes, and on the challenging Something-Something dataset, involving interactions between a human and other objects [52].

There are not many available video datasets that require modeling of difficult object interactions. Improvements are often made by averaging the final predictions over space and time [38]. The complex interactions and the structure of the space-time world still seem to escape the modeling capabilities. For this reason, and to better understand the role played by each component of our model in relation to some very strong baselines, we introduce a novel dataset, named SyncMNIST. We make several MNIST digits move in complex ways. We designed the dataset such that the relationships involved are challenging in both space and time. The dataset contains 600 K videos showing multiple digits, where all of them move randomly, apart from a pair of digits that moves synchronously—that specific pair determines the class of the activity pattern, for a total of 45 unique digit pairs (classes) plus one extra class (no pair is synchronous). In order to recognize the pattern, a given model has to reason about the location in space of each digit and track them across the entire time in order to learn that the class label is associated with a pair that moves synchronously. The data has 18×18 size digits moving on a black 64×64 background for 10 frames. In Fig. 8.3, we present frames from three different videos used in our experiments. We trained and evaluated our models first on an easier three digits (3SyncMNIST) dataset and then, only the best models were trained and tested on the harder five digits dataset (5SyncMNIST).

We compared against four strong baseline models that are often used on video understanding tasks. For all tested models, we used a convolutional network as a backbone for a larger model. It is a small CNN with three layers, pretrained to classify a digit randomly placed in a frame of the video. It is important that the ranking on SyncMNIST of published models, such as MeanPooling+LSTM, Conv+LSTM, I3D, and Non-Local, correlates with the ranking of the same models on other datasets, such as UCF-101 [53], HMDB-51 [54], Kinetics (see Carreira and Zisserman [46]), and Something-Something (see Wang and Gupta [36]). Also important is that the performance of the different models seems to be well correlated with the ability

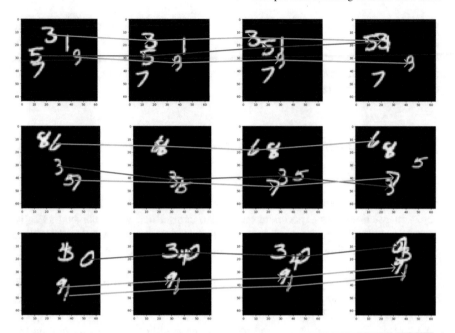

Fig. 8.3 On each row we present frames from videos of 5SyncMNIST dataset. In each video sequence, two digits follow the exact same pattern of movement. The correct classes: "3–9" "6–7", and "9–1"

of a specific model to compute over time. This aspect, combined with the fact that by design on SyncMNIST the temporal dimension is important, makes the tests on SyncMNIST relevant.

Mean pooling + LSTM: Use backbone for feature extraction, spatial mean pool, and temporally aggregate them using an LSTM. This model is capable of processing information from distant time steps but it has poor understanding of spatial information.

ConvNet + LSTM: Replace the mean pooling with convolutional layers that are able to capture fine spatial relationships between different parts of the scene. Thus, it is fully capable of analyzing the entire video, both in space and in time.

I3D: We adapt the I3D model [46] with a smaller ResNet [55] backbone to maintain the number of parameters comparable to our model. 3D convolutions are capable of capturing some of the longer range relationships both spatially and temporally.

Non-local: We used the previous I3D architecture as a backbone for a non-local [35] model. We obtained best results with one non-local block in the second residual block.

Table 8.1 Accuracy on SyncMNIST dataset, showing the capabilities of different parts of our model

Model	3 SyncMNIST	5 SyncMNIST
Mean + LSTM	77.0	–
Conv + LSTM	95.0	39.7
I3D	–	90.6
Non-local	–	93.5
RSTG: space-Only	61.3	–
RSTG: time-Only	89.7	–
RSTG: homogenous	95.7	58.3
RSTG: 1-temp-stage	97.0	74.1
RSTG: all-temp-stages	**98.9**	94.5
RSTG: positional All-temp	–	**97.2**

8.2.3.1 Implementation Details for RSTG

Our Recurrent Neural Graph Model (RSTG) uses the initial three-layer CNN as backbone, an LSTM with 512 hidden state size for the f_{time}, and RSTG-to-vec as aggregation. We use three scales with 1×1, 2×2 and 3×3 grids with nodes of dimension 512. We implement our model in Tensorflow framework [56]. We use cross-entropy as loss function and trained the model end to end with SGD with Nesterov Momentum with value 0.9 for momentum, starting from a learning rate of 0.0001 and decreasing by a factor of 10 when performance saturates.

In Table 8.1, results show that RSTG is significantly more powerful than the competitors. Note that the graph model runs on single-image-based features, without any temporal processing at the backbone level. The only temporal information is transmitted between nodes at the higher graph level.

8.2.3.2 Ablation Study

Solving the moving digits task requires a model capable of capturing pairwise interactions both in space and time. RSTG is able to accomplish that through spatial connections between nodes and the temporal updates of their state. In order to prove the benefits of each element, we perform experiments that show the contributions brought by each one and present them in Table 8.1. We observed the efficient transfer capabilities of our model between the two versions of the SyncMNIST dataset. When pretrained on 3SyncMNIST, our best model RSTG-all-temp-stages achieves 90% of its maximum performance in a number of steps in which an uninitialized model only attains 17% of its maximum performance.

Space-only RSTG: We create this model in order to prove the necessity of having powerful time modeling. It performs the space-processing stage on each frame, but ignores the temporal sequence, replacing the recurrence with an average pool across time dimension, applied for each node. As expected, this model obtains the worst results because the task is based on the movement of each digit, an information that could not be inferred only from spatial exploration.

Time-only RSTG: This model performs just the time-processing stage, without any message passing between nodes. The features used in the recurrent step are the initial features extracted from the backbone neural network, which takes as input single frames.

Homogeneous space-time RSTG: This model allows the graph to interact both spatially and temporally, but learn the same set of parameters for the MLPs that compute messages in time and space. Thus, time and space are computed in the same way.

Heterogeneous space-time RSTG: We developed different schedulers for our spatial and temporal stages. In the first scheduler, used in the **1-temp RSTG** model, for each time step, we performed three successive spatial iteration, followed by a single final temporal update. The second scheduler, the **all-temp RSTG** model, alternates between the spatial and temporal stages (as presented in Algorithm 8.1). We use one time-processing stage before each of the three space-processing stages, and a last time-processing stage to obtain the final node's representation.

Positional all-temp RSTG: This is the previous all-temp RSTG model but enriched with positional embeddings used in f_{send} function as explained in Sect. 8.2.2. This model, which is our best and final model, is also able to reason about global locations of the entities.

8.2.4 Experiments: Learning Complex Human-Object Interactions

In order to evaluate our method in a real-world scenario involving complex interactions and activities, we use the Something-Something-v1 dataset [52]. It consists of a collection of 108499 videos with 86017, 11522, and 10960 videos for train, validation, and test splits, respectively. It has 174 classes for fine-grained interactions between humans and objects. It is designed such that classes can be discriminated not by some global context or background but from the actual specific interactions.

For this task, we investigate the performance of our graph model combined with two backbones, a 2D convolutional one (C2D), based on ResNet-50 architecture, and an I3D [46] model inflated also from the ResNet-50. We start with backbones pretrained on Kinetics-400 [46] dataset as provided by Wang et al. [35] and train the whole model end to end.

We analyze both of our aggregation types, described in Sect. 8.2.2.4. For RSTG-to-vec, we use the last convolutional features given by the I3D backbone as input to our graph model and obtain a vector representation. To facilitate the optimization process we use residual connections in RSTG, by adding the results of the graph processing to the pooled features of the backbone. For the second case, we use

Table 8.2 Ablation study showing where to place the graph inside the I3D backbone

Model	Top-1	Top-5
RSTG-to-vec	47.7	77.9
RSTG-to-map res2	46.9	76.8
RSTG-to-map res3	47.7	77.8
RSTG-to-map res4	48.4	78.1
RSTG-to-map res3-4	**49.2**	**78.8**

intermediate features of I3D as input to the graph and also add them to the graph output by a residual connection and continue the I3D model. For this purpose, we need both the input and the output of the graph to have the same dimension. Thus, we use RSTG-to-map to obtain a 3D map at each time step.

Training and evaluation. For training we uniformly sample 32 frames from each video resized such that the height is 256, preserving the aspect ratio and randomly cropped to a 224 × 224 clip. For inference we apply the backbone fully convolutional on a 256 × 256 crop with the graph taking features from larger activation maps. We use 11 square clips uniformly sampled on the width of the frames for covering the entire spatial size of the video and use 2 samplings along the time dimension. We mean pool the clips output for the final prediction.

Results. We analyze how our graph model could be used to improve I3D by applying RSTG-to-map at different layers in the backbone and RSTG-to-vec after the last convolutional layer. In all cases, the model achieves competitive results, and the best performance is obtained using the graph in the res3 and res4 blocks of the I3D as shown in Table 8.2. We compare against recent methods on the Something-Something-v1 dataset and show the results in Table 8.3. Among the models using 2D ConvNet backbones, ours obtains the best results (with a significant improvement of more than 8% over all methods using a 2D backbone, for the Top-1 setup). When using the I3D backbone, RSTG reaches state-of-the-art results, with 1% improvement over all methods (Top-1 case) and 3.1% improvement over top methods (Top-1 case) with the same 3D-ResNet-50 backbone.

8.2.4.1 Concluding Remarks

In this section, we presented the Recurrent Space-time Graph (RSTG) neural network model, which is designed to learn efficiently in space and time. The graph, at each moment in time, starts by receiving local space-time information from features produced by a certain backbone network. Then it moves towards global understanding by passing messages over space between different locations and scales and recurrently in time, by having a different past memory for each space-time iteration. Our model is unique in the literature in the way it processes space and time, with

Table 8.3 Top-1 and Top-5 accuracy on Something-Something-v1

Model	Backbone	Val Top-1	Val Top-5
C2D	2D ResNet-50	31.7	64.7
TRN [45]	2D Inception	34.4	–
Ours C2D + RSTG	2D ResNet-50	**42.8**	**73.6**
MFNet-C50 [57]	3D ResNet-50	40.3	70.9
I3D [36]	3D ResNet-50	41.6	72.2
NL I3D [36]	3D ResNet-50	44.4	76.0
NL I3D + Joint GCN [36]	3D ResNet-50	46.1	76.8
ECO$_{Lite-16F}$ [58]	2D Inc+3D Res-18	42.2	–
MFNet-C101 [57]	3D ResNet-101	43.9	73.1
I3D [42]	3D Inception	45.8	76.5
S3D-G [42]	3D Inception	48.2	78.7
Ours I3D + RSTG	3D ResNet-50	**49.2**	**78.8**

several main contributions: (1) it treats space and time differently; (2) it factorizes them and uses recurrent connections within a unified graph model, with relatively low computational complexity; (3) it is flexible and general, being relatively easy to adapt to various learning tasks in the spatiotemporal domain; and (4) our ablation study justifies the structure and different components of our model, which obtains state-of-the-art results on the challenging Something-Something dataset. For future work, we plan to further study and extend our model to other higher level tasks such as semantic segmentation in spatiotemporal data and vision-to-language translation. We will also study how to use and integrate the RSTG graph model, as a building block of the Visual Story Network concept which we present next, in the final part of our book.

8.3   Putting Things Together

The material presented in this book evolves around a few key ideas which state and prove, through various experiments and computational models, that unsupervised learning is possible in the real world if we take advantage of the rare alignments, co-occurrences, and agreements between many different cues, features, and classifiers. Such agreements are extremely unlikely at random and when they happen it is because of the presence of a certain visual category in the space-time vicinity. In every chapter, we showed how this idea could be used for unsupervised learning as it applies to different tasks, such as matching features using higher order geometric constraints, selecting discriminative groups of features (that may be weak by themselves but much more powerful in combination), recognizing visual object categories as well as discovering and segmenting objects in images and videos. While each solution proved efficient when applied to a particular problem, we have not yet discussed the possibility of a unified model, which can put all ideas together in a single, general model for unsupervised learning in space and time. We know that such a system is possible and already exists, since biological brains can learn effortlessly, in an almost completely unsupervised way, to visually recognize and understand the natural world.

In this final section, we present a novel model, the Visual Story Graph Neural Network (in short, the Visual Story Network), inspired from an earlier concept [59], which is based on the core ideas and models proposed in this book. The Visual Story Network could handle unsupervised learning from a general perspective. The proposed Visual Story system is meant to learn in an unsupervised fashion a full visual understanding of the visual scene through a self-supervised scheme that exploits the mutual consensus between its multiple nodes and neural pathways.

We go beyond the conceptual level and discuss the possibility of creating the Visual Story Net in practice using the ideas, principles, and tools developed in this book. Starting from the recursive graph neural network presented in the previous section and equipped with the necessary computational models and experimental results detailed in the previous chapters, we can now start imagining, both at a pure conceptual as well as implementation levels, how to create the Visual Story Net. Before getting deeper into the details, we should first synthesize and relate together the relevant key ideas and results presented in the book, which will help us better understand how such a general unsupervised learning approach might become possible.

Later, in this section, we present a unified view of the Visual Story Graph Neural Network based on ideas presented throughout the book. Before we get to that point, we show how the material from the previous chapters relates to the general model that will follow.

8.3.1 Agreements at the Geometric Level

In the first two chapters dedicated to solving and learning efficiently for graph matching and clustering, the methods introduced were based on the statistical significance of accidental alignments in unsupervised learning, at the lower level of geometry, shape, and relatively simple local appearance. The main idea was that accidental alignments are rare. When such alignments happen it is typically not by accident. There is an underlying cause: the presence of an object groups together such appearance and geometric symmetries and alignments. Thus, such agreements in form and appearance become strong cues for grouping and can be used reliably for finding correspondences between different images and views of the same object. In Chaps. 1, 2, and 3, we presented efficient algorithms for matching, clustering, and learning, such as the spectral matching and integer projected fixed point method which fully exploit the statistics of alignments in real-world data and transform the exponentially expensive problems of clustering and matching into ones that can be efficiently solved in practice. Then, in Chaps. 4, 5, and 6, we show how these methods immediately apply to other formulations and tasks such as feature selection, classifier learning, and object discovery in video sequences.

8.3.2   Agreements at the Semantic Level

In Chap. 4, we move towards the higher level of agreements among object category-level classifiers. Such agreements are not based on fine geometrical alignments but on co-occurrences of classifier positive firings. These classifiers are trained to respond positively to the presence of different object categories in an image. We showed that often classifiers that are unrelated at a semantic level are positively correlated with a given class that is very different semantically from the class they were trained to respond to. For example, in Chap. 4, Fig. 4.1 pretrained classifiers for classes library, steel arch bridge, freight car, mobile home, and trolley are consistently selected as relevant for the class "train", as they share common appearance or contextual features with the "train" class, even though they might mean completely different things. Then, in Fig. 4.15 (Chap. 4), we present several other such interesting examples. Such groups of co-firing classifiers can be discovered in an unsupervised way and they usually signal the presence of a new, previously unseen, category. Their co-firing is often strong and precise enough such that their group could constitute the basis for learning, in an unsupervised way (based entirely on the co-occurrence of their positive outputs), a novel classifier for the new class. That novel classifier is, in fact, the average output of the group and we showed that this average is, in fact, the optimal solution for a specific formulation of the learning task that relates to the material from Chaps. 2 and 3. We show how the unsupervised learning task relates to unsupervised clustering (Chap. 3) and feature selection and adapt the integer projected fixed point method (IPFP) from Chaps. 2 and 3 to this new task.

Even though the task of classifier learning is apparently very different from the problem of feature matching (for which IPFP was first introduced), they are linked in very strong and interesting ways in Chaps. 2 , 3, and 4. As also discussed in Chap. 3, from a mathematical formulation and optimization point of view, the methods presented in the second chapter, namely, Spectral Matching (SM) and Integer Projected Fixed Point (IPFP) algorithm, are immediately related to the clustering formulation in Chap. 3, where the IPFP method is applied, as is, to the novel clustering task. The formulation of that task is interesting as it produces discrete solutions at the optimum, exactly as needed by the clustering problem. A strong property of IPFP is that while it optimizes within a continuous domain, it returns optimal solutions in the desired, discrete domain. Then, the classifier learning in Chap. 4 can be formulated as a clustering problem too, with the same discrete solution, which transforms classifier learning into feature selection and learning of classifier ensembles.

At a higher level, what unites feature matching and classifier learning, beyond the mathematical formulations and methods, is the principle of mutual agreements: geometric alignments (in the case of matching) or co-occurrences of positive classifier outputs (in the classification case). Then, when such agreements happen they are indication of strong positive features, which could be used as supervisory signal for descendant components of the unsupervised system. Such unlikely co-occurrences taking place in the real world strongly indicate the presence of the positive class, therefore our unsupervised learning system must capture them and select their corresponding features in order to start learning about that new class.

8.3.3 Agreements as Highly Probable Positive (HPP) Features

In Chap. 5, we generalize the concept of learning from unlikely agreements and introduce the concept of Highly Probable Positive (HPP) features. We show that such agreements are a case of HPP features and present a method that discovers objects in the spatiotemporal domain of video sequences using such strategy of learning with HPP features, over several stages in which we grow from simpler to more powerful classifiers. Chapter 5 also marks an important step towards taking advantage of the consistency and coherence in the spatiotemporal domain. We also move from the simpler task of deciding the presence or absence of a given category in a video (Chap. 4) to the finer task of producing actual object segmentations for each frame in the video.

We first observe that objects occupy relatively small regions in space that display unique color distributions. In other words, when very simple pixels that tend to have certain colors co-occur in the same region of space that represents Highly Probable Positive (HPP) signal that often indicates the presence of an object. We then could use such HPP patterns reliably in order to discover the full extent of the object in the image and start learning about it. In the first part of Chap. 5, we demonstrate the power of learning from simple colors at the level of pixels, when such colors represent HPP signal. We present a simple algorithm for object mask discovery and lay down the intuitive basis for learning with HPP features.

In the second part of Chap. 5, we show how to learn with HPP features over multiple iterations, in order to discover high-quality masks of foreground objects in video sequences. We compare our method to others in the literature and demonstrate state-of-the-art results. Learning from HPP features over several iterations is possible in the following way: first we start with some very simple classifiers that can only "see" pixel colors in order to detect object regions with high precision, even though the recall might be very low. That is the same idea as in the first part of Chap. 5. For example, we could only be able to detect, initially, 15% of the object area, while leaving the rest in the background. However, if that initial 15% of the area is detected with high precision (meaning very low false positive rate: very few positively labeled pixels belong to the background) then we could use it to learn simple object and background color distributions, by considering that small area as the region of positive samples and all the rest as the region of negative ones. These simple distributions then can be used to discover a larger part of the object, with significantly better recall while the precision could still be kept high. At the next iteration, based on regions that are more accurately labeled as positive or negative, we can train more powerful patch-based classifiers (instead of single-pixel-based) that take as input more information (patch vs. pixel) and have less corrupted training sets. The output of that second iteration model is expected to be even more accurate so, at the third iteration, it can be used as ground truth to train an even more powerful classifier: in that case we use in our experiments a U-Net neural network model that is known to perform well on the segmentation task. Note that at each step we apply supervised learning techniques in order to actually train the classifier. What makes the process unsupervised is not the technique for training (e.g., gradient descent with backpropagation to optimize a classification cost that considers certain labels, which are, in fact, noisy, as ground truth). What makes the process unsupervised is the fact that at the core the supervisory signal is given by the HPP signal not by human annotators. Besides generalizing the usage of unlikely agreements for unsupervised learning introduced in the first chapters, in Chap. 5, we make another connection to the previous material when we use IPFP to learn, in an unsupervised fashion, compact patch-based descriptors based on patterns of color co-occurrences that best separate the foreground object from the background. It is interesting to see how such graph-based optimization techniques (e.g., spectral matching and IPFP), which were developed initially for graph matching (Chap. 2), can be successfully applied to many different tasks that are involved in unsupervised learning. Another such case is discussed in Chap. 6, when the same spectral method is used to discover objects as clusters of motion and appearance patterns through space and time. The fact of the matter is that these techniques optimize by discovering a cluster in the graph: the meaning of the cluster differs from task to task, but the same core idea of clustering (which is another name for discovering agreements among many entities) is seen again and again in the world of unsupervised learning.

We develop further the idea (introduced in Chap. 5 in its basic form) of learning over multiple teacher-student generations in Chap. 7. There we discuss in detail what conditions must be met in order for this process to continue learning by itself, which we summarize here: (1) the outputs of several classifiers have to agree consistently within local regions of space and time. This is the main idea behind classifier co-occurrences and geometric alignments; (2) we need the possibility to select with high precision good quality answers on which groups of classifiers agree. These are HPP features to be used as supervisory signal; (3) then we need to introduce, at the next generation, more powerful classifier models that will be trained on such selected HPP outputs. Thus, we can go from simple to more complex and powerful classifiers and evolve during the course of several generations of teachers and students; and (4) at every generation we have to have access to new, larger, and potentially more challenging sets of unlabeled or iteratively self-labeled training data. That can sufficiently ensure the diversity for avoiding over-fitting when training more and more powerful classifiers. The very essence of classifier evolution includes not only the evolution of the classifier complexity and generalization power but also access to the training data that can match that complexity.

8.3.4 Motion Patterns as HPP Features

In Chap. 6, we start to exploit more the aspect of motion by considering the flow between the pixels from one frame to the next, in the video. Thus, objects could be seen as clusters of motion at the pixel level and also of appearance, through lower or higher level features, within the space-time domain of the video. Motion is seen as a complementary dimension to the appearance one: the two dimensions are capturing independent views of the world. They could be put in agreement and re-enforce each other, in an unsupervised learning scheme such that predictions based on motion patterns agree with predictions based on appearance features. In Chap. 6, we present our approach to discover objects in video as clusters in space and time, based on finding the eigenvector of a specific Feature-Motion matrix that is associated with the video. There is a (column, row) element in the matrix for pair of pixels in the video, which corresponds to a potential link in space and time between the two. The matrix is symmetric, positive semi-definite, and represents the adjacency matrix of graph in which each node corresponds to a pixel in the video. Its principal eigenvector reflects the foreground object as its main space-time cluster, and completely defines the object masks in every frame. The matrix could be huge in theory but extremely sparse in practice. It is used only at a pure conceptual, mathematical level and never built explicitly. The method accesses directly matrix' adjacency lists (non-zero elements) without creating the full matrix. The final result is indeed its exact principal eigenvector. Thus, the algorithm is directly related to spectral matching, as it is based on the same power iteration for computing the eigenvector. It is only the actual meaning of the matrix that differs.

8.3.5 Learning over Multiple Teacher-Student Generations

In Chaps. 2–6, we addressed unsupervised learning based on unlikely agreements among multiple sources of information, at different levels of semantics (geometric matching versus object category recognition versus foreground object segmentation) and using different cues, features, or classifiers. The material in Chap. 7 focuses mostly on how we can learn over multiple generations such that learning never truly stops.

If we ever want to develop a working, practical system that learns by itself then we should have a solution that enables continuous learning. In Chap. 7, we propose the teacher-students paradigm that learns over multiple generations. At each generation, we have a teacher composed of several students from the previous generation, which vary in structure and tend to make independent mistakes. They function as an ensemble and their united output is expected to be more powerful than each individual one. The principle of learning from unlikely agreements is again seen, now at the level of powerful classifiers that are trained for semantic object segmentation. Their agreements are also used in order to train, in an unsupervised fashion, an automatic selection mechanism that filters out lower quality outputs from the ensemble of students that form the teacher. Thus, only the good quality segmentations that pass through the selector becomes supervisory signal for the next generation. These good quality segmentations that are selected play the role of Highly Probable Positive (HPP) samples, and thus the connection to the material in Chap. 5 is immediate. The same ideas developed throughout the book are seen at work and make possible the more ambitious case of continuous learning over multiple generations of teachers and students, who become themselves teachers at the next iteration. Each generation brings more diverse and powerful students and, accordingly, it receives more varied and challenging automatically teacher-labeled training data. That ensures the possibility of acquiring new more complex knowledge and achieve superiority and better generalization power over previous generations. Thus, by training more powerful students at the next generation, we also ensure more powerful teachers at the generation that will follow after, within a cycle that is capable of ensuring continuous teacher-student learning.

8.3.6 Building Blocks of the Visual Story Network

The elements introduced so far, different models and methods developed, all having at the core the same key ideas, could make possible the creation of the more general Visual Story model. We are by now equipped not only with the theoretical concepts, principles, and intuitions but also with the actual algorithms and more practical solutions that could establish a basis for VSGNN. The recurrent space-time graph neural net model, presented at the beginning of this chapter, is a real example of a complex system that is capable to learn, with minimal manual supervision, to recognize high-level concepts in space and time. In the following section, we present the building blocks for creating the unified story model, which synthesize the fundamental ideas discussed so far.

8.4 The Dawn of the Visual Story Graph Neural Network

The ideas, models, algorithms, and experiments in this book, all strongly suggest that many classifiers can offer context to each other and create together, at a higher level, a unified system that is capable to learn from unlabeled data, by itself.

Such a society of many classifiers, offering context and re-enforcing each other, could ensure a more solid and robust performance at inference time, due to the possibility of reaching a consensus through multiple pathways.

The multiple processing pathways could also be used within a self-supervised learning scenario, in which in order to train a single classifier, supervisory signal could come from the remaining ones, which could form its context.

As we discuss next, the better connected they are through multiple pathways within a common graph of classifiers, the more compact and harmonious the group of classifiers could be, each with its dedicated class having a well-defined role in the larger context of the space-time story. A poor connectivity imposes fewer constraints, since there are fewer pathways to decide on a given category, situation or reach a certain conclusion. Fewer constraints bring larger and more frequent errors, increased fragility and decreased capacity to learn efficiently by itself.

8.4.1 Classifiers Should Be Highly Interconnected

Classifiers in the Visual Story network should be interconnected through multiple processing pathways in order to provide strong context to each other. The group of classifiers should form a consistent, coherent and robust whole such that for each classifier, the remaining ones provide complementary context.

The more connected are the classes used to explain and give an interpretation of the world scene, the stronger the mutual context is. That helps both inference and training. We want to be able to predict the output of a given classifier from the output of the rest. The better we could do that, the stronger the agreement between classifiers. We would have a strong ensemble at inference and a strong and robust supervisory signal at training time. Through such mutual context, VSN can then provide self-supervisory labels to each classifier, for cases where there is strong agreement along pathways from the context classifiers. We will provide specific details and examples in the next section, which discusses potential implementation approaches.

Now we are ready to propose our eighth principle for unsupervised learning, which captures one of the two key insights of the Visual Story Network. While this principle is not demonstrated in practice yet, we state it here as it constitutes a foundation for our proposed universal unsupervised learning concept:

Principle 8

A universal unsupervised learning machine would need many complementary interpretation layers of the same scene, which should be highly interconnected in order to provide strong context to each other. For any processing pathway reaching a given output layer, the rest of the pathways reaching the same layer should provide sufficient HPP agreements in order to obtain an acceptably robust supervisory signal.

8.4.2 Relation to Adaptive Resonance Theory

Our proposed unsupervised learning principle, which states that in order to learn about the world by itself a system would need strong agreements among multiple processing pathways, is related to Grossberg's classical Adaptive Resonance Theory in neuroscience [60–66], which refers to these agreement as "resonance". The idea about the existence of multiple pathways for processing the same type of visual inputs and concepts (which in our case means predicting the same interpretation layer of the world through many neural pathways) is also related to the classical work of Lashley in the 1950s [67], who concluded that memory and learning is distributed across the brain and there is not a confined single local part of the brain representing an engram that performs processing related to a specific memory. Our own conclusions, while coming seven decades later and in the context of an engineering field, in which we create artificial things instead of studying the brain, seem to agree with the classical neuroscience work: visual concepts should be connected to many parts of the "brain" and they involve many neural networks for predicting the same abstract interpretation layer or concept. Moreover, Grossberg [66] goes on to say that for something to be consciously seen there needs to be an agreement, or resonance, between the features generated by the bottom-up view (in our case: the local processing pathway) and the once produced by the top-down view (in our case: multiple pathways coming from the overall context provided distributively by the graph). Grossberg suggests that all conscious states are resonant, in other words, when seeing happens at the conscious level, then all these views, top-down and bottom-up, are in harmony. These ideas also relate to the initial version of our Visual Story Graph concept [59], when there were fewer computational solutions and less experimental evidence that engineering such a system is within our reach.

Today there are engineering solutions, which show experimentally that unsupervised learning and discovery greatly benefits from maximizing consensus or resonance among multiple hypotheses or classifier outputs. However, these solutions are specifically designed for limited problem domains such as: 3D computer vision problems [68], feature selection [69] or vision-to-language translation [70]. What we aim in this book is to propose a relatively simple and general computational prototype for unsupervised learning, sufficiently complete and realistic. We leverage consensus and use it as self-supervisory signal within an ecosystem with many interpretation layers, interconnected through multiple neural pathways, in order to put together everything that is sensed, acted upon and predicted. Our claim is that the more round and global the system is and the more interconnected its interpretation layers are the higher the quality of the consensus: harder to achieve but stronger and more robust when it is achieved.

8.4.3 Multiple Layers of Interpretation: Depth, Motion, and Meaning

The world scene is interpreted in different ways. We could have one layer of interpretation for each way in which we see the world. One layer could be interpreting the world at the level of depth and 3D geometry. Another layer could be at the level of motion: how the camera or objects move in the scene. Other layers could address different levels of interpretation. For example, one such layer could segment the spatiotemporal scene into objects. Another layer which could take into consideration longer chunks of time could segment it into different regions in which different activities take place.

The main idea that we propose (Fig. 8.4), and which is in line with recent self-supervised approaches, is to have one layer of interpretation, which could be low level (depth or motion) or high level (semantic segmentation) predict another layer, through a neural network connection with input from the predictive layer and output to the second, predicted layer. We could then see these layers as nodes in the graph and the nets which connect them as edges. If the graph is strongly connected then there will be multiple pathways between different layers. We would also have multiple pathways that reach the same layer, which itself would function as input to predict one or several other layers. The more connected the graph, the more pathways we could find to connect different nodes in different ways.

For example, we could have a pathway that goes from a low-level RGB layer to a depth layer, which could then predict motion. Depth and motion could then predict segmentation. Another pathway could take us directly from RGB to segmentation. Then yet another pathway could take us from a semantic segmentation layer of the full image to a layer segmenting only moving objects in the scene. That layer could also be predicted from depth or from RGB. Any one of these layers could then provide input over larger volumes of space and time for layers at higher levels of interpretation, such as could be layers for actions or complex activities. We could often have as input, several layers from different types of interpretation layers (nodes), as well. However, every time there is a single node at the destination, which is sufficient to implement the mutually supervised learning scheme (which is self-supervised w.r.t to the VSN itself, but unsupervised w.r.t outside world). A specific segment in space and time could correspond to a specific activity that takes place in that region, e.g., "two people are walking together" in that specific region of space and time, which is segmented in each frame on that specific interpretation layer.

Predictive pathways could also come backwards from a larger context in space and time to lower levels of interpretation. Thus, we could predict what an object is both from its local appearance information (RGB) as it is the usual case in semantic segmentation, as well as from its role played in the larger contextual story (action or activity). Several predictive pathways could also come from the level of object interactions, activities, and beyond, from the realm of sound, language, and abstract thought to predict the existence in time and space of a specific object category.

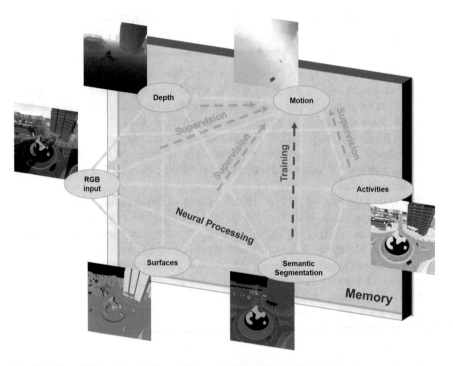

Fig. 8.4 Visual Story Graph Neural Network (VSGNN or VSN) Schematic overview of our proposed Visual Story Network. Each view of the world scene has one or several layers (nodes in the Visual Story Graph) of interpretation. Each layer is essentially an image in which values, which could have pixel-to-pixel correspondences with the input image or be a global map of the world, have certain meanings. Thus, we could have a layer of motion which capture how things are moving in the scene or layers for higher level semantic segmentation capturing different objects or classes. We could also have layers that capture interactions and activities over space and time, such that each pixel in the image is given a complex activity or simpler action label. Our idea is that the layers should learn to predict each other whenever possible, through deep neural pathways represented here by edges. Thus, many pathways from different sources always reach a single node. Unsupervised learning will then become possible by a mutually supervised approach in which each oriented edge (which could be modeled by a deep neural net—an example is shown here with a red dotted arrow) receives as supervisory signal at the end node the predictions from other pathways that reach the same node (shown as green dotted arrows)—but only when those predictions are highly probable correct (HPP cases). This type of learning will encourage consensus, reduce noisy fluctuations, and overall improve generalization and robustness. It will also make learning in an unsupervised way (or with minimal supervision) possible. Note that only very few possible interpretation layers (nodes) are shown in the figure. A biological "brain", for example, could have thousands of such layers and could include more abstract processing and descriptions in natural language. Also note that each node has access to memory, and therefore memory states could also participate in providing supervisory signal. Memory states could also be re-trained and updated through a similar process, if needed

For example, at the lower level of appearance, *I know there is a car on the road because there is an object that looks perfectly like a car.* At the same time, at the higher level of semantic interpretation *I also know that there is a car on the road because I know that I am watching a car race.* And, *I also know there must be a car there because my friend, sitting next to me, just told me to look at "how fast that car is."* Then, *I also know there is a car on that particular place on the road because I know it was a car in that neighborhood a fraction of a second ago and based on its predicted motion it should be at that new location now.*

As we can see, several layers of interpretation at different scales in space and time and at different levels of semantics could exist simultaneously for a given image. Also, they should be in agreement if we want to obtain a coherent understanding of the scene. As mentioned before, each layer can be seen as a node in a graph with multiple pathways being active between any two nodes. Thus, layers that contain static semantic regions, depth, motion layers, moving objects layers or layers that contain regions where a specific activity takes place could all be interconnected such that there is predictive processing and information going along each pathway. We could imagine how all this can be modeled and implemented. For example, the interpretation layers could be binary image representations or soft feature maps, having different meanings and the same size as the input image. They would represent the nodes of the graph. Then a directed edge between two nodes could be modeled as a neural network that takes as input the source layer (node) and predicts the other, destination layer. We could also have memory nodes, in which we learn to retain previous states which could also serve as sources or even destinations (in case we want to re-interpret the past based on current interpretations).

Let us consider the scenario in which we have layers for motion (trajectories of objects describing how things move over the next couple of seconds), semantic segmentation (what category each segmented object belongs to), a depth layer (that captures the 3D structure of the scene), a camera motion layer, a semantic layer of activities (what objects do in the scene), and, of course, the RGB input. We could immediately imagine how we could form edges (or hyperedges) between pairs (or tuples) in this group of layers, such that one layer or several could predict another. Note that a layer contains both spatial and temporal information, which could be captured by memory states (nodes) with recurrent connections. Then, from the RGB layer, for example, we could predict in principle all the others with direct connections. We could also form links that predict motion from depth, and vice versa, predict current depth from past depth and motion. We could also go back to the RGB channel and predict it from depth, motion, and past RGB. The layer of activities could be predicted from RGB directly, as well, or from the others that consider physical properties such as depth and motion or semantic information such as the segmentation layer. The activities layer could itself then be used as a source to predict the segmentation or even depth and motion.

Note that none of these pathways has perfect confidence or accuracy. However, the paths together form a very strong graph in terms of predictive power. When several such paths meet at a node and are in agreement it is very likely that they are correct. The fact that they are together and completely describe all the views of the world, locally (RGB and local objects) or more globally (complex activities) over space and time, make it possible to have it produce coherent predictions along multiple pathways.

8.4.4 Local Objects and Their Global Roles in the Story

Such coherence makes it possible for the graph to represent, what we call, a full visual story, with different actors having specific appearance locally and roles they play at the more global level in space and time. Then the local pathways which are based on appearance should agree with the global pathways that consider more complex spatiotemporal interactions on the actor's identity. The actor will appear on many different interpretation layers, from the local ones considering appearance to mid-level ones that capture motions and local actions to the higher level ones that interpret the actor by its role in the story. There should be a strong correlation between these layers such that the system should be able to predict any layer from many different subgroups of other layers.

Next-generation classifiers should predict locally what context predicts globally: Our initial VSN model is based on different interpretation layers which are implemented as activation maps, or images which represent either a semantic segmentation of the input image itself or a semantic segmentation of the world map w.r.t to a global or local system of coordinates. We consider that the semantic segmentation paradigm could be applied to most visual recognition problems and could represent the output of virtually any task that involves a combination of "what" (semantic class) and "where" (positional values in some space—which do not have to be related to the input image itself). The VSN will consist of many such interpretation layers and there will be many different networks and analytical systems that are able to predict one layer from many different sources of input. In our view, the notion of context is relative to a given classifier (processing pathway) and refers to all the other pathways which are able to predict (with some margin of error) the output interpretation layer of our given classifier. Therefore, if a specific classifier is more local in space and time (it has access only to local or limited information—and most classifiers are of this type since they inherently have a limited receptive field w.r.t the entire input in space and time) its context is by definition global, as it includes everything else which collectively uses global information to produce the same output layer. As stated before, in our VSN, all classifiers are interconnected and aim to be able to recognize everything there is to recognize in the world model.

This context, as defined above, is a powerful resource in space and time. At that level, the context is the one providing the story and the local classifiers that are associated with physical objects receive roles in this story. What we want is to be able to predict from the local appearance and behavior in space and time, the role which a specific object plays in the overall story, for example, is it a certain bus that takes kids to school? Or is it a building where movies are played, or is he a mailman? We ask for a lot from the local classifier and we hope that the next-generation classifier will be able to figure out, from limited space-time information, the class that best fits within the larger story-like context.

8.4.5 Unsupervised Learning in the Visual Story Network

The initial nodes (layers) in the VSN can be defined from the start. A minimum amount of supervised training could also be done in order to initialize the networks that form the edges of VSN. The amount of supervision needed at the beginning for pretraining needs to be decided after sufficient experiments and validation. However, the first question now is how to continue the learning process in an unsupervised fashion. We claim that this could be done in a self-training manner, from the point of view of the net as a society of classifiers that provide supervisory context to each other. However, from the outside it is a case of unsupervised learning in which space-time data passes through the net and is interpreted by each layer in turn. Most layers should have a direct path from input RGB to their output. Direct paths could also have memory and could be implemented with various architectures, similar or identical to the recurrent graph neural network presented earlier in this chapter. We would like a neural net that is suitable for layer-to-layer translation (e.g., RGB to semantic segmentation) and for that U-Nets are very effective. We also want to add memory nodes and recurrent links, to take advantage of the space and time continuum.

Once the interpretation layers are activated they start sending predictive messages through different pathways, such that each pathway will predict the corresponding output layer, which is the layer at the end of the pathway. It is clear that when the network is not yet fully trained there will be many times when such predictions will contradict each other. We can then train each pathway, by passing the gradient through it from output to input and by freezing everything else and using the information flow through other contextual pathways as supervisory signal.

At this point, it should be clear that we, in fact, are using all the main ideas that were demonstrated through the book such as learning with HPP signal from Chap. 5 (use as supervisory signal only contextual outputs that strongly agree), using ensemble of students as teacher (the contextual teacher formed by several student pathways) and their agreement as selection (Chap. 7), and using apparently very different and unrelated classes as context for others (see Chap. 4). The material in Chap. 6 could be used to provide ways to construct motion layers, in which objects are discovered in an unsupervised way as strong clusters in their space-time neighborhoods. The Recurrent Space-time Graph Neural Net (RSGT), presented at the beginning of this chapter, could also provide ideas about how to combine neural networks with graph processing and how to pass messages between nodes at different scales in space and time and between present nodes and memory ones. The recurrent space-time graph neural net could be the building block for connecting different interpretation layers and also for relating RGB directly to that specific interpretation layer. RSGT net could be, in fact, the network used along edges and a much better choice in space and time than the image-based U-Net model.

8.4.6   Learning Evolution over Multiple Generations

The Visual Story Net could evolve over generations by adding new and more powerful pathways (neural networks) for each existing category in order to predict and generalize better the consensus of the rest of the pathways. For example, as it was the case in Chap. 7, a more powerful net could predict from less data over time the output of the graph from larger chunks of data in space and time. The end result would be increased prediction accuracy as well as smoother and more consistent graph behavior with fewer noisy fluctuations. One can show that by using several pathways as supervisory signal for a single edge in the visual story graph, it will result in decreased output variance. That happens because the expected output of pathways towards a specific layer over the whole graph is used as ground truth for a single one-edge path. Thus, the difference between the single output of any edge and the expected output (approximated with output from the rest of the context) will decrease, which means (almost by definition) that variance is expected to decrease. This fact is expected to lead to learning a more robust, less fluctuating, and overall more intelligent neural story graph.

8.4.7   Learning New Categories

New categories with new layers will be added in the graph as potential candidates, whenever there will be a suspicion of unlikely alignments and co-occurrences that are not immediately linked to known classes or categories. For example, if several layers produce a positive output (fire) over a given region in time and space and there is no class (associated to an existing interpretation layer) associated with that firing, we will add a new candidate layer with an associated "anonymous" candidate class. The name given to the class is irrelevant—what matters is that the class will receive a candidate layer for itself with new pathways between its layer and the others and a direct link between the RGB input and that layer. Thus, its layer will enter the game of contextual prediction together with other layers: it will attempt to predict other layers and also be predicted and learn from the mutual consensus of others.

However, the "new class" will only be considered as a candidate class. Its "true existence" and "usefulness" will be monitored for a while. The performance of a new class (or old class for that matter) should be its ability to help predict and be predicted by others. In that way, it can participate in improving the consistency and smooth behavior of the graph over time. If a class is generally well predicted from its local and global pathways (if there is a large amount of agreement between the multiple pathways that are supposed to predict it) then those pathways that are in agreement improve the overall consensus within the VSN. This automatically decreases variance in overall prediction that is mainly due to noise. Reduced variance in overall output will also reduce errors and the corresponding costs coming with those errors (costs related to energy, time, safety, for example). Ideally we want disagreements between different pathways to appear at the right time, when new knowledge needs to be learned. Such disagreements and surprises are healthy and absolutely needed to learn new things and move towards a different point of equilibrium.

Consensus becomes a reward in itself because it brings so many real benefits in practice, as it makes robust inference and unsupervised learning possible. It is a universal reward in the Visual Story World, as it applies to the functioning and learning of the whole Visual Story Graph, not just to a particular class or situation. Consensus is not easy to obtain, especially when different paths use different types of information that are very different from each other, such as color appearance, depth, motion, and eventually the many possible complex ways to attach semantic meaning. While semantic layers could be more subjective and belong to deeper levels inside the graph, the lower levels of color appearance, motion, and depth are much more "physical" and could come directly from the outside world. Mistakes at those lower level layers could be very costly and such mistakes could and also should be quantitatively measured and evaluated.

8.5 Visual Stories Towards Language and Beyond

Entities of the spatiotemporal world can be seen in a dual way: each could be predicted and interpreted using its own local information in space and time that belongs mostly to its physical presence, movement, and immediate action. A car, for example, could be recognized based solely on the color (appearance) information that comes from its pixels. At the same time, the semantic entity could be recognized based on its role played in the larger story. In the case of the car, its role could include *transporting people from one point to another, drive on recognizable roads, follow certain traffic rules, go through several landmark places, and usually be driven by a person.* Interestingly enough, note that when we look for the definition of a "word" or "entity" in the dictionary we usually find its role played in the larger context. That role carries more meaning than that object's looks or how that entity is formed physically. However, all the information, both from the local level of appearance and from the role played in the larger story and its path through space and time, define that car and give it meaning. That complete information tells about what the car does, what is its purpose, who uses it, where and how, how it came into existence, how it looks, who designed it and then manufactured it, who owned it so far, where does it come from, and where it goes. Note that answers to these types of questions are nothing but destinations along different processing pathways in the Visual Story Net, which take input from each other in order to achieve consensus and build a coherent story about that car.

"That car" is just a very simple example. Entities in the visual story graph could belong to higher level categories and have more complex roles in the overall story. But let us see what is "the story" and what is "the role" in the Visual Story Graph?

> The role of an entity on its associated interpretation layer is defined by the multiple pathways that come from higher or lower levels of interpretation in space and time and achieve consensus with respect to that entity. It also includes the pathways that start from that entity and reach consensus with respect to other entities on other interpretation layers. The entity together with all its incoming and outgoing pathways, which find agreements, completely describe the coherent role that entity "plays" in the Visual Story Net in terms of appearance, movement, physical properties, interactions, and higher level complex activities. Then, the overall story is the totality of such roles that are in consensus together within the Visual Story Network. In other words, if the role is a node with all its connections in space and time that reach consensus with other connections and nodes, then the story is the maximal subgraph in which all nodes and connections are in agreement.

For example, if a layer says that *This morning John is going to work* through one pathway, it should also say the same thing through a different pathway. One could be more local and completely described by the path that John takes on that morning. The other could be defined by the larger context, which could include John's usual habits and specific tasks scheduled for that day or the things he said on that morning or the SMS messages he received before. The fact that he goes to work should be clear no matter what is the pathway taken in the graph and that fact along with all its incoming and outgoing pathways that eventually find consensus in the Visual Story *is the role* played by that piece of knowledge in the overall story in space and time. We could immediately see that once the statement *This morning John goes to work* (which itself could be represented on an interpretation layer that is dedicated to such activities) is accepted as truth, it immediately becomes context for lower level classifiers that are supposed to recognize entities such as "person", "John", "walk", "goes to work", "work". At the same time, those more atomic classifiers are, in turn, fundamental in deciding that "John indeed goes to work," within the larger picture. That piece of information now could be passed through other more complex pathways to a larger network that goes even beyond vision towards a global understanding and more abstract reasoning about the world. It could be used, for example, to figure out that *John is not at home because he goes to work*. We could also use it to infer that *very soon John will be at work*, so if we want to talk to him we should look for him there. At the same time, it should come as a surprise if *John did not reach work* during that morning even if *he was for sure going towards work*. From that surprising finding, which represents, in fact, a disagreement between some pathways and others, we could be able to learn new things that we did not know before. For example, we could learn that *John, on his way to work, met an old friend he did not see in years and took the day off to play golf.* And that could be the first time we learn that *John likes to play golf.* We see how, almost imperceptibly, we go from vision to logical reasoning and language through the many different channels that try to put the pieces together.

It is again, consensus that puts everything together and creates a stable, solid, and unified view or story about the world. In that story all actors and actions and places are as connected as possible in order to form a single whole. That unified view is painted on a single dome on which all entities find their place in space and time. When one moves, changes, or does something, all the other pay attention and could potentially be affected. When a new entity comes in, more space is made, the dome gets larger, and more connections are added, which could increase complexity but could also simplify by making agreements stronger and getting rid of noise.

It is also true that disagreements along established paths bring the possibility of learning new things and adapting to the continuously changing world, by accordingly changing the weights along those paths in order to reach the desired agreement. The future is fundamentally unknown so there will always be some inconsistencies between what is predicted through a more complex pathway and context and what is seen at the local spatiotemporal level.

The Visual Story Network in space and time also reveals the need for language, which is the one that expresses the story beyond what is seen here and now at the local level. The story is more about the context and to put the local pieces together than about a single particular piece in space and time. It is more about the role played by actors than about how they look. Many researchers in artificial intelligence are asking themselves about "What is context?". In the Visual Story Net, the context with respect to a given actor is everything from the outside that relates to that actor— everything that predicts its appearance and actions and everything that is influenced and can be predicted by them.

Language could express in coherent form the content of such a visual story graph, conditioned on a specific problem or goal. Language does not need to say everything, the whole story, but what it says should make sense and be coherent. It should "go through" at least some pathways in space and time that find consensus in the graph. The important aspect is that language does not talk about what we see now and here, necessarily, but more about how what we see now and here makes sense within the overall story and how it is connected to the other interpretation layers—which could be at higher levels of semantics and abstraction and less connected to the current physical world of here and now. So, from this point of view, we can expect language to describe a world that is not always true or that is not necessarily accurate. There should be no surprise that each person might have a different story with respect to a situation or a sequence of events. The story is more about layers that are deeper in the graph and less about lower level layers that come from sensors. However, it is only by finding agreements among deeper layers that consider larger chunks of space and time and interpret the world in more semantic and meaningful way that we could predict and also find agreements at the level of what we see here, now and measure with sensors.

8.5.1   *Learning from Language*

There is clearly a very close connection between vision and language. The two are deeply linked but research at their intersection is still in its infancy. As we have discussed so far, it is clear that vision is not only about physical objects that are seen in the scene. Vision creates a coherent story about the world in space and time which includes not only the physical objects but also how they interact, what they do over shorter or longer periods of time. The link between perception and abstract thought is very subtle, in fact, the two might be the two sides of the same thing since every conscious thought is perceived and also understood in relation to other perceptions. It is not uncommon to say that *In the park I saw a group of people talking about how Lionel Messi almost magically dribbled three opponents and scored a magnificent goal last night* and the moment we saw those people to actually imagine how Messi scored the goal. There would be a mix of perceptions at different levels, of object, interactions, and events, a whole story, caused not only by real-time visual and sound input coming to our brain but, in fact, mostly by our revived memory and past experiences. They would all be put together into a single coherent visual and perceptual story, which we would then describe in language to others. What is very interesting and most important here is that there would be almost no real barrier between what we see then, on the spot (people) and our interpretation (talking about a game) and story construction (what happened at the event). The transition between physical objects and semantic thought is so gradual and subtle that, from the mind's point of view, it could simply be of one and the same type. When listening carefully to someone speak or ourselves speak, we will soon realize that in our language there is no clear barrier between descriptions of the physical world and interpretations of what happens in the world and the other people's minds. Language and vision are very deeply linked and they seem to be put together in order to create a coherent story of the world, which aims to tell the truth but often, due to lack of strong evidence, it is better at being coherent than accurate.

In Fig. 8.5a, we show an image of people, three children and an adult, in an outdoor scene. It seems pretty easy to see what objects are in the image, where they are, and what they are probably doing. At least at the level of physical objects, things should be all very clear. However, at a closer inspection (Fig. 8.6), we notice that, in fact, we cannot see the smaller objects if only we have access to information from inside their bounding boxes. This should come as a surprise, since they are clearly seen *in the image*. This example proves once more time our point that context plays a central role in vision and that local as well as global processing has to come together and reach a consensus for us to actually "see". Basically, every object in the scene says something about the other objects and about the whole scene, with its interactions and activities in the virtual space and time imagined. All these views of the scene, all its objects, actions, and interactions then try to predict each other in order to obtain a coherent story. It is only after our minds reach such consensus among sufficiently many pathways that we finally decide, subconsciously, to see.

Fig. 8.5 What does it mean to "see"? When looking at an image we want to know where and what are all the objects, what they do, infer what happened in the immediate past and predict what happens in the future. We mentally construct a story describing what takes place in a coherent way, such that all pathways of thought reach a consensus

The language of thought takes things a bit further as it attempts to put together the physical world seen in space and time and build a coherent story of events (most of it only imagined) that could offer a plausible and probable explanation of what happens in space and time. Written or spoken language is then only reflecting, as concise, accurate, and coherent that language of thought, combining perceptions, logical connections, recognition, and imagination at different levels of semantic abstractions.

To prove this point, we have shown a group of Masters' students in artificial intelligence at the University Politehnica of Bucharest the picture shown in Fig. 8.5 and asked them to describe it. We gave them five minutes to write about what they see in the image, without giving them any other instruction. The descriptions are shown below, in the gray box. We immediately note that, while all descriptions are plausible and generally agree at a high level, they also differ significantly when addressing different concrete aspects, such as the questions regarding the adult (shown in red), the activity of the children (shown in blue), or the time of the day (shown in green). Moreover, descriptions often include details about what actors speak or even think (shown in pink).

A. Many objects cannot be recognized by their
local appearance alone

B. Often, to „see" objects we need their context

Leg

Hand

Shoe

Food

Fig. 8.6 **A** We present patches containing the bounding boxes around different objects in the scene. Notice that they are not recognized from the visual information inside their bounding box alone, even though they are very well seen when put in the context of the image (**B**). This again proves our point that local and global information and processing come together along multiple pathways to reach a consensus and form an image of the world scene that is both seen and understood at a semantic level. Since the act of seeing depends a lot on previously learned experience there should be no surprise if different people create different coherent stories in their minds

Between vision and language: Here we present descriptions of the image, in natural language, given by eight Masters' students in Artificial Intelligence, of similar ages and backgrounds:

Student A: *A lady teaches some children how to prepare food for the next day. They are all getting ready to spend the night in the tent. It seems that the kids are in a camp, where they learn how to survive in nature.*

Student B: *A group of 4 people are in the woods, on the shore of a lake. I believe it is a family on vacation or on a trip. The father shows the others how to cook on the grill and they look and give their opinion on what he is doing there.*

Student C: *A family is camping. The family consists of 3 children and a mother. They are in a forest near a lake. The forest looks reddish because of the sun at sunset.* The mother prepares something to eat *while the children look at her. Next to them is a tent.*

Student D: *The image represents a natural scenery of a forest. Where an adult blonde Caucasian female prepares a grill for 3 children of different ages and race.* They are camping near a lake *in the afternoon sun. Weather appears to be warm, probably during summer.*

Student E: *The image denotes a quiet atmosphere at the day's dawn on a sky without clouds.*

Student F: *The photo is the scene of a lake campsite in the middle of summer. Judging by the position of the sun, it is afternoon,* and *the characters are most likely involved in the preparation of the meal.*

Student G: *A family is camping. The family consists of 3 children and a mother. They are in a forest near a lake. The forest looks reddish because of the sun at sunset.*

Student H: *A family with 3 children, or just a mother with the three children, spent a night with the tent in the nature near a lake/river. Now, they've been awake for a while and are cooking something on the grill. Mother teaches kids how to take care of the barbeque. At the same time, she teaches them to appreciate everything that happens ... the scent of burned wood, the beautiful light that surrounds them. We have 4 people, a tent, a barbecue, fir trees, water, food, grilling utensils.*

8.5.2 Unsupervised Learning by Surprise

On a day like every other, various actors and objects are in the scene and do what they are supposed to do. The scene looks as usual: a sunny day on the busy street, with people coming and going, cars finding their way through traffic. Everybody seems to be on their usual path. But somewhere, in a less noticeable place, something new happens. In fact, if we watch closely enough, there is always something new taking place, everywhere. The scene is not quite as expected and people's behavior is not really that predictable. There is everywhere a bit of the unexpected. After every corner there is a small surprise and the world shows its true face again, novel again and not completely predictable.

However, through its multiple active pathways the Visual Story Graph is not easily fooled. Through its many layers of interpretations and robust design optimized for leveraging consensus, it manages to figure out when a new type of object or action appears. The context then, very strong due to its multiple processing pathways, provides the right supervisory signal. If a new unseen car model appears on the street, the global context unmistakably recognizes it as a car and takes the new and initially unrecognized local appearance as a new training case. It could be a very interesting training case, a difficult one for which new learning resources are allocated. Immediately after, in the middle of a park a group of people perform a certain pattern of moves that is not recognized at all, not even by its context. It is only recognized as a group of people doing something, but not as a specific activity, ritual, dance, or sport. There are many classifiers that trigger at the same time and it is unlikely that nothing new happens there. The HPP feature detector is in full functioning and it will catch that pattern for which it will allocate a new potential candidate class. It is still not perfectly clear that the pattern of people moving and behaving in certain ways is indeed a novel category that will be seen again. It might be just a random pattern of movements or an *ad hoc* game that those people have just designed on the spot and are playing now. Nevertheless, that pattern has now its allocated new class and layers of interpretation, linked to all the current group of local or global contextual classifiers that initially triggered it. The few instances seen now are used to learn initial pathways between it and its context. However, nobody knows if this will be recognized later, if it will ever be seen again.

The mind-like Visual Story (VSN) system might not even pay too much attention to the fact that a new class is born until it is seen again. Many such new candidate classes could be born and then fade away as they are never seen again. From the point of view of the mind-like VSN, the process of learning has always a supervised form, a self-supervised form in which context of multiple pathways is teacher for the learning student. From the point of view of the outside world, the system is always unsupervised, since it can never predict how data will come and when new classes will be born. It is only through observations of grouping and rare alignments and co-occurrences that the system catches the possible presence of a new unknown entity. And that information comes completely from the outside, it is at the mercy of Mother Nature. From that point of view, the system and its world should be in agreement if the system wants any chance of survival. It needs to be sufficiently good at self-supervision in order to survive the coming into existence of the truly unpredictable, the totally new classes and situations. It needs to use all its potential resources for detecting groupings and alignments, to maintain its balance when put in novel situations. For that it needs to learn as quickly as possible many potential new classes most of which will be completely dropped later, as simply noise. Others, however, will be crucial to keep to be recognized again and thus strengthen its understanding of the world and improve its equilibrium.

We now state the ninth principle of unsupervised learning, which deals with the capability of the system to learn from disagreements.

> **Principle 9**
> The Visual Story Network will gain fundamentally new knowledge through disagreements along different pathways, but only when there would still be sufficient agreements along other pathways in order to maintain its equilibrium state in space and time.

By the equilibrium state we mean that the system will still agree on sufficiently many interpretation layers so that it could connect the new class (layer) to the ones which are in equilibrium (agreement). Without that solid agreement on other layers, the new class could not be linked to the old ones and could not be possibly learned and maintained in time, based on our context-supervised scheme.

8.5.3   Discover Itself

As the system discovers new classes, it will soon discover the presence of a unique new class of completely different kind. As the robotic system observes but is also capable to move and act in the world it will very soon learn about the possibility of its own existence, which will come up as other new classes learned before, also as a new class. As it can move and act by itself, observations will be directly linked to its own actions and movements. Agreements between such observations, interpretations, and actions will immediately trigger agreements and co-occurrences which will be labeled as HPP signal and trigger the creation of a new candidate class, that of *its own self*. Note that in the following paragraphs the words "self" and "I" are used in symbolic form, only to denote that special, new class learned, which will correspond to the system's own existence.

Now that it has discovered and learned about its own existence and also deeply inclined to keep consensus among its different pathways, the system will naturally have a tendency towards seeing, understanding, planning, and acting towards keeping the "self" class strong and in consensus with the others. It will very soon be clear that this "class" plays a central role in the system, most vital for maintaining the entire consensus and coherent functioning of the whole Visual Story. There is no class, no entity, no interpretation layer of the world, no action or semantic understanding that does not depend on the "self" class. Preserving and protecting this class will protect all the others and will again improve consensus, which the system is designed to do (according to the unsupervised learning principles). It would not be at all unexpected if the system would eventually learn best ways to survive, as a way to protect the "self" class, which in turn maintains general consensus in the Visual Story Network, over space and time.

In this view, survival becomes an efficient way to optimize consensus within a structure in which the "self" class becomes central, with all the others being dependent on it. The main role of this class would be to be involved in predicting all the others and also be predicted by them: *I make this action, therefore the world will change in this following expected way and therefore I will continue to exist. Or, if I make this other action the world will change in this other way and then I might not exist anymore.* Preserving the class "I" becomes a top priority since it becomes pretty soon clear that the best way to maintain consensus is to keep "I exist" in strong agreement from as many pathways as possible. If consensus in space and time is a priority, then "I exist" is an equally important one. However, ultimately the system will need all the others and the world in order to predict its own existence. It will also need its own existence in order to predict the world. It would soon become clear that it is as much about "I" as it is about the rest of the seen classes in the world. The same universal need for consensus will drive it towards evolving the most intelligent means, layers of interpretation and actions in order to achieve and maintain an even better consensus for an even smoother and longer existence.

Sometimes, in order to achieve a temporary consensus related to the current state of understanding the world, the system may damage a more global consensus not yet understood. As its knowledge expands the system can self-reflect on the past actions in order to improve its expectancy of a longer term consensus and equilibrium between the best prediction of "I" and its external world. It will also learn by trial and error. Its trials, however, are not random, but drawn from distributions that represent its current best understanding of itself in relation to the outside world.

We are now ready to propose our final principle of unsupervised learning:

Principle 10
The Visual Story Network could become self-aware (learn a new class equivalent to "I exist") only when it will be allowed to make actions that change the world. Thus, its own decisions, which will cause observed changes of the outside world, will also start being observed and constitute input to pathways that will ultimately reach the "I exist" layer, in which different semantic properties with respect to the world (e.g., pose) could be given to the new "I exist" class. This class will play a central role in maintaining consensus over time, between actions and their consequences (observations).

We see how the Visual Story Network learns to optimize overall consensus, reduce variance, while remaining in agreement with the past, present and future sensor readings. By doing so, it also improves its own chances of survival and the prediction of the "self" class, which becomes central to maintaining such overall consensus. This should remind us of homeostasis [71], which is the innate tendency of living organisms towards a relatively stable equilibrium between their many interdependent elements, in order to maintain their life.

8.5.4 Dreams of Tomorrow

We started the story of this book in the first chapter by laying down several fundamental ideas that we believe are key to create working, real-world unsupervised learning systems. Then, during the next chapters, we showed those ideas at work. We applied them to various tasks and proposed different solutions. We presented extensive experimental validations for each problem and each algorithm and also showed how they all relate at a higher level and could be used to imagine a unified unsupervised learning model.

In the final chapter, we showed how the two main computational approaches discussed in the book, namely, graphical models and neural networks could be effectively put together into a single recursive graph neural network to learn to understand data at high semantic levels over space and time.

Then, in the very last part, we adventured into the unknown world of the future and created in our imagination, while staying deeply rooted in the key principles and models presented so far, the ambitious idea of a Visual Story Net. If we can make the VSN system possible then it would quickly grow beyond the realm of vision, in its need to obtain a full and coherent understanding of its surrounding world. If we further equip VSN with robotic capabilities of sensing, moving, and acting in the world it would soon discover the new class of "self" and naturally develop a need for survival (e.g., consistently predict the class "I exist", all the time) in order to keep the overall consensus among all other classes alive, with the self playing a central role. However, the importance of the "self" class will be well balanced by the other classes, since the existence of the "self" will depend on the consensus from others. Thus, they all ultimately depend on the interaction with the outside world. While the system is self-trained and self-centered from its own point of view, it remains unsupervised and strongly dependent on spatiotemporal data coming from the outside, larger context.

While we cross the border of science and reach the realm of fiction with no clear experiments to prove the feasibility of the Visual Story Net, we argue that it is only through fiction and imagination that we could ultimately start from a dream and then make it come true. As the field of artificial intelligence is growing incredibly fast, borders between fiction and science will start being crossed more and more often, as fiction becomes science and then science becomes the next starting point for fiction. We, as researchers, scientists, and engineers, should be prepared and get ready to both dream and act. Now, more than ever in history, it is the right moment to have a vision: imagine the pieces, put them together in a coherent way, and then make it happen. We are also self-supervised with respect to our own knowledge, while being fundamentally unsupervised with respect to the outside world. So we too should be looking for consensus through as many pathways as possible, in order to succeed in making our dreams come true.

References

1. Maurer D, Lewis TL (2018) Visual systems. In: The neurobiology of brain and behavioral development. Elsevier, pp 213–233
2. Yue-Hei Ng J, Hausknecht M, Vijayanarasimhan S, Vinyals O, Monga R, Toderici G (2015) Beyond short snippets: deep networks for video classification. In: Proceedings of the IEEE conference on computer vision and pattern recognition, pp 4694–4702
3. Karpathy A, Toderici G, Shetty S, Leung T, Sukthankar R, Fei-Fei L (2014) Large-scale video classification with convolutional neural networks. In: Proceedings of the IEEE conference on computer vision and pattern recognition, pp 1725–1732
4. Battaglia PW, Hamrick JB, Bapst V, Sanchez-Gonzalez A, Zambaldi V, Malinowski M, Tacchetti A, Raposo D, Santoro A, Faulkner R, et al (2018) Relational inductive biases, deep learning, and graph networks. arXiv:180601261
5. Gilmer J, Schoenholz SS, Riley PF, Vinyals O, Dahl GE (2017) Neural message passing for quantum chemistry. In: Precup D, Teh YW (eds) Proceedings of the 34th international conference on machine learning, proceedings of machine learning research, vol 70, pp 1263–1272
6. Besag J (1986) On the statistical analysis of dirty pictures. J R Stat Soc Ser B (Methodological) 259–302
7. Hummel RA, Zucker SW (1983) On the foundations of relaxation labeling processes. IEEE Trans Pattern Anal Mach Intell 3:267–287
8. Geman S, Geman D (1984) Stochastic relaxation, gibbs distributions, and the bayesian restoration of images. IEEE Trans Pattern Anal Mach Intell 6:721–741
9. Geman S, Graffigne C (1986) Markov random field image models and their applications to computer vision. In: Proceedings of the international congress of mathematicians, Berkeley, CA, vol 1, p 2
10. Lafferty J, McCallum A, Pereira FC (2001) Conditional random fields: probabilistic models for segmenting and labeling sequence data
11. Kumar S, Hebert M (2006) Discriminative random fields. Int J Comput Vis 68(2):179–201
12. Pearl J (2014) Probabilistic reasoning in intelligent systems: networks of plausible inference. Elsevier
13. Ravikumar P, Lafferty J (2006) Quadratic programming relaxations for metric labeling and markov random field map estimation. In: Proceedings of the 23rd international conference on machine learning. ACM, pp 737–744
14. Schaeffer SE (2007) Graph clustering. Comput Sci Rev 1(1):27–64
15. Leordeanu M, Sukthankar R, Hebert M (2012) Unsupervised learning for graph matching. Int J Comput Vis 96:28–45
16. Ng AY, Jordan MI, Weiss Y (2002) On spectral clustering: analysis and an algorithm. In: Advances in neural information processing systems, pp 849–856
17. Bruna J, Zaremba W, Szlam A, LeCun Y (2013) Spectral networks and locally connected networks on graphs. arXiv:1312.6203
18. Henaff M, Bruna J, LeCun Y (2015) Deep convolutional networks on graph-structured data. arXiv:1506.05163

19. Defferrard M, Bresson X, Vandergheynst P (2016) Convolutional neural networks on graphs with fast localized spectral filtering. In: Advances in neural information processing systems, pp 3844–3852
20. Kipf TN, Welling M (2017) Semi-supervised classification with graph convolutional networks. In: International conference on learning representations (ICLR)
21. Duvenaud DK, Maclaurin D, Iparraguirre J, Bombarell R, Hirzel T, Aspuru-Guzik A, Adams RP (2015) Convolutional networks on graphs for learning molecular fingerprints. In: Advances in neural information processing systems, pp 2224–2232
22. Battaglia P, Pascanu R, Lai M, Rezende DJ et al (2016) Interaction networks for learning about objects, relations and physics. In: Advances in neural information processing systems, pp 4502–4510
23. Xu K, Hu W, Leskovec J, Jegelka S (2019) How powerful are graph neural networks? In: International conference on learning representations. https://openreview.net/forum?id=ryGs6iA5Km
24. Velikovi P, Cucurull G, Casanova A, Romero A, Li P, Bengio Y (2018) Graph attention networks. In: International conference on learning representations. https://openreview.net/forum?id=rJXMpikCZ
25. Li Y, Tarlow D, Brockschmidt M, Zemel R (2016) Gated graph sequence neural networks. In: International conference on learning representations (ICLR)
26. Jain A, Zamir AR, Savarese S, Saxena A (2016) Structural-rnn: deep learning on spatio-temporal graphs. In: Proceedings of the IEEE conference on computer vision and pattern recognition, pp 5308–5317
27. Dehghani M, Gouws S, Vinyals O, Uszkoreit J, Kaiser L (2019) Universal transformers. In: International conference on learning representations. https://openreview.net/forum?id=HyzdRiR9Y7
28. Santoro A, Faulkner R, Raposo D, Rae J, Chrzanowski M, Weber T, Wierstra D, Vinyals O, Pascanu R, Lillicrap T (2018) Relational recurrent neural networks. In: Bengio S, Wallach H, Larochelle H, Grauman K, Cesa-Bianchi N, Garnett R (eds) Advances in neural information processing systems, vol 31. Curran Associates, Inc., pp 7310–7321
29. Felzenszwalb PF, Huttenlocher DP (2005) Pictorial structures for object recognition. Int J Comput Vis 61(1):55–79
30. Lazebnik S, Schmid C, Ponce J (2006) Beyond bags of features: spatial pyramid matching for recognizing natural scene categories. In: CVPR
31. He K, Zhang X, Ren S, Sun J (2015) Spatial pyramid pooling in deep convolutional networks for visual recognition. IEEE Trans Pattern Anal Mach Intell 37(9):1904–1916
32. Vaswani A, Shazeer N, Parmar N, Uszkoreit J, Jones L, Gomez AN, Kaiser Ł, Polosukhin I (2017) Attention is all you need. In: Advances in neural information processing systems, pp 5998–6008
33. Hochreiter S, Schmidhuber J (1997) Long short-term memory. Neural Comput 9(8):1735–1780
34. Santoro A, Raposo D, Barrett DG, Malinowski M, Pascanu R, Battaglia P, Lillicrap T (2017) A simple neural network module for relational reasoning. In: Guyon I, Luxburg UV, Bengio S, Wallach H, Fergus R, Vishwanathan S, Garnett R (eds) Advances in neural information processing systems, vol 30. Curran Associates, Inc., pp 4967–4976
35. Wang X, Girshick R, Gupta A, He K (2018) Non-local neural networks. In: The IEEE conference on computer vision and pattern recognition (CVPR), vol 1, p 4
36. Wang X, Gupta A (2018) Videos as space-time region graphs. In: Proceedings of the European conference on computer vision (ECCV), pp 399–417
37. Baradel F, Neverova N, Wolf C, Mille J, Mori G (2018) Object level visual reasoning in videos. In: ECCV
38. Chen Y, Kalantidis Y, Li J, Yan S, Feng J (2018) A2-nets: double attention networks. In: Advances in neural information processing systems, pp 350–359
39. Szegedy C, Vanhoucke V, Ioffe S, Shlens J, Wojna Z (2016) Rethinking the inception architecture for computer vision. In: Proceedings of the IEEE conference on computer vision and pattern recognition, pp 2818–2826

40. Chollet F (2017) Xception: deep learning with depthwise separable convolutions, pp 1610–02,357
41. Sun L, Jia K, Yeung DY, Shi BE (2015) Human action recognition using factorized spatio-temporal convolutional networks. In: Proceedings of the IEEE international conference on computer vision, pp 4597–4605
42. Xie S, Sun C, Huang J, Tu Z, Murphy K (2018) Rethinking spatiotemporal feature learning: speed-accuracy trade-offs in video classification. In: Proceedings of the European conference on computer vision (ECCV), pp 305–321
43. Tran D, Wang H, Torresani L, Ray J, LeCun Y, Paluri M (2018) A closer look at spatiotemporal convolutions for action recognition. In: Proceedings of the IEEE conference on computer vision and pattern recognition, pp 6450–6459
44. Donahue J, Anne Hendricks L, Guadarrama S, Rohrbach M, Venugopalan S, Saenko K, Darrell T (2015) Long-term recurrent convolutional networks for visual recognition and description. In: Proceedings of the IEEE conference on computer vision and pattern recognition, pp 2625–2634
45. Zhou B, Andonian A, Oliva A, Torralba A (2018) Temporal relational reasoning in videos. In: Proceedings of the european conference on computer vision (ECCV), pp 803–818
46. Carreira J, Zisserman A (2017) Quo vadis, action recognition? A new model and the kinetics dataset. In: 2017 IEEE conference on computer vision and pattern recognition (CVPR). IEEE, pp 4724–4733
47. Simonyan K, Zisserman A (2014) Two-stream convolutional networks for action recognition in videos. In: Advances in neural information processing systems, pp 568–576
48. Zhao Y, Xiong Y, Lin D (2018) Trajectory convolution for action recognition. In: Bengio S, Wallach H, Larochelle H, Grauman K, Cesa-Bianchi N, Garnett R (eds) Advances in neural information processing systems, vol 31. Curran Associates, Inc., pp 2204–2215. http://papers.nips.cc/paper/7489-trajectory-convolution-for-action-recognition.pdf
49. Shi X, Chen Z, Wang H, Yeung DY, Wong WK, Chun Woo W (2015) Convolutional LSTM network: a machine learning approach for precipitation nowcasting. In: NIPS
50. Wang Y, Long M, Wang J, Gao Z, Yu PS (2017) Predrnn: recurrent neural networks for predictive learning using spatiotemporal LSTMS. In: NIPS
51. Wang Y, Jiang L, Yang MH, Li LJ, Long M, Fei-Fei L (2019) Eidetic 3d LSTM: a model for video prediction and beyond. In: International conference on learning representations. https://openreview.net/forum?id=B1lKS2AqtX
52. Goyal R, Kahou SE, Michalski V, Materzynska J, Westphal S, Kim H, Haenel V, Fruend I, Yianilos P, Mueller-Freitag M et al (2017) The "something something" video database for learning and evaluating visual common sense. In: ICCV, vol 1, p 3
53. Soomro K, Zamir AR, Shah M (2012) Ucf101: a dataset of 101 human actions classes from videos in the wild. arXiv:12120402
54. Kuehne H, Jhuang H, Garrote E, Poggio T, Serre T (2011) Hmdb: a large video database for human motion recognition. In: 2011 international conference on computer vision. IEEE, pp 2556–2563
55. He K, Zhang X, Ren S, Sun J (2016) Deep residual learning for image recognition. In: 2016 IEEE conference on computer vision and pattern recognition (CVPR), pp 770–778
56. Abadi M, Agarwal A, Barham P, Brevdo E, Chen Z, Citro C, Corrado GS, Davis A, Dean J, Devin M, Ghemawat S, Goodfellow I, Harp A, Irving G, Isard M, Jia Y, Jozefowicz R, Kaiser L, Kudlur M, Levenberg J, Mané D, Monga R, Moore S, Murray D, Olah C, Schuster M, Shlens J, Steiner B, Sutskever I, Talwar K, Tucker P, Vanhoucke V, Vasudevan V, Viégas F, Vinyals O, Warden P, Wattenberg M, Wicke M, YY, Zheng X (2015) TensorFlow: large-scale machine learning on heterogeneous systems. https://www.tensorflow.org/, software available from tensorflow.org
57. Lee M, Lee S, Son SJ, Park G, Kwak N (2018) Motion feature network: fixed motion filter for action recognition. In: ECCV
58. Zolfaghari M, Singh K, Brox T (2018) Eco: efficient convolutional network for online video understanding. In: Proceedings of the European conference on computer vision (ECCV), pp 695–712

59. Leordeanu M, Sukthankar R (2017) Towards a visual story network using multiple views for object recognition at different levels of spatiotemporal context. In: The physics of the mind and brain disorders. Springer, pp 573–610

60. Carpenter GA, Grossberg S (1987) A massively parallel architecture for a self-organizing neural pattern recognition machine. Comput Vis Graph Image Process 37(1):54–115

61. Chang HC, Grossberg S, Cao Y (2014) Wheres waldo? How perceptual, cognitive, and emotional brain processes cooperate during learning to categorize and find desired objects in a cluttered scene. Front Integr Neurosci 8:43

62. Fazl A, Grossberg S, Mingolla E (2009) View-invariant object category learning, recognition, and search: how spatial and object attention are coordinated using surface-based attentional shrouds. Cogn Psychol 58(1):1–48

63. Grossberg S (1976) Adaptive pattern classification and universal recoding: I. parallel development and coding of neural feature detectors. Biol Cybern 23(3):121–134

64. Grossberg S (2000) The complementary brain: unifying brain dynamics and modularity. Trends Cogn Sci 4(6):233–246

65. Grossberg S (2013) Adaptive resonance theory: how a brain learns to consciously attend, learn, and recognize a changing world. Neural Netw 37:1–47

66. Grossberg S (2015) From brain synapses to systems for learning and memory: object recognition, spatial navigation, timed conditioning, and movement control. Brain Res 1621:270–293

67. Lashley K (1950) In search of the engram. In: Symposia. Society of experimental biology, vol 4, pp 454–482

68. Probst T, Paudel DP, Chhatkuli A, Gool LV (2019) Unsupervised learning of consensus maximization for 3d vision problems. In: The IEEE conference on computer vision and pattern recognition (CVPR)

69. Tang C, Chen J, Liu X, Li M, Wang P, Wang M, Lu P (2018) Consensus learning guided multi-view unsupervised feature selection. Knowl-Based Syst 160:49–60

70. Duta I, Liviu Nicolicioiu A, Bogolin SV, Leordeanu M (2018) Mining for meaning: from vision to language through multiple networks consensus. In: British machine vision conference

71. Betts Gordon J, et al (2014) Anatomy and physiology

Index

© Springer Nature Switzerland AG 2020
M. Leordeanu, *Unsupervised Learning in Space and Time*,
Advances in Computer Vision and Pattern Recognition,
https://doi.org/10.1007/978-3-030-42128-1

Printed in the United States
by Baker & Taylor Publisher Services